Lecture Notes in Computer Science   11039

Commenced Publication in 1973
Founding and Former Series Editors:
Gerhard Goos, Juris Hartmanis, and Jan van Leeuwen

More information about this series at http://www.springer.com/series/7412

Danail Stoyanov · Zeike Taylor
Francesco Ciompi · Yanwu Xu et al. (Eds.)

Computational Pathology and Ophthalmic Medical Image Analysis

First International Workshop, COMPAY 2018
and 5th International Workshop, OMIA 2018
Held in Conjunction with MICCAI 2018
Granada, Spain, September 16–20, 2018
Proceedings

 Springer

Editors
Danail Stoyanov
University College London
London, UK

Francesco Ciompi 🄳
Radboud University Medical Center
Nijmegen, The Netherlands

Zeike Taylor
University of Leeds
Leeds, UK

Yanwu Xu 🄳
Baidu
Beijing, China

Additional Workshop Editors *see next page*

ISSN 0302-9743   ISSN 1611-3349   (electronic)
Lecture Notes in Computer Science
ISBN 978-3-030-00948-9   ISBN 978-3-030-00949-6   (eBook)
https://doi.org/10.1007/978-3-030-00949-6

Library of Congress Control Number: 2018955277

LNCS Sublibrary: SL6 – Image Processing, Computer Vision, Pattern Recognition, and Graphics

This Springer imprint is published by the registered company Springer Nature Switzerland AG
The registered company address is: Gewerbestrasse 11, 6330 Cham, Switzerland

Additional Workshop Editors

Tutorial and Educational Chair

Anne Martel
University of Toronto
Toronto, ON
Canada

Workshop and Challenge Co-chair

Lena Maier-Hein
German Cancer Research Center (DKFZ)
Heidelberg
Germany

First International Workshop on Computational Pathology, COMPAY 2018

Nasir Rajpoot(iD)
University of Warwick
Coventry
UK

Jeroen Van der Laak(iD)
Radboud University Medical Center
Nijmegen
The Netherlands

Mitko Veta(iD)
Eindhoven University of Technology
Eindhoven
The Netherlands

Stephen McKenna(iD)
University of Dundee
Dundee
UK

David Snead
University Hospitals Coventry
  and Warwickshire
Coventry
UK

5th International Workshop on Ophthalmic Medical Image Analysis, OMIA 2018

Emanuele Trucco
University of Dundee
Dundee
UK

Xin Jan Chen
Soochow University
Suzhou
China

Mona K. Garvin
University of Iowa
Iowa City, IA
USA

Hrvoje Bogunovic
Medical University of Vienna
Vienna
Austria

COMPAY 2018 Preface

We were very excited to host the first MICCAI COMPAY workshop in the rapidly emerging area of computational pathology, the study of disease using computational analysis of digitized images of tissue slides. We believe this first event on computational pathology and its synergy with advanced image analysis and deep learning provided a space for researchers in the MICCAI community to meet, discuss, and share their advances in these fields. The MICCAI conference was the perfect venue and *it was* the best time for this to happen. The aim of COMPAY was to bring together scientific researchers, medical experts, and industry partners working in the field of computational pathology, in order to push further innovative and clinically relevant solutions for digital pathology. We strived to provide a platform for scientific discussion on computational pathology with a focus on artificial intelligence and deep learning, which can help foster cooperative projects at an international level. We hope that you will find the contributions on the state of the art computational pathology stimulating and enjoyable. We are grateful to the MICCAI organizers for giving us this opportunity. We also extend our sincere gratitude to all the reviewers who helped ensure the high quality of papers presented at COMPAY 2018, the first of hopefully a series of workshops at MICCAI.

August 2018

Francesco Ciompi
Jeroen van der Laak
Nasir Rajpoot
Stephen McKenna
Mitko Veta
David Snead

OMIA 2018 Preface

Age-related macular degeneration, diabetic retinopathy, and glaucoma are main causes of blindness. Oftentimes blindness can be avoided by early intervention, making computer-assisted early diagnosis of retinal diseases a research priority. Related research is exploring retinal biomarkers for systemic conditions such as dementia, cardiovascular disease, and complications of diabetes. Significant challenges remain, including reliability and validation, effective multimodal analysis (e.g., fundus photography, optical coherence tomography, and scanning laser ophthalmoscopy), more powerful imaging technologies, and the effective deployment of cutting-edge computer vision and machine learning techniques. The 4th International Workshop on Ophthalmic Medical Image Analysis (OMIA5) addressed all these aspects and more, this year in collaboration with the ReTOUCH retinal image challenge.

August 2018

Yanwu Xu
Emanuele Trucco
Mona K. Garvin
Xinjian Chen
Hrvoje Bogunović

Organization

COMPAY 2018 Organizing Committee

Francesco Ciompi   Radboud University Medical Center, The Netherlands
Jeroen van der Laak   Radboud University Medical Center, The Netherlands
Nasir Rajpoot   University of Warwick, UK
Stephen McKenna   University of Dundee, UK
Mitko Veta   Eindhoven University of Technology, The Netherlands
David Snead   University Hospitals Coventry and Warwickshire NHS Trust, UK

OMIA 2018 Organizing Committee

Hrvoje Bogunović   Medical University of Vienna, Austria
Xinjian Chen   Soochow University, China
Mona K. Garvin   University of Iowa, USA
Emanuele Trucco   University of Dundee, UK
Yanwu Xu   Institute for Infocomm Research, Singapore

Contents

5th International Workshop on Ophthalmic Medical Image Analysis, OMIA 2018

First International Workshop on Computational Pathology, COMPAY 2018

Improving Accuracy of Nuclei Segmentation by Reducing Histological Image Variability

Yusuf H. Roohani[1,2($\boxtimes$)] (iD) and Eric G. Kiss[1]

[1] Stanford University School of Medicine,
291 Campus Drive, Stanford, CA 94305, USA
yusuf.x.roohani@gsk.com
[2] GlaxoSmithKline, 200 Cambridgepark Drive, Cambridge, MA 02140, USA

Abstract. Histological analyses of tissue biopsies is an essential component in the diagnosis of several diseases including cancer. In the past, evaluation of tissue samples was done manually, but to improve efficiency and ensure consistent quality, there has been a push to evaluate these algorithmically. One important task in histological analysis is the segmentation and evaluation of nuclei. Nuclear morphology is important to understand the grade and progression of disease. However, implementing automated methods at scale across histological datasets is challenging due to differences in stain, slide preparation and slide storage. This paper evaluates the impact of four stain normalization methods on the performance of nuclei segmentation algorithms. The goal is to highlight the critical role of stain normalization in improving the usability of learning-based models (such as convolutional neural networks (CNNs)) for this task. Using stain normalization, the baseline segmentation accuracy across distinct training and test datasets was improved by more than 50% of its base value as measured by the AUC and Recall. We believe this is the first study to perform a comparative analysis of four stain normalization approaches (histogram equalization, Reinhart, Macenko, spline mapping) on segmentation accuracy of CNNs.

Keywords: Histology · Stain normalization · Nuclei segmentation
Convolutional neural networks · Machine learning

1 Introduction

Diagnoses made by pathologists using tissue biopsy images are central for many tasks such as the detection of cancer and estimation of its current stage [2]. One routine yet important step within histological analyses is the segmentation of nuclei. Nuclear morphology is an important indicator of the grade of cancer and the stage of its progression [3]. It has also been shown to be a predictor of cancer outcome [4]. Currently, histological analysis such as these are done manually, with pathologists counting and evaluating cells by inspection. Developing

© Springer Nature Switzerland AG 2018
D. Stoyanov et al. (Eds.): COMPAY 2018/OMIA 2018, LNCS 11039, pp. 3–10, 2018.
https://doi.org/10.1007/978-3-030-00949-6_1

automated methods to perform this analysis will help pathologists maintain consistent quality, allow for greater use of histological analysis by reducing cost and throughput.

However, automating nuclei detection is not a trivial task and can be challenging for a number of reasons - one important challenge is the lack of stain standardization. Stain manufacturing and aging can lead to differences in applied color. It could also be the result of variation in tissue preparation (dye concentration, evenness of the cut, presence of foreign artifacts or damage to the tissue sample), stain reactivity or image acquisition (image compression artifacts, presence of digital noise, specific features of the slide scanner). Each stain has different absorption characteristics(sometimes overlapping) which impact the resulting slide color. Finally, storage of the slide samples can have aging effects that alter the color content [2,5]. Radiologists have established standards (such as DICOM) to ensure consistency between scans from different origins and time. Ideally, histopathology would also work within a framework like DICOM where images can be standardized against experimental conditions to ensure consistency across datasets.

Recently, there has been considerable interest in the application of novel machine learning tools such as deep learning to aid in routine tasks such as segmentation. These models generally work on raw pixel values, but could achieve greater accuracy through reducing the variance contributed by slide and stain specific variables. However, the approach must not be too general or else false positives will occur through altering the image signal [1,2].

The aim of this project is to address the impact of variability in histological images on the accuracy of deep learning based algorithms for segmentation. A Convolutional Neural Network (CNN) was trained to perform nuclei segmentation and tested to get a baseline. Four stain normalization techniques, histogram equalization, Reinhard, Macenko, and spline mapping were then applied as means to reduce color variability of the raw images. The CNN was trained and tested again for each of these normalization conditions to get segmentation accuracy in each case. This paper is unique in that it employs a wide variety of normalization methods, uses deep learning based nuclei segmentation accuracy as a metric, and tests the model on a different dataset to understand model generalizability.

2   Stain Color Normalization

Stain normalization techniques involve transforming image pixel values. There are a wide array of techniques in literature, but most involve statistical transformations of images in various color spaces. Below we provide an overview of the four techniques used. Since our goal was mainly to highlight the impact of stain normalization as opposed to finding the best approach, we acknowledge that there is scope for further expanding the following list (e.g.: Automatic Color Equalization [12], HSV channel shifting, adding random Gaussian noise etc.). We chose the following four because they were reasonably diverse and were easily implementable using University of Warwick's stain normalization toolbox [10]. For methods requiring stain vector estimation, this was done using an image from the training set in both cases of training and testing.

- **Histogram equalization:** Histogram equalization is a commonly used image processing technique that transforms one histogram by spreading out its distribution to increase image contrast. In this analysis, histogram equalization was performed on each RGB channel in Matlab, effectively normalizing the color intensities frequencies between two images [9].
- **Macenko color normalization:** The Macenko color normalization method transforms the images to a stain color space by estimating the stain vectors then normalizes the stain intensities. Quantifying the stain color vectors (the RGB contribution of each stain) provides a more robust means of manipulating color information [8].
- **Reinhard color normalization:** Reinhard color normalization aims to make one image 'feel' like another by transforming one image distribution to be closer to another. Reinhard transforms the target and source images into L" color space. This was created to minimize the correlation between channels. After the source and target images are in this colorspace, descriptive statistics are used to transform the target image's colorspace as described in [7]. Finally, the average channel values of the source are added back to the data points and it is transformed back to RGB colorspace.
- **Spline mapping:** Conceptually, the spline mapping technique is similar to the Macenko technique in that it estimates the stain vectors, deconvolves the image, maps the stain intensity to a target image before reconstructing back in RGB colorspace. Khan makes contributions in automatic stain vector calculation using a classifier with global and local pixel value information, and a non-linear stain normalization method [5].

3   Image Segmentation Using CNNs

This section describes the methods used to generate deep learning based image segmentation models. The goal was to train a model using one dataset and to test using another. We aimed to perform this approach using different normalization strategies to narrow down on an approach that best reduced variability and improved performance.

3.1   Model Selection

We first trained and validated the model on the same dataset to make sure that our training procedure was working correctly. For this we used breast tissue slices from [3]. We randomly split the dataset into 70% train and 30% validation. After experimenting with different network architectures, the final architecture that was chosen after the validation procedure was (Conv-BNorm-ReLU)x6 - (Fully Convolutional) - Softmax [3].

We chose to use a fully convolutional network (FCN) instead of a regular fully connected network so as to enhance throughput and quickly return the network based on results [6]. This meant that, at the time of inference, our model could simply process an entire image in one pass instead of needing it to be broken down into pixelwise patches. However, for training the network, we used a fully connected ultimate layer, and fed patches instead of a whole image as input. This allowed us to have greater control over the size and composition of the training classes, given their skewed distribution in the training set. This is also why we chose FCN over more recent architectures that work better with whole images such as U-Net [13] and Mask-RCNN [14]. We also realize that there are deeper architectures that could be used with FCN for further improving pixel level accuracy but our main focus was on showing the value of stain normalization as opposed to finding the optimal architecture for segmentation. We used the Caffe deep learning framework to design these models.[1]

3.2   Dataset

The training dataset consisted of 143 histological sections of breast tissue from the Case Western Reserve digital pathology dataset. Each RGB image was 2000×2000 pixels, $20\times$ magnification and was H&E stained. Manually annotated masks were provided for over 12000 nuclei. We found that, for training, a patch size of 64×64 (87.8% baseline validation accuracy) worked better for training than 32×32 (82% accuracy). A total of 400,000 unique patches were generated for each scenario.

However, it was not sufficient to randomly sample from non-nuclear regions as defined by the hand annotations. There was a significant probability of sampling unannotated nuclei while developing negative patches for the training set. To address this problem, we used the approach outlined by [3]. Nuclei are known to absorb greater levels of the eosin (red) stain and so the red channel in the images was enhanced. A negative mask was thus generated defining regions outside of these enhanced red zones that were deemed safe for non-nuclei class patch selection. We also made sure to allocated a third of the non-nuclei patches to boundaries around the nuclei so that these would be clearly demarcated in the output. Moreover, positive and negative samples were equal in number even after accounting for these changes. The model prediction accuracy was found to benefit from these approaches.

The test set was composed also of breast tissue slices from a hand annotated dataset provided by the BioImaging lab at UCSB [11]. Referred to as the BioSegmentation benchmark, these were 58 H&E stained images at a much smaller resolution (896×768). This dataset proved to be ideal for model testing because the images were quite different from our training set both in terms of image quality and resolution and also in terms of the staining used (more eosin content). Patching was not required for the test set because we were using a fully convolutional network.

---

[1] Code: https://github.com/yhr91/NucleiSegmentation/tree/master/BMI-260.

3.3 Training

Once our model architecture and dataset generation approach had been finalized, we began to train separate model for each of the normalization scenarios as shown in Fig. 1. We used a batch size of 1000 because that could fit comfortably in our memory (P100 GPU 16 GB × 2).

There were four models - these corresponded to the four techniques outlined previously: Histogram Equalization (SH), Macenko (MM), Reinhard (RH), Spline Mapping (SM). There was a also a model for the unnormalized (Unnorm) case. Model performance would generally begin to plateau around 5–10 epochs. We did not notice overfitting in any model until 25 epochs except in SM. However, we could not continue training much beyond that point due to time constraints.

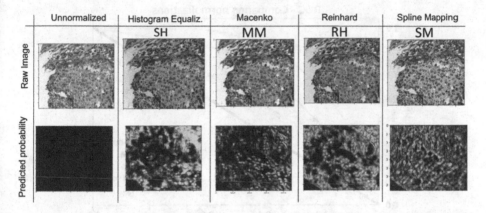

Fig. 1. The first row shows the test image as fed into the model after stain normalization (labelled using acronym). The normalization applied on the test image was the same as that applied on the training dataset in that case. The bottom row shows the model predicted output on the test images.

4 Results

4.1 Visual Inspection

The top row in Fig. 1 shows the original images after being transformed using the four different stain normalization approaches. We can see that all four images appear different in some respect. For example, HE and RH, which involve stain normalization through working directly with the color values show a noticeable blue tint. This is more pronounced in HE, where non-nuclear regions in the top right of the cell get quite heavily stained with hematoxylin. On the other hand, SM and MM, which both use stain vectors to map an input image to a target space, don't show a blue-ish tint and provide a much more robust transformation that is true to the semantic information of the parent image (e.g.: treating nuclei and background regions distinctly).

The bottom row looks at the class probability for nuclear regions as predicted by models trained on datasets that were each stain normalized differently. Clearly all four normalized sets perform far better than the unnormalized dataset where almost no nuclei were detected due to the drastic change in staining as compared to what the model had been trained on. HE does pick up most of the nuclei but also a lot of background noise due to its inability to differentiate clearly between different types of staining. RH is also more sensitive to noise but does a better and clearer detection of nuclei as is visible in the clear boundaries. SM clearly performs the best at segmenting nuclei while also being most robust to false positives.

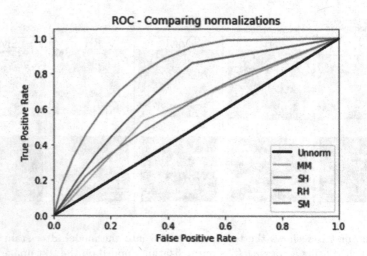

Fig. 2. ROC curve for models trained using different stain normalization schemes

4.2   Quantitative Assessment

To perform a more rigorous quantitative assessment, we looked at metrics calculated over a randomly selected set of 15 test images, using an output threshold of 0.5 for binarization (see Table 1). Simply calculating classification accuracy would be insufficient for this sort of segmentation problem. For instance, even if a classifier were to only classify pixels as non-nuclear regions it would still be around 85–90% accurate because the vast majority of pixels don't lie within nuclei.

Given the set of all nuclear pixels in the set, recall tells us what fraction of those were picked up by the model. Clearly SM does a great job in this area. SH and RH also do well but when we look at their precision values they are not as high as those for SM. Precision measures how many of the positives that you picked up were actually relevant. This indicates the tendency of SH and

RH to pick up more false positives than SM. This trade-off between true and false positives is best captured by the ROC curve (Fig. 2). Here, we see that the unnormalized case doesn't add any value at all while all the normalization scenarios show improved prediction accuracy. SM is the clear winner showing an excellent range of operation at a TPR of >80/90% while only allowing an FPR of 50%. This is very impressive considering how the model was trained on a staining visually very different from the one in the test data. This difference is quantitatively captured by the AUC. Finally, the F-score is another attempt to capture segmentation accuracy without getting bogged down by all the true negatives. It calculates the intersection of pixels that have been classified as nuclei in both the prediction and the ground truth and it divides that over the union of all pixels classified as nuclei by either set. Again, SM is seen to be the best at improving accuracy of the algorithm.

Table 1. Quantitative comparison of model performance under different forms of stain normalization

Normalization	Precision	Recall	F-score	AUC	Epochs
None	0.006	0.00	0.00	0.50	25
Histogram equalization (SH)	0.025	0.52	0.05	0.61	25
Macenko (MM)	0.026	0.18	0.05	0.61	25
Reinhard (RH)	0.04	0.55	0.07	0.71	12.5
Spline mapping (SM)	0.05	0.70	0.09	0.83	6

5   Discussion

Through this study, we have explored several stain normalization approaches that were all shown to reduce inter slide variability. The results (particularly AUC, F-score) clearly indicate that using a stain normalization approach increases the performance of the deep learning based segmentation algorithm. We found that SM performed better than all other approaches. We believe this is because it use a non-linear mapping function that is more accurate than the other approaches. It is able to delineate between different regions and map them appropriately to the target space.

We also noticed that the model seems to perform more poorly in case of normalizations that are biased more towards the eosin channel. In future, it may make sense to normalize the stain of the training dataset using two different approaches. This would push the model to become robust to these subtle changes and be less dependent on any one channel. Moreover, stain normalization could also be looked at as a regularization approach to enhance generalizability of deep learning based models in this space and prevent overfitting. On the other hand, we must remain conscious of the fact that staining color is a very valuable source of information in histological analyses and adopt a balanced approach towards stain normalization.

6   Conclusion

In this study, we looked at the impact of stain normalization as a means of improving the accuracy of segmentation algorithms across datasets. To the best of our knowledge, this is the first study that compares the chosen four stain normalization techniques through assessing their usability in the context of deep learning based segmentation models. There is scope for expanding upon this work with a deeper analysis of why certain normalization approaches or model architectures are better suited for this task.

References

1. Ghaznavi, F.: Digital imaging in pathology: whole-slide imaging and beyond. Annu. Rev. Pathol.: Mech. Dis. **8**, 331–359 (2013)
2. Irshad, H.: Methods for nuclei detection, segmentation, and classification in digital histopathology: a review' current status and future potential. IEEE Rev. Biomed. Eng. **7**, 97–114 (2014)
3. Janowczyk, A., Madabhushi, A.: Deep learning for digital pathology image analysis: a comprehensive tutorial with selected use cases. J. Pathol. Inform. (2016)
4. Basavanhally, A., Feldman, M., Shih, N.: Multi-field-of-view strategy for image-based outcome prediction of multi-parametric estrogen receptor-positive breast cancer histopathology: comparison to oncotype DX. J. Pathol. Inform. **2**, S1 (2011). https://doi.org/10.4103/2153-3539.92027
5. Khan, K.M., et al.: A nonlinear mapping approach to stain normalization in digital histopathology images using image-specific color deconvolution. IEEE Trans. Biomed. Eng. **61**(6), 1729–1738 (2014)
6. Long, J., Shelhamer, E., Darrell, T.: Fully convolutional networks for semantic segmentation. In: Proceedings of the IEEE Conference on Computer Vision and Pattern Recognition, pp. 3431–3440 (2015)
7. Reinhard, E., Ashikhmin, M., Gooch, B., Shirley, P.: Color transfer between images. IEEE Comput. Graph. Appl. **21**(5), 34–41 (2001)
8. Macenko, M., et al.: A method for normalizing histology slides for quantitative analysis. In: ISBI, vol. 9, pp. 1107–1110, June 2009
9. https://www.math.uci.edu/icamp/courses/math77c/demos/hist_eq.pdf. Accessed 24 Jul 2018
10. http://www2.warwick.ac.uk/fac/sci/dcs/research/tia/software/sntoolbox. Accessed 24 Jul 2018
11. http://bioimage.ucsb.edu/research/bio-segmentation . Accessed 24 Jul 2018
12. Rizzi, A., Gatta, C., Marini, D.: From retinex to automatic color equalization: issues in developing a new algorithm for unsupervised color equalization. J. Electron. Imaging **13**(1), 75–85 (2004)
13. Ronneberger, O., Fischer, P., Brox, T.: U-Net: convolutional networks for biomedical image segmentation. In: Navab, N., Hornegger, J., Wells, W.M., Frangi, A.F. (eds.) MICCAI 2015. LNCS, vol. 9351, pp. 234–241. Springer, Cham (2015). https://doi.org/10.1007/978-3-319-24574-4_28
14. He, K., Gkioxari, G., Dollár, P., Girshick, R.: Mask R-CNN. In: 2017 IEEE International Conference on Computer Vision (ICCV), pp. 2980–2988. IEEE, October 2017

Multi-resolution Networks for Semantic Segmentation in Whole Slide Images

Feng Gu(✉), Nikolay Burlutskiy, Mats Andersson, and Lena Kajland Wilén

ContextVision AB, Linköping, Sweden
feng.gu@contextvision.se

Abstract. Digital pathology provides an excellent opportunity for applying fully convolutional networks (FCNs) to tasks, such as semantic segmentation of whole slide images (WSIs). However, standard FCNs face challenges with respect to multi-resolution, inherited from the pyramid arrangement of WSIs. As a result, networks specifically designed to learn and aggregate information at different levels are desired. In this paper, we propose two novel multi-resolution networks based on the popular 'U-Net' architecture, which are evaluated on a benchmark dataset for binary semantic segmentation in WSIs. The proposed methods outperform the U-Net, demonstrating superior learning and generalization capabilities.

Keywords: Deep learning · Digital pathology · Whole slide images

1 Introduction

The working pattern of an experienced pathologist is frequently characterized by a repeated zooming in and zooming out motion while moving over the tissue to be graded. This behavior is similar if a microscope is used or if a whole slide image (WSI) is observed on a screen. The human visual system needs these multiple perspectives to be able to grade the slide. Only in rare cases can a local neighborhood of a slide be safely graded, independent from the surroundings. The heart of the matter is that the slide only represents a 2D cut out of a complex 3D structure. A glandular tissue in 3D resembles the structure of a cauliflower. Depending on the position of the 2D cut, the size and shape of the glands on the slide may vary significantly. To assess if a deviation in size or shape is due to the position of the cut or to a lesion, a multi-resolution view is crucial. The structure of the surrounding glands must be accounted for when the current gland is being investigated at a higher resolution.

Digital pathology opens the possibility to support pathologists by using fully convolutional networks (FCNs) [11] for semantic segmentation of WSIs. Standard FCNs do however face the same challenge with respect to multi-resolution. It can be argued that a network like U-Net [13] can to some extent handle multi-resolution, since such a structure is inherent in the network. However, patches

© Springer Nature Switzerland AG 2018
D. Stoyanov et al. (Eds.): COMPAY 2018/OMIA 2018, LNCS 11039, pp. 11–18, 2018.
https://doi.org/10.1007/978-3-030-00949-6_2

(image regions of a WSI at a given resolution) with the finest details should probably be extracted at the highest resolution, to utilize such a capability. It may also require the patch size to be considerably larger, making it infeasible for the VRAM of a modern GPU to fully explore the multi-resolution. As a result, approaches capable of learning from data and aggregating information efficiently and effectively at multiple resolutions are desired.

In this paper, two novel multi-resolution networks are proposed to learn from input patches extracted at multiple levels. These patches share the same centroid and shape (size in pixels), but with an octave based increase of the pixel size, micrometers per pixel (mpp). Only the central high resolution patch is segmented at the output. The proposed methods are evaluated and compared with the standard U-Net on a benchmark dataset of WSIs.

2   Related Work

Semantic segmentation problems were initially solved by traditional machine learning approaches, where hand crafted features were engineered [15]. Researchers applied methods, such as predictive sparse decomposition and spatial pyramid matching, to extract features of histopathological tissues [4]. However, deep learning approaches based on FCNs [11] showed significantly higher performance and eventually have substituted them [7]. To overcome the so called 'checkerboard artifacts' of transposed convolutions, several approaches have been proposed, e.g. SegNet [1], DeepLab-CRF [5], and upscaling using dilated convolutions [18]. To increase localization of learned features, high resolution features from the downsampling path can be aggregated with the upsampled output. Such an operation is known as 'skip connections', which enables a successive convolution layer to learn and assemble a more precise output based on the aggregated information. Several researchers successfully demonstrated that architectures with skip connections can result in better performance. Such networks include U-Net, densely connected convolutional networks [9], and highway networks with skip connections [16]. On overall, U-Net has proved to be one of the most popular networks for biomedical segmentation tasks [3].

One limitation of standard FCNs is the fact that the networks are composed of convolution layers with a set of filters that have the same receptive field size. The receptive field size corresponds to the context that a network can learn from, and eventually influences the network performance. Grais *et al.* [8] proposed a multi-resolution FCN with different receptive field sizes for each layer, for the audio source separation problem. Such a design allowed to extract features of the same input at multiple perspectives (determined by the receptive field sizes), and thus to capture global and local details from the input. Fu *et al.* introduced a multi-scale M-Net [6] to tackle the problem of joint optic disc and cup segmentation, where the same image contents of different input shapes or scales are passed through the network. However, both methods were designed to handle the same input audio or image content, while learning features from multiple perspectives by either employing encoders with varied respective fields or taking inputs with multiple scales. Roullier *et al.* [14] proposed multi-resolution graph-based analysis of whole slide images for mitotic cell segmentation. The approach is based on domain specific knowledge, which cannot be easily transferred to another problem domain. Recently, an approach of using multi-resolution information in FCN was described in [12]. However, the fusion of multi-resolution inputs was performed before encoders, instead of within the network. In addition, the approach could only be applied to a subset of small regions of interests, rather than WSIs.

Networks that incorporate inputs extracted from different resolutions with respect to the same corresponding tissue area in WSIs are desired, to tackle the challenge of multi-resolution effectively. In addition, the networks should be scalable in terms of resolutions and VRAM efficient for training and prediction. These motivated us to develop the multi-resolution networks in this work.

3   Algorithmic Formulation

A common practice of handling a WSI with deep learning is to divide it into multiple equally sized patches [17]. Here the deep learning task is formulated as a binary semantic segmentation problem of patches, where each patch is considered an image example. At prediction, a trained model first predicts each patch individually, and then stitches predictions of all the patches, to form the prediction of the entire slide (a probabilistic map indicating the probability of each pixel belonging to the class of interest).

3.1   Learning and Inference

Let $(\mathbf{x}, y) \in \mathbf{X} \times \mathbf{Y}$ be a patch or an example of a given dataset, where $\mathbf{X} \subseteq \mathbb{R}^{N \times D \times 3}$ and $\mathbf{Y} \subseteq \mathbb{N}^{N \times D}$. The value of N is equal to the number of examples (or patches), and D is the dimensionality of the feature vector (i.e. the product of height and width of the patch $h \times w$). So $\mathbf{x}$ can be a RGB image extracted from a slide, and y can be a binary ground truth mask associated with the RGB image. We can formulate a deep network as a function $f(\mathbf{x}; \mathbf{W})$, where $\mathbf{W}$ is a collection of weights of all the parametrized layers. The learning task is a process of searching for the optimal set of parameters $\hat{\mathbf{W}}$ that minimizes a loss function $\mathcal{L}(y, f(\mathbf{x}; \mathbf{W}))$. The output of the function f can be transformed to a probabilistic value in the range of $[0, 1]$ via a sigmoid function. A commonly used loss function for binary semantic segmentation is the binary cross entropy loss.

To counter over-fitting and improve the generalization capability of a trained model, a regularization term $\mathcal{R}(\cdot)$ is often added to the objective function as

$$\mathcal{E}_{\mathbf{W}} = \sum_{i=1}^{N} \mathcal{L}\left(y, f(\mathbf{x}_i; \mathbf{W})\right) + \lambda \mathcal{R}(\mathbf{W}) \tag{1}$$

where the scalar λ determines the weighting between two terms. One popular regularization function is ℓ_2-regularization, such that $\mathcal{R}(\mathbf{W}) = \|\mathbf{W}\|_2^2$. Search of

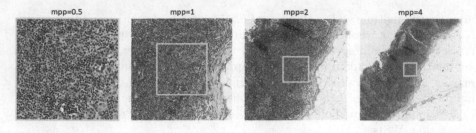

Fig. 1. From left to right are the patches with the same central coordinates, where mpp = 0.5 is equivalent to 20x and so forth. The increase of mpp values corresponds to the zooming out action to enlarge the field of view, and the yellow squares represent the effective tissue ares at different magnifications. (Color figure online)

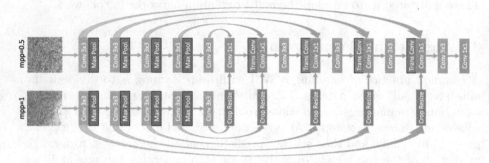

Fig. 2. An illustration of the proposed MRN methods when two resolutions are involved. The dark blue boxes represent stacks of two 3 × 3 convolution layers with ReLU activations; the red boxes are 2 × 2 max pooling layers; the light blue boxes are 1 × 1 convolution layers with identity activations; the green boxes are 2 × 2 transposed convolution layers with stride = 2 and ReLU activations. (Color figure online)

the optimal set of parameters $\hat{\mathbf{W}} = \arg\min_{\mathbf{W}} \mathcal{E}_{\mathbf{W}}$ for the objective function is known as optimization in machine learning. Popular optimizers include stochastic gradient descent (SGD), adaptive gradient (AdaGrad), and root mean square propagation (RMSProp). Recently, adaptive moment estimation (Adam) [10] has become a particularly popular method for optimizing deep networks.

3.2 Multi-resolution Networks

Here we propose two multi-resolution networks (MRN) that are based on the architecture of U-Net [13]. A standard U-Net can be seen as two parts, an 'encoder' for downsampling and a 'decoder' for upsampling. The downsampled feature maps are concatenated with the corresponding layers of the decoder in the upsampling pathway. The proposed MRN employ multiple encoders corresponding to different resolutions that are structurally identical for downsampling, and one single decoder for upsampling.

The input shapes of all resolutions are identical, and the examples share the common central coordinates and effectively cover tissue areas in a pyramid manner, as in Fig. 1. Let $(\mathbf{x}, y)$ be an example, where $\mathbf{x}_j = [\mathbf{x}_1, \mathbf{x}_2, \ldots, \mathbf{x}_J]$ and $y = y_1$, where the resolutions are in a descending order. The shapes of $\mathbf{x}$ and y are there $h \times w \times 3 \times J$ and $h \times w \times 1$ respectively. The rationale behind such an arrangement is that the pixel correspondence is more cumbersome compared to a standard U-Net. A key issue is to enable a sufficient receptive field for the low resolution branches of the network to successfully convey the information from the peripheral regions into the central parts.

To preserve the information relevant to the area of interest (i.e. the central part) at a lower resolution, we center crop the output feature maps of each encoder unit and then resize them back to the original resolutions via upscaling. We can defined a nested function $u \circ v$ such that

$$u : \mathbb{R}^{w \times h \times c} \to \mathbb{R}^{\lfloor \frac{w}{\gamma} \rfloor \times \lfloor \frac{h}{\gamma} \rfloor \times c} \quad \text{and} \quad v : \mathbb{R}^{\lfloor \frac{w}{\gamma} \rfloor \times \lfloor \frac{h}{\gamma} \rfloor \times c} \to \mathbb{R}^{w \times h \times c} \tag{2}$$

where the cropping factor is $\gamma = 2^N$, since resolutions at different levels of a WSI are usually downsampled by a factor 2 in both height and width. On one hand, the function u center crops a real-valued tensor of shape $h \times w \times c$ (height, width, and channels) to the shape of $\lfloor \frac{w}{\gamma} \rfloor \times \lfloor \frac{h}{\gamma} \rfloor \times c$. On the other hand, the function v upscales the output of u to the original shape. For upscaling, we present two options, namely 'MRN-bilinear' via bilinear interpolation and 'MRN-transposed' through transposed convolution.

The outputs of $u \circ v$ are concatenated with the convoluted feature maps of the corresponding layers in the encoder of the highest resolution. The concatenated feature maps are then passed though a 1×1 convolution layer with an identity activation, before being combined with layers in the decoder. The 1×1 convolution acts as a weighted sum to aggregate feature maps from all the resolutions, while keeping the number feature maps in the decoder constant despite the number of resolutions involved. Figure 2 illustrates an example of such networks when two resolutions are involved, and this is easily expandable with more resolutions.

4   Experimental Conditions

4.1   Implementation Details

We implemented all the networks in TensorFlow, with 'SAME' padding. Batch normalization and ℓ_2-regularization with $\lambda = 0.005$ were applied to all the convolution and transposed convolution layers, to improve convergence rates and counter over-fitting. We employ the Adam optimizer with default parameters ($\eta = 0.001$, $\beta_1 = 0.9$, $\beta_2 = 0.999$, and $\epsilon = 10^{-8}$). The input shape is 512 in height and width, and the collection of resolutions are mpp $\in \{0.5, 1, 2, 4\}$, where the U-Net deals with one of the resolutions at each time and the MRN methods handles all the resolutions simultaneously. The batch size is equal to 16, which is limited by the VRAM of an NVIDIA Titan XP. The number maximum epochs is set to 500 for the training to be terminated.

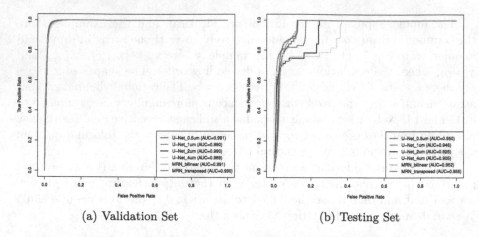

(a) Validation Set   (b) Testing Set

Fig. 3. Comparisons of the standard U-Net and MRNs on CAMELYON16, when different thresholds are applied to the predictions.

4.2 Segmentation Experiments

CAMELYON datasets [2] are the only few publicly available WSI datasets with pixel-level annotations. In particular, the CAMELYON16 dataset has both the training and testing sets available, and is one of the most popular benchmark datasets in the field of digital pathology. As a result, it was chosen for evaluating the proposed methods against the standard U-Net, for binary semantic segmentation of 'normal' and 'tumor' classes in WSIs[1]. There are 269 slides in the training set, 159 of which are normal and the remaining 110 are tumor. As pointed out in [17], 18 tumor slides have non-exhaustive annotations and thus are excluded from the experiments. The training set is then randomly divided into 'training' (80%) and 'validation' (20%), where the validation set is used to select the best model with respect to lowest validation losses. The testing set has 130 slides, 80 of which are normal and the rest are tumor. We excluded 2 tumor slides, due to non-exhaustive annotations.

5 Results and Analysis

In this section, we compare the methods from both quantitative and qualitative perspectives. Quantitatively, we intend to evaluate their learning abilities on the training data, and more importantly the generalization capabilities on unseen data in the testing set. Therefore, we evaluated on both the validation set and the testing set, the ROC curves are displayed in Fig. 3. On the validation set, the results are rather identical, and MRN-transposed is marginally better. This indicates that all methods are able to learn from the given data.

---

[1] Note we have no intention to tackle the CAMELYON16 tasks of slide-based or lesion-based classifications, or the CAMELYON17 task of determining the pN-stage for a patient. Those tasks are beyond the scope of this work.

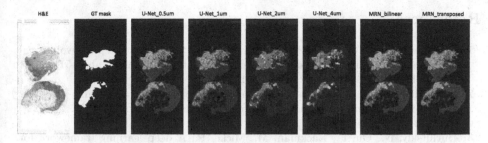

Fig. 4. A qualitative comparison of all methods on 'test_090' slide.

Results of the testing set vary more significantly. First of all, the standard U-Net performance decreases as the mpp value increases, while the proposed networks both outperform the U-Net variants. The reason for the better performance of MRN-transposed can be that it has a higher capacity than MRN-bilinear, since transposed convolutions are parameterized and bilinear interpolations are not.

To understand the results qualitatively, we plot the original H&E slide, the annotation mask, and the predictions of trained models, as shown in Fig. 4. As the mpp value goes up, predictions of the U-Net variants become increasingly sparser and less confident (implied by darker colors). However, the predictions of both MRN-bilinear and MRN-transposed contain sufficient amount of details and are relatively more confident. This explains why they produce the best performance when evaluating at the pixel level.

6   Conclusions and Future Work

In this paper, we proposed two novel multiple resolution networks, to learn from and infer on WSIs at different resolutions. The proposed methods produce state-of-the-art results and outperform the standard U-Net on a benchmark dataset, for binary semantic segmentation. These results demonstrate their superior learning and generalization capabilities. In addition, the proposed methods are memory efficient, since constant input shapes of different resolutions make the increase in VRAM linear for training and prediction. Furthermore, we can now train one model for all resolutions of interest, instead of training one model for each.

As for the future work, we would like to apply the proposed methods to other more challenging problems, e.g. multi-class semantic segmentation. Other network architectures can also be transformed to be multi-resolution capable, following the same principles proposed in this work. In addition, we will experiment with other building blocks of semantic segmentation networks, to develop methods with higher capacities.

Acknowledgements. The authors would like to thank ContextVision AB, Sweden for supporting the research, and the organizers of CAMELYON challenges for making the datasets available to the community.

References

1. Badrinarayanan, V., Kendall, A., Cipolla, R.: SegNet: a deep convolutional encoder-decoder architecture for image segmentation. CoRR abs/1511.00561 (2015). http://arxiv.org/abs/1511.00561
2. Bejnordi, B.E., et al.: Diagnostic assessment of deep learning algorithms for detection of lymph node metastases in women with breast cancer. JAMA **318**(22), 2199–2210 (2017)
3. Burlutskiy, N., Gu, F., Backman, M., Micke, P.: A deep learning framework for automatic diagnosis in lung cancer. In: International Conference on Medical Imaging with Deep Learning (2018)
4. Chang, H., Nayak, N., Spellman, P.T., Parvin, B.: Characterization of tissue histopathology via predictive sparse decomposition and spatial pyramid matching. In: MICCAI, pp. 91–98 (2013)
5. Chen, L., Papandreou, G., Kokkinos, I., Murphy, K., Yuille, A.L.: Semantic image segmentation with deep convolutional nets and fully connected CRFs. CoRR abs/1412.7062 (2014). http://arxiv.org/abs/1412.7062
6. Fu, H., Cheng, J., Xu, Y., Wong, D.W.K., Liu, J., Cao, X.: Joint optic disc and cup segmentation based on multi-label deep network and polar transformation. IEEE Trans. Med. Imaging (2018)
7. Garcia-Garcia, A., Orts-Escolano, S., Oprea, S., Villena-Martinez, V., Rodríguez, J.G.: A review on deep learning techniques applied to semantic segmentation. CoRR abs/1704.06857 (2017). http://arxiv.org/abs/1704.06857
8. Grais, E.M., Wierstorf, H., Ward, D., Plumbley, M.D.: Multi-resolution fully convolutional neural networks for monaural audio source separation. CoRR abs/1710.11473 (2017). http://arxiv.org/abs/1710.11473
9. Huang, G., Liu, Z., Weinberger, K.Q.: Densely connected convolutional networks. CoRR abs/1608.06993 (2016). http://arxiv.org/abs/1608.06993
10. Kingma, D.P., Ba, J.L.: Adam: a method for stochastic optimization. In: International Conference on Learning Representations (2015)
11. Long, J., Shelhamer, E., Darrell, T.: Fully convolutional networks for semantic segmentation. In: Conference on Computer Vision and Pattern Recognition (2015)
12. Mehta, S., Mercan, E., Bartlett, J., Weaver, D.L., Elmore, J.G., Shapiro, L.G.: Learning to segment breast biopsy whole slide images. CoRR abs/1709.02554 (2017). http://arxiv.org/abs/1709.02554
13. Ronneberger, O., Fischer, P., Brox, T.: U-Net: convolutional networks for biomedical image segmentation. In: MCCAI (2015)
14. Roullier, V., Lezoray, O., Ta, V.T., Elmoataz, A.: Multi-resolution graph-based analysis of histopathological whole slide images: application to mitotic cell extraction and visualization. Comput. Med. Imaging Graph. **35**(7), 603–615 (2011)
15. Sood, A., Sharma, S.: Image segmentation and object recognition using machine learning. In: Unal, A., Nayak, M., Mishra, D.K., Singh, D., Joshi, A. (eds.) SmartCom 2016. CCIS, vol. 628, pp. 204–210. Springer, Singapore (2016). https://doi.org/10.1007/978-981-10-3433-6_25
16. Srivastava, R.K., Greff, K., Schmidhuber, J.: Highway networks. CoRR abs/1505.00387 (2015). http://arxiv.org/abs/1505.00387
17. Wang, D., Khosla, A., Gargeya, R., Irshad, H., Beck, A.H.: Deep learning for identifying metastatic breast cancer (2016). arXiv:1606.05718v1
18. Yu, F., Koltun, V., Funkhouser, T.: Dilated residual networks. In: Conference on Computer Vision and Pattern Recognition (CVPR) (2017)

Improving High Resolution Histology Image Classification with Deep Spatial Fusion Network

Yongxiang Huang[(⊠)] and Albert Chi-Shing Chung

Lo Kwee-Seong Medical Image Analysis Laboratory, Department of Computer Science and Engineering, The Hong Kong University of Science and Technology, Clear Water Bay, Hong Kong
{yhuangch,achung}@cse.ust.hk

Abstract. Histology imaging is an essential diagnosis method to finalize the grade and stage of cancer of different tissues, especially for breast cancer diagnosis. Specialists often disagree on the final diagnosis on biopsy tissue due to the complex morphological variety. Although convolutional neural networks (CNN) have advantages in extracting discriminative features in image classification, directly training a CNN on high resolution histology images is computationally infeasible currently. Besides, inconsistent discriminative features often distribute over the whole histology image, which incurs challenges in patch-based CNN classification method. In this paper, we propose a novel architecture for automatic classification of high resolution histology images. First, an adapted residual network is employed to explore hierarchical features without attenuation. Second, we develop a robust deep fusion network to utilize the spatial relationship between patches and learn to correct the prediction bias generated from inconsistent discriminative feature distribution. The proposed method is evaluated using 10-fold cross-validation on 400 high resolution breast histology images with balanced labels and reports 95% accuracy on 4-class classification and 98.5% accuracy, 99.6% AUC on 2-class classification (carcinoma and non-carcinoma), which substantially outperforms previous methods and close to pathologist performance.

Keywords: Histology imaging · Computer-aided diagnosis
Image classification · Deep learning

1 Introduction

Histology imaging on tissue slice is a critical method for pathology analysis, indicating further targeted therapies. Pathologists perform histological analysis and morphological assessment on microscopic structure and tissue organization to diagnose and grade the cancer type [13]. On histology images, discriminative features of a certain cancer type can be observed at nuclei-level, ductal-level, cellular-level and overall tissue organization [6]. The diagnosis of biopsy tissue is

© Springer Nature Switzerland AG 2018
D. Stoyanov et al. (Eds.): COMPAY 2018/OMIA 2018, LNCS 11039, pp. 19–26, 2018.
https://doi.org/10.1007/978-3-030-00949-6_3

tedious and non-trivial. In-observer disagreement often exists between pathologists due to the complex diversity and distribution of discriminative features [4].
Therefore, developing an accurate computer-aided diagnosis (CAD) system to
automatically classify histology images can greatly improve diagnosis efficiency
and provides valuable diagnosis reference to pathologists with dissensions [15].

In recent years, deep convolutional networks have achieved state-of-the-art
performance on a large number of visual classification tasks [7]. The success
of deep CNN relies on the large available training set that is well labeled and
is limited to the size of input image considering the high computational cost.
However, for classification problems in biomedical images, the input is often high
resolution images, such as breast cancer histology images.

The challenges of developing a CNN for high resolution histology images classification include: (1) the distribution of discriminative features over a histology
image is complex and one patch on a histology image does not necessarily contain
discriminative features consistent with the image-wise label; (2) In most cases,
only the image-wise ground truth label is given due to the high cost of annotation
on high resolution images, which complicates the problem; (3) Dramatic downsampling leads to the loss of discriminative details at nuclei-level and ductal-level,
thus training a CNN on whole histology images is usually inappropriate. [3] proposed to divide a high resolution breast histology image into patches and train a
VGG-like patch-wise CNN, from which the image-wise label can be inferred by
voting on patch-wise predictions. [12] utilized deep CNNs for feature extraction
and gradient boosted trees for classification, which achieved better performance.
[3] developed a patch-based CNN and utilized a linear regression fusion model
to predict image-wise labels.

Compared to previous methods, in this paper, we propose to use a deep spatial fusion network to model the complex distribution of discriminative features
over patches. We also investigate a more effective method to extract hierarchical
discriminative features on patches. The proposed method outperforms previous
state-of-the-art solutions and it is the first work that utilizes the spatial relationship between patches to improve image-wise prediction for breast cancer
histology image classification.

2   Methods

2.1   Architecture

The architecture of the proposed method is depicted in Fig. 1. The input to the
network is high resolution histology images. As discussed in Sect. 1, only image-
wise ground truth labels are available and discriminative features may distribute
sparsely on the whole image, which indicates that all patches are not necessarily
consistent with image-wise label. We propose a spatial fusion network to model
this fact and produce a robust image-wise prediction.

The architecture is composed of two principal components: (1) an adapted
deep residual network trained to discover hierarchical discriminative features and

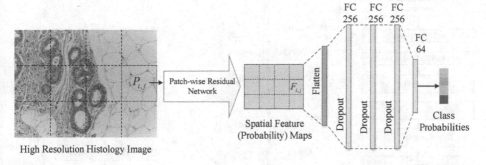

High Resolution Histology Image

Fig. 1. A schematic view of the proposed Spatially Fused Residual Network. The high resolution histology image is sampled into patches by a non-overlap sliding widow. F_{ij} represents deep CNN features for a patch P_{ij}, where i, j are the patch row and column index, respectively.

predict the probabilities of different cancer type for local image patches. Compared to VGG-like convolutional network [14], the skip connection structure of residual network reduces the vanishing gradient phenomenon in backpropagation and thus have a better performance on extracting critical visual feature with a deeper convolutional network. (2) a deep spatial fusion network, which is designed to utilize the spatial relationship between patches with the input of spatial feature maps. For the sake of simplicity and generalization, patch-wise probability vector is adopted as the base unit of the spatial feature maps as shown in Fig. 1. The fusion model learns to correct the bias of patch-wise predictions and yields robust image-wise prediction compared to typical fusion methods, which will be discussed in the paper.

2.2 Patch-Wise Residual Network

Residual neural networks (ResNets) [8] are adopted in our proposed architecture instead of plain feedforward deep convolutional neural network. Compared to plain CNN, residual networks mitigate the difficulty of training deep network using shortcut connections and residual learning [8]. Additionally, the identity shortcut connections enable flow of information across layers without attenuation caused by non-linear transformations. Therefore, hierarchical features from the low level to higher level are combined to make the final prediction, which is very useful in histology image classification considering discriminative features are distributed in the image from the cellular level to tissue level.

We have adjusted the original residual network developed for ILSVRC2015 classification task so that it works more appropriately on histology images classification. The modified residual network architecture is shown in Fig. 2. (1) The input layer is adjusted to receives normalized image patches of size 512×512, sampled from whole histology images. (2) The depth of the network is chosen to be 18 layers with 4 block units for fully exploring regional patterns in different scale. The receptive fields of the four block groups are of size 19×19 to

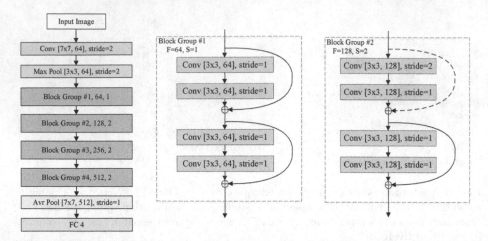

Fig. 2. The structure of the adapted residual network. Parameters in green boxes indicate "[kernel size, channels], stride". Each convolution layer follows by batch normalization and ReLU for regularization and non-linearity. Each block group consists of two building blocks [8]. The parameters in blue boxes indicate "#index, the number of output feature map in each convolutional layer of the same group, stride of the first convolutional layer". In block group #1, only identity shortcut connections are used because the input and the output are of the same dimensions. In block group #2, both projection shortcut (dotted line) and identity shortcut are used. The projection shortcut matches the dimensions of the input and the output by 1×1 convolution. Block group #3 and #4 share structure with block group #2. (Color figure online)

$43 \times 43, 51 \times 51$ to $99 \times 99, 115 \times 115$ to 211×211, and 243×243 to 435×435 pixel respectively, which effectively respond to region patterns in nuclei, nuclei organization, structure and tissue organization [3].

2.3   Deep Spatial Fusion Network

The aim of fusion model is to predict the image-wise label $\hat{y}$ among K classes $\mathbf{C} = \{C_1, C_2, .., C_K\}$, given all patch-wise probability (feature) maps $\mathbf{F}$ output by the proposed residual network. The image-wise label prediction is defined by MAP estimate [5] as follows,

$$\hat{y} = \operatorname{argmax}_{y \in \mathbf{C}} P(y|\mathbf{F}). \tag{1}$$

Suppose the whole high resolution image is divided into $M \times N$ patches. We first organize all patch-wise probability maps in spatial order, such that:

$$\mathbf{F} = \begin{pmatrix} F_{11} & F_{12} & \ldots & F_{1N} \\ F_{21} & F_{22} & \ldots & F_{2N} \\ \vdots & \vdots & \ddots & \vdots \\ F_{M1} & F_{M2} & \ddots & F_{MN} \end{pmatrix}. \tag{2}$$

A deep neural network (DNN) is applied to utilize the spatial relationship between patches. As shown in Fig. 1, the proposed fusion model consists of 4 fully-connected layers, each of which follows by ReLU activation function [11]. The deep multi-layer perceptron(MLP) learns to transform the spatial distribution of local probability maps to a global class probabilities vector during image-wise training. To increase the robustness of the model and avoid overfitting, we insert one dropout layer before each hidden layer. Notice that dropout layer is also inserted between the flatten probability maps and the first hidden layer. Dropping out half of the probability maps, the models tend to yield an image-wise prediction with half information of patches through minimizing the cross-entropy loss in training.

3   Experiments

3.1   Dataset and Preprocessing

We validate the proposed method on two public histological breast cancer images dataset, the Bio-imaging Challenge 2015 Breast Histology dataset (BIC) [1], and the BACH 2018 dataset [2]. Both datasets consist of Hematoxylin and Eosin (H&E) stained microscope histology images on breast tissue biopsy. The images are annotated by two pathologists and classified into 4 classes: (1) normal tissue, (2) benign lesion, (3) in situ carcinoma, (4) invasive carcinoma, according to the predominant cancer type in each image. Among the 4 classes, in situ carcinoma and invasive carcinoma fall to malignant carcinoma. As some works focus on analysis for malignant-benign classification, we also evaluate the performance of the proposed model on 2-class classification in our experiments.

The BIC dataset consists of 286 high resolution images of size 2048 × 1536 pixels, split into 249 for training and 36 for testing. The BACH dataset consists of 400 images of the same size, split into 360 for training and 40 for testing. For both dataset, the 4 class labels are evenly distributed, hence it is fair to use accuracy as the evaluation metric. To avoid overfitting due to the small training dataset, we perform strong data augmentation as described in the next subsection. Before augmentation, to reduce the variance incurred by H&E staining, the images are normalized using the method proposed in [10].

3.2   Network Training

We first extract 512 × 512 pixel patches with overlapping from the high resolution images, as the input for the patch-based deep model. As patch-wise labels are not given in the training dataset, we initially assume the patch labels are consistent with the image-wise ground truth. It may incur bias during patch-based training and reduce the patch-wise classification accuracy. However, the bias will be alleviated during image-based training in the second stage under the supervised learning of image-wise labels. Due to the limited number of training samples, to prevent overfitting, we perform three kinds of image augmentation in each iteration: (1) random rotation; (2) horizontal flipping; (3) random enhancement of contrast and brightness. Thus, we generate 201,600 patches from the BACH dataset and 140,000 patches from the BIC dataset respectively. The residual network is trained on 32-sized mini-batches to minimize the cross-entropy cost function using Adam Optimization [9] with learning rate 10^{-3} for 50 epoch. After training, the patch-wise network encodes a 512×512 patch to 10×10 feature maps and 4-class probabilities.

To train the spatial fusion network, we perform similar data augmentation and generate 5,760 high resolution images from the BACH training dataset and 3,984 from the BIC training dataset. After augmentation, each high resolution image is divide into 12 non-overlapping patches of size 512×512 pixels. Each patch is fed into the residual network separately and output 512 feature maps of size 10×10 and a class probability vector of size 1×4. Probabilistic vectors of patches in the same image are then combined into a probabilistic map following their spatial order, which becomes the input for the spatial fusion network. The generated probability map can be seen as a high-level feature map that encodes all the patch-wise discriminative features and the image-wise spatial context-aware features. Supervised by the image-wise ground truth, the weights of the spatial fusion network are learned by using mini-batch gradient descent (batch size 32) with Adam optimization. During training, to minimize the cross-entropy loss, the spatial fusion model learns to encode the biased probabilistic map into a k-class vector approximating the image-wise ground truth ($k = 4$). By utilizing the spatial context-aware feature hidden in the probabilistic map, the image-based classification accuracy can be effectively improved.

3.3   Results

We first evaluated the performance of the patch-based residual network and then focused on the effectiveness of the proposed spatial fusion network by conducting multiple comparison experiments on the two datasets. In the first experiment, we reimplemented the published state-of-the-art framework [3] on the BIC dataset as the baseline method, which is based on a patch-based plain CNN architecture followed by multiple vote-based fusion strategies (named Baseline). For a fair comparison, we only replaced the plain CNN architecture with the proposed patch-wise residual network (named ResNet + Vote) and evaluated the two methods using the same dataset setting. The results are shown in Table 1 Residual + Spatial Network is our proposed method, which is evaluated on two dataset for a comprehensive comparison. All methods are evaluated with stratified ten-fold cross-validation on the same released dataset respectively.

On the BIC dataset, the proposed method reports an accuracy of 86.1% for 4-class classification, which outperforms the baseline method [3] by 8.31%. The proposed patch-wise residual network brings an improvement of 3.8% by replacing the plain CNN in the baseline method. The deep spatial fusion network further improves the ResNet + Vote method by 4.51% further, which demonstrates utilizing the spatial context-aware feature map is more effective than using multiple voting strategies for patch-wise result fusion. On the BACH dataset, the

Table 1. Quantitative comparisons on two public datasets.

Datasets	Methods	4-class ACC	2-class ACC	STD
BIC	Baseline	0.778	0.833	-
BIC	Residual + Vote	0.816	0.850	-
BIC	Residual + Spatial Network	**0.861**	0.889	-
BACH	CNNs + GDT	0.872	0.938	0.026
BACH	Residual + Spatial Network	**0.950**	0.985	0.022

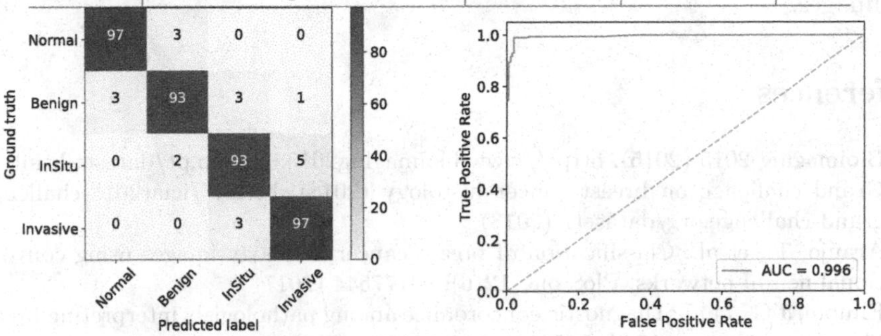

(a) Confusion matrix   (b) ROC

Fig. 3. (a) Confusion matrix without normalization, representing the 10-fold cross-validation result on 4-class classification of 400 high resolution histology images. (b) Performance of 2-class classification (non-carcinoma and carcinoma) in terms of AUC

proposed method reported 95.0% accuracy on 4-class classification and 98.5% accuracy, 99.6% AUC on 2-class classification (carcinoma and non-carcinoma). As a comparison, CNNs + GDT [12] is a published state-of-the-art method on the BACH dataset, which adopted several deep CNNs (ResNet50, InceptionV3, and VGG16) by model ensemble and used gradient boosted trees classifier to extract features at different scales. The proposed spatial fusion network outperforms [12] by 7.8% on 4-class classification without using any ensemble technique. The classification performance in terms of confusion matrix and receiver operating characteristic curve (ROC) are shown in Fig.3.

All experiments were implemented using PyTorch and performed on a NVIDIA 1080Ti GPU. The training of spatial residual network took approximately 30 min for each iteration. No ensemble technique is used in testing. The test time for classifying a single high resolution histology image took roughly 80 ms.

4   Conclusion

In this paper, we propose a deep spatial fusion network that models the complex construction of discriminative features over patches and learns to correct the patch-wise prediction bias on high resolution histology image. Also, we propose an adapted patch-wise residual network that effectively extract hierarchical visual features from cellular-level to overall tissue organization. Unlike previous patch-based CNN methods, the proposed architecture explores the spatial relationship between patches. Experiment results show that a substantially better performance than previous work even without using the ensemble.

In future work, we plan to extend the current work by: (1) incorporating concise patch-wise feature maps on spatially organized probability maps, (2) employing the deep spatial fusion model to annotate malignant patches to assist diagnosis, and (3) transferring the proposed model to other high resolution medical images.

References

1. Bioimaging 2015 (2015). http://www.bioimaging2015.ineb.up.pt/dataset.html
2. Grand challenge on breast cancer histology (2018). https://iciar2018-challenge.grand-challenge.org/dataset/ (2018)
3. Araújo, T., et al.: Classification of breast cancer histology images using convolutional neural networks. PloS one **12**(6), e0177544 (2017)
4. Elmore, J.G., et al.: Diagnostic concordance among pathologists interpreting breast biopsy specimens. JAMA **313**(11), 1122–1132 (2015)
5. Greig, D.M., Porteous, B.T., Seheult, A.H.: Exact maximum a posteriori estimation for binary images. J. R. Stat. Soc. Ser. B (Methodol.) **51**, 271–279 (1989)
6. Gurcan, M.N., Boucheron, L.E., Can, A., Madabhushi, A., Rajpoot, N.M., Yener, B.: Histopathological image analysis: a review. IEEE Rev. Biomed. Eng. **2**, 147–171 (2009)
7. He, K., Zhang, X., Ren, S., Sun, J.: Delving deep into rectifiers: surpassing human-level performance on imagenet classification. In: Proceedings of the IEEE International Conference on Computer Vision, pp. 1026–1034 (2015)
8. He, K., Zhang, X., Ren, S., Sun, J.: Deep residual learning for image recognition. In: Proceedings of the IEEE Conference on Computer Vision and Pattern Recognition, pp. 770–778 (2016)
9. Kingma, D.P., Ba, J.: Adam: a method for stochastic optimization. arXiv preprint arXiv:1412.6980 (2014)
10. Macenko, M., et al.: A method for normalizing histology slides for quantitative analysis. In: IEEE International Symposium on Biomedical Imaging: From Nano to Macro, ISBI 2009, pp. 1107–1110. IEEE (2009)
11. Nair, V., Hinton, G.E.: Rectified linear units improve restricted Boltzmann machines. In: Proceedings of the 27th International Conference on Machine Learning (ICML 2010), pp. 807–814 (2010)
12. Rakhlin, A., Shvets, A., Iglovikov, V., Kalinin, A.A.: Deep convolutional neural networks for breast cancer histology image analysis. arXiv preprint arXiv:1802.00752 (2018)
13. Rosen, P.P.: Rosen's Breast Pathology. Lippincott Williams & Wilkins, Philadelphia (2001)
14. Simonyan, K., Zisserman, A.: Very deep convolutional networks for large-scale image recognition. arXiv preprint arXiv:1409.1556 (2014)
15. Yamada, S., Komatsu, K., Ema, T.: Computer-aided diagnosis system for medical use, October 1993. US Patent 5,235,510

Construction of a Generative Model of H&E Stained Pathology Images of Pancreas Tumors Conditioned by a Voxel Value of MRI Image

Tomoshige Shimomura[1](✉), Kugler Mauricio[1], Tatsuya Yokota[1],
Chika Iwamoto[2], Kenoki Ohuchida[2], Makoto Hashizume[2],
and Hidekata Hontani[1]

[1] Nagoya Institute of Technology, Gokiso-cho, Showa-ku,
Nagoya, Aichi 466-8555, Japan
simomura@iu.nitech.ac.jp, {kugler.mauricio,t.yokota,hontani}@nitech.ac.jp
[2] Kyushu University, 744 Motooka Nishi-ku, Fukuoka 819-0395, Japan

Abstract. In this paper, we propose a method for constructing a multi-scale model of pancreas tumor of a KrasLSL.G12D/+; p53R172H/+; PdxCretg/+ (KPC) mouse that is a genetically engineered mouse model of pancreas tumor. The model represents the correlation between the value at each voxel in the MRI image of the tumor and the pathology image patches that are observed at each portion corresponds to the location of the voxel in the MRI image. The model is represented by a cascade of image generators trained by a Laplacian Pyramid of Generative Adversarial Network (LAPGAN). When some voxel in a pancreas tumor region in an MRI image is selected, the cascade of generators outputs patches of the pathology images that can be observed at the location corresponds to the selected voxel. We trained the generators by using an MRI image and a 3D pathology image, the latter was first reconstructed from a spatial series of the 2D pathology images and was then registered to the MRI image.

1   Introduction

Modelling the correlations between pathology images and MRI images has been investigated (e.g. [9]). The former images can be used for definitive diagnosis and the latter images can be obtained non-invasively. Models that represent the correlations of these images would improve the confidence of diagnosis and can be used for predicting histopathological status from the corresponding MRI images. In this study, we construct a non-parametric model that represents the correlation between the voxel value of an MRI image and corresponding histopathological images of a pancreas tumor of a KPC mouse. This model plays an important role in a system that acquires various kinds of information currently obtained from histopathological images with probability information from MR images.

© Springer Nature Switzerland AG 2018
D. Stoyanov et al. (Eds.): COMPAY 2018/OMIA 2018, LNCS 11039, pp. 27–34, 2018.
https://doi.org/10.1007/978-3-030-00949-6_4

For constructing such a model, we employ a conditional Laplacian Pyramid of Generative Adversarial Network (LAPGAN) [1].

A Generative Adversarial Network (GAN) [4] can construct a sufficiently representative latent model of target images, while simultaneously learning a generator and a discriminator: The generator can create sample images that are intended to come from the same distribution with the training data and the discriminator examines samples to determine whether they are real or fake. The latent model of target images can be represented by a manifold, from which the generator can create the fake images by sampling [10]. In many cases, a noise signal, z, is input to the generator and the noise signal, z, can be used as the local coordinate system on the manifold [2]. Generators learned by conditional GANs take not only noise signals but also other condition signals as inputs, where the condition signals restrict the output sample images. The condition signals also can be used as the local coordinate system on the manifold. When the input values of the condition signals are fixed, the generator creates fake images that correspond to a sub-manifold restricted by the conditions. The manifolds with the local coordinate systems can represent the correlations between the target images and the condition signals [10]. In this study, we construct a generator that takes noise signals and a voxel value of an MRI image as input and outputs corresponding pathology image patches sampled from the sub-manifold determined by the MRI voxel value.

The spatial resolution of an MRI image and that of a microscope pathology image are largely different. Each single voxel of an MRI image corresponds to a large image patch of the pathology image. We hence assume that each voxel value correlates with low-resolution features of the pathology images and employ a LAPGAN, which can generate a cascade of image generators, each of which creates a sample fake image that represents the difference between a high-resolution image and the corresponding low-resolution input image [1].

2   Method

2.1   Outline of the Proposed Method

Figure 1 shows the outline of the construction of a multi-scale pancreas tumor model from an MRI image and the corresponding pathology images of a KPC mouse. The MRI image of a whole body of a KPC mouse was captured just before the pancreas tumor was extracted. A 3D pathology image of the tumor was reconstructed from a spatial series of 2D microscope images of the tumor and the 3D pathology image was non-rigidly registered to the tumor region in the MRI image in order to obtain a set of training data for the LAPGAN, in which each datum is a pair of the voxel value of the MRI image and the corresponding image patch in the microscope image. Applying conditional LAPGAN to the training data, we construct a cascade of generators, which can generate a pathology patch image that corresponds to the input voxel value of the MRI image.

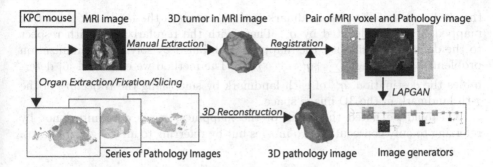

Fig. 1. Construction of a multiscale model of pancreas tumor using an MRI image and the corresponding pathology images

2.2 Images Used in the Experiments

The training images for the conditional LAPGAN was obtained as follows. An MRI image of the whole body of the KPC mouse was captured just before the organs including the whole part of the tumor was extracted. The spatial resolution of the MRI image was $0.1536\,\mathrm{mm} \times 0.1536\,\mathrm{mm} \times 0.5\,\mathrm{mm}$. The tumor was spherical and its diameter was about $2\,\mathrm{cm}$. Two MRI images of the extracted organs were obtained before and after the organs were formaline-fixed. Registering the two MRI images, we found that the organs around the tumor shrunk but the tumor itself did not deform in the formaline fixation.

The extracted organs were paraffin-embedded after the formaline-fixation. We first cut the paraffin block into five small blocks, of which the thickness was about $5\,\mathrm{mm}$, and one of the small blocks, which contained the center portion of the tumor, was sliced into a spatial series of about 800 thin sections. The thickness of the section was set to $4\,\mu\mathrm{m}$. The number of the sections obtained from the $5\,\mathrm{mm} = 5000\,\mu\mathrm{m}$ thick block was less than $1250 = 5000/4$ because of the loss generated by the slicing. We dyed the thin sections by the Hematoxylin and Eosin (H&E) stain. The microscopy images of these stained sections were then captured with the spatial resolution, $0.22\,\mu\mathrm{m} \times 0.22\,\mu\mathrm{m}$.

2.3 Reconstruction of a 3D Microscope Image

For the 3D reconstruction of the microscope image, we employed the non-rigid registration method proposed in [6]. Let the given 2D microscope images be denoted by $I_1(\boldsymbol{x}), I_2(\boldsymbol{x}), \ldots, I_M(\boldsymbol{x})$, where M is the total number of the 2D microscope images and $\boldsymbol{x} = [x, y]^\top$ denotes the 2D image coordinates. It is assumed that the given images are roughly aligned, for example, by a rigid registration method. Let the deformation mapping computed for $I_i(\boldsymbol{x})$ be denoted by ϕ_i $(i = 1, 2, \ldots, M)$. Then, the 3D image, $J(x, y, z)$, is reconstructed as $J(x, y, z = i) = I_i(\phi_i^{-1} \circ \boldsymbol{x})$. The mapping, ϕ_i, is computed from a set of landmarks located in $I_i(\cdot)$. Let $\boldsymbol{p}_i^j$ denote the 2D image coordinates of the j-th landmark $(j = 1, 2, \ldots, N)$ in the i-th image, $I_i(\cdot)$ and let the coordinates of the *destination* of the j-th landmark in $I_i(\cdot)$, to which the landmark should be mapped by ϕ_i, be denoted by $\boldsymbol{q}_i^j$. Then, with the regularization with respect to the deformation rigidity, the mapping is obtained by solving a minimization problem: $\phi_i = \arg\min_\phi \sum_j \|\boldsymbol{q}_i^j - \phi \circ \boldsymbol{p}_i^j\|^2$. The method we employed [6] determines the destination, $\boldsymbol{q}_i^j$, of each landmark by smoothing the *trajectory* of the j-th landmark in the 3D image space.

It should be noted that each of the mappings, ϕ_i, is determined not by referring to only consecutive two images but by referring to all the given images.

2.4   Registration Between MRI Image and Pathology Image

The tumor region in the MRI image and the tumor region in the reconstructed 3D microscope image were registered. A mutual information based non-rigid registration method [8] was employed. Assuming that the deformation of the tumor in the 3D microscope image mainly occurred when each thin section of the tumor specimen was placed on the slide glass, we restricted the movement of the control points to a plane where the z-coordinate is constant: Let the mapping to be computed be denoted by ψ and let $\boldsymbol{X}' = \psi \circ \boldsymbol{X}$, where the three-vectors, $\boldsymbol{X}' = [x', y', z']^\top$ and $\boldsymbol{X} = [x, y, z]^\top$, denote the 3D coordinates in the reconstructed microscope image, $J(\cdot)$. We computed ψ that maximizes the mutual information under a condition that $z' = \psi \circ z = z$ is satisfied. The mapping, ψ, then keeps the plane, $z = i$, corresponding to each microscope image, $I_i(\cdot)$, flat.

As the result of the registration described above, each voxel in the MRI image is corresponded to a specific region in the 3D microscope image as shown in Fig. 2.

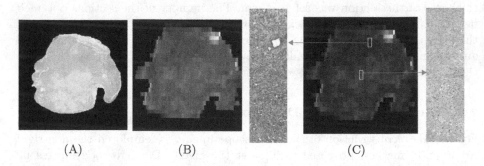

(A)   (B)   (C)

Fig. 2. Registration between the tumor region in the MRI image and the reconstructed 3D pathology image. (A): A pathology image of the tumor region. The pale pink portion includes necrosis region. (B): The corresponding MRI slice image. As shown in (C), the registration makes the correspondence between each voxel in the MRI image and a portion in the pathology image. (Color figure online)

2.5   Construction of Training Image Data for LAPGAN

Let the index of the voxel in the MRI image be denoted by $m \in \mathbb{N}$ and let the 3D region in the reconstructed 3D microscope image, $J(\psi^{-1} \circ X)$, that corresponds to the m-th voxel of the MRI image be denoted by Γ_m. Let the portion in the plane, $z = i$, included in Γ_m be denoted by Γ_{mi}. The region, Γ_{mi}, in the deformed i-th microscope image, $J(x', y', z' = i)$, corresponds to the m-th voxel in the MRI image. Let the value of the m-th voxel of the MRI image be denoted by v_m and let a set of 256×256 image patches included in Γ_{mi} be denoted by $\{\mathcal{I}_{mis}(x)|s = 1, 2, \ldots S_{mi}\}$, where S_{mi} denotes the number of patches sampled from Γ_{mi}. We sample the patches only from the H&E stained images and augmented the patches by applying random rotation to $\mathcal{I}_{mis}$ for increasing the number of training data. Then, we obtain a set of training data,

$$\mathcal{D} = \{(v_m, \mathcal{I}_{mis})|m, i, s \in \mathbb{N}\}, \tag{1}$$

in which each datum is a pair of the voxel value of the MRI image and the corresponding patch in H&E stained microscope images.

For avoiding the mode collapse, we quantize the MRI voxel value, v_m, by applying a K-means clustering method so that the variety of the data that have same condition is increased. Let the number of the clusters be denoted by K and let the clusters of voxel values be denoted by $\mathcal{C}_k$ $(k = 1, 2, \ldots, K)$. Then, we can obtain K sets, $\mathcal{D}_k$, of training image patches from $\mathcal{D}$ such that

$$\mathcal{D}_k = \{\mathcal{I}_{mis}|v_m \in \mathcal{C}_k; m, i, s \in \mathbb{N}\}, \tag{2}$$

where the image patches included in $\mathcal{D}_k$ correspond to the MRI voxel values that are included in $\mathcal{C}_k$. We use the index, k, as the condition for the LAPGAN.

The LAPGAN constructs a series of image generative models within a Laplacian pyramid framework. In the Laplacian pyramid framework, an image is represented in a coarse-to-fine fashion, that is, by a series of band-passed images plus a low-frequency residual. Let $D_{\downarrow}$ denote a downsampling operation with a factor, two (2): When the size of an input image, I, is $W \times W$, then $D_{\downarrow} \circ I$ is a new image of size $W/2 \times W/2$. Following the paper [1], we first built a Gaussian pyramid, $g(\mathcal{I}_{mis})$ from each image patch, $\mathcal{I}_{mis}$, such that $g(\mathcal{I}_{mis}) = [\mathcal{I}^0_{mis}, \mathcal{I}^1_{mis}, \ldots, \mathcal{I}^L_{mis}]$, where

$$\mathcal{I}^{l+1}_{mis} = D_{\downarrow} \circ G_{\sigma} \circ \mathcal{I}^l_{mis}, \tag{3}$$

G_{σ} denotes a Gaussian smoothing with the variance, σ^2, $\mathcal{I}^0_{mis} = \mathcal{I}_{mis}$, L denotes the number of the levels in the pyramid, and $l = 0, 1, \ldots, L - 1$ denotes the level. From the Gaussian pyramid, $g(\mathcal{I}_{mis})$, we then constructed the series of the band-passed images, $\mathcal{B}^l_{mis}$, which can be computed from $\mathcal{I}^l_{mis}$.

The LAPGAN we implemented constructs a series of $L+1$ image generators: The l-th generator $(l = 0, 1, \ldots, L - 1)$ generates the band-passed image, $\mathcal{B}^l_{mis}$, from $\mathcal{I}^{l+1}_{mis}$ and the last (the L-th) generator can generate the lowest-frequency image residual, $\mathcal{I}^L_{mis}$, from a Gaussian noise image. The cascade of the $L + 1$ generators, in the descending order of l, can generate the original image, $\mathcal{I}_{mis}$.

The l-th generator ($l = 0, 1, \ldots, L - 1$) is constructed from sets of the training data, $\{\mathcal{D}_k^l | k = 1, 2, \ldots, K\}$, where $\mathcal{D}_k^l = \{(\mathcal{I}_{mis}^{l+1}, \mathcal{B}_{mis}^l) | v_m \in \mathcal{C}_k, \; m, i, s \in \mathbb{N}\}$. The dataset, $\mathcal{D}_k^l$, consists of the pairs of the downsampled image of the low-frequency residual, $\mathcal{I}_{mis}^{l+1}$, and the band-passed image, $\mathcal{B}_{mis}^l$, both are obtained from the microscope images that correspond to the voxel value, $v_m \in \mathcal{C}_k$, of the MRI image. The last (the L-th) image generator is constructed from the sets of the images, $\{\bar{\mathcal{D}}_{mis}^L | k = 1, 2, \ldots, K\}$, where $\bar{\mathcal{D}}_k^L = \{\mathcal{I}_{mis}^L | v_m \in \mathcal{C}_k, \; m, i, s \in \mathbb{N}\}$. The last generator does not need the band-passed images for the training.

2.6   Conditional LAPGAN

We constructed $L+1$ image generators, $\mathsf{G}^0, \mathsf{G}^1, \ldots, \mathsf{G}^L$, by using the LAPGAN. The inputs of the last generator, G^L, are the Gaussian noise image, z, and the index, k, of the class of the corresponding MRI voxel value. The output is a residual image of the lowest-frequency, $\mathcal{I}^L$, which is indistinguishable from the training images, $\mathcal{I}_{mis}^L \in \bar{\mathcal{D}}_k^L$. One can generate variety of such the indistinguishable images by changing the input Gaussian noise images. The inputs of the other generators, G^l ($l = 0, 1, \ldots, L - 1$) are the lower-frequency image, $\mathcal{I}^{l+1}$, the Gaussian noise image, z, and the index, k. The output of G^l is a band-passed image, $\mathcal{B}^l$, that can generate the higher-frequency image, $\mathcal{I}^l$, from the input lower-frequency image, $\mathcal{I}^{l+1}$ as $\mathcal{I}^l = \mathcal{B}^l + O_{sm} \circ U_\uparrow \circ \mathcal{I}^{l+1}$, where the resultant image, $\mathcal{I}^l$, is indistinguishable from the training images, $\mathcal{I}_{mis}^l \in \mathcal{D}_k^l$. Changing the input Gaussian noise image, one can generate variety of images that are indistinguishable from the training data for the discriminator.

Let the discriminators corresponding to G^l be denoted by D^l. Given the dataset, $\{\bar{\mathcal{D}}_k^L | k = 1, 2, \ldots, K\}$, the LAPGAN constructs the L-th generator, G^L, and discriminator, D^L, by solving the problem, $\min_\mathsf{G} \max_\mathsf{D} F^L(\mathsf{G}, \mathsf{D})$, where (some superscripts, L, are abbreviated)

$$F^L(\mathsf{G}, \mathsf{D}) = \mathbb{E}_{\mathcal{I}, k \sim P_{\text{data}}(\mathcal{I}, k)} [\log \mathsf{D}(\mathcal{I}, k)]$$
$$+ \mathbb{E}_{z \sim P_{\text{noise}}(z), k \sim P_{\text{data}}(k)} [\log(1 - \mathsf{D}(\mathsf{G}(z, k), k))]. \tag{4}$$

Given the dataset, $\{\mathcal{D}_k^l | k = 1, 2, \ldots, K\}$, the LAPGAN constructs the l-th generator, G^l for $l = 0, 1, \ldots, L - 1$, by solving the problem, $\min_\mathsf{G} \max_\mathsf{D} F^l(\mathsf{G}, \mathsf{D})$, where (again, some superscripts are abbreviated):

$$F^l(\mathsf{G}, \mathsf{D}) = \mathbb{E}_{\mathcal{B}, k \sim P_{\text{data}}(\mathcal{B}, k)} [\log \mathsf{D}(\mathcal{B}, k)]$$
$$+ \mathbb{E}_{\mathcal{I}, k \sim P_{\text{data}}(\mathcal{I}, k), z \sim P_{\text{noise}}(z)} [\log(1 - \mathsf{D}(\mathsf{G}(\mathcal{I}, z, k), k))]. \tag{5}$$

We employed CNNs for the generators and discriminators. We initialized the networks by using the method proposed in [3] and employed Adam [7] and the batch normalization [5] for the stochastic optimization.

3   Results

Setting $K = 4$, we divided voxel values of the MRI image into four (4) clusters. Examples of the pathology images, $\mathcal{I}_{mis}$, included in $\mathcal{D}_1$ (the brightest portion) and $\mathcal{D}_4$ (the darkest portion) are shown in Fig. 3. The dataset included about one million image patches. The distribution of the patch patterns are different among the clusters, $\mathcal{D}_k$. Setting the number of the cascade $L = 3$, we constructed $L + 1$ generators. The cascade of the generators, $\mathsf{G}^L, \mathsf{G}^{L-1}, \ldots, \mathsf{G}^0$, can output a fake pathology image from a given Gaussian noise, z, and the condition, $k \in [1, K]$. Figure 4 shows examples of the randomly chosen generated images. Corresponding to the bright voxel values of the MRI image, the cascade of the generators sampled patches similar to those in the necrosis portion with a higher probability.

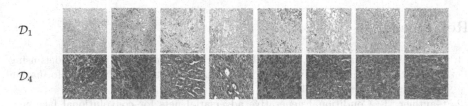

Fig. 3. Examples of the training pathology images, $\mathcal{I}_{mis}$, included in $\mathcal{D}_1$ and $\mathcal{D}_4$

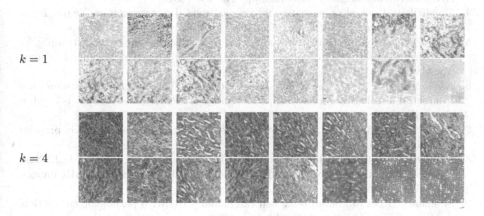

Fig. 4. Examples of the fake pathology images generated by the cascade of the generators for the condition $k = 1$ and $k = 4$.

4   Discussion and Conclusion

We constructed a multiscale model of pancreas tumor that can generate a H&E stained pathology image from a voxel value of the corresponding MRI image. For obtaining a set of training images, we first reconstructed a 3D pathology image of the pancreas tumor and then registered it to the tumor region in the MRI image. For the construction of the multiscale model, we employed the conditional LAPGAN. The resultant generators output pathology image patches that look like those in the necrosis region with a high probability when the brighter voxel value of the MRI image is input. We constructed the model from a partial region of only one pancreas tumor observed from one KPC mouse. The future works include to annotate histopathological/genetic information to each portion of the histopathology image and to construct a model that can derive the histopathological/genetic information with confidence from each voxel of a given MR image.

References

1. Denton, E.L., Chintala, S., Fergus, R., et al.: Deep generative image models using a Laplacian pyramid of adversarial networks. In: Advances in Neural Information Processing Systems, pp. 1486–1494 (2015)
2. Gauthier, J.: Conditional generative adversarial nets for convolutional face generation. Class Project for Stanford CS231N: Convolutional Neural Networks for Visual Recognition, Winter semester 2014(5), 2 (2014)
3. Glorot, X., Bengio, Y.: Understanding the difficulty of training deep feedforward neural networks. In: Proceedings of the Thirteenth International Conference on Artificial Intelligence and Statistics, pp. 249–256 (2010)
4. Goodfellow, I., et al.: Generative adversarial nets. In: Advances in Neural Information Processing Systems, pp. 2672–2680 (2014)
5. Ioffe, S., Szegedy, C.: Batch normalization: accelerating deep network training by reducing internal covariate shift. In: International Conference on Machine Learning, pp. 448–456 (2015)
6. Kawamura, N., et al.: Landmark-based reconstruction of 3D smooth structures from serial histological sections. In: Proceedings of SPIE, vol. 10581, pp. 1–7. International Society for Optics and Photonics (2018)
7. Kingma, D.P., Ba, J.: Adam: a method for stochastic optimization. arXiv preprint arXiv:1412.6980 (2014)
8. Rueckert, D., Sonoda, L.I., Hayes, C., Hill, D.L., Leach, M.O., Hawkes, D.J.: Nonrigid registration using free-form deformations: application to breast MR images. IEEE Trans. Med. Imaging 18(8), 712–721 (1999)
9. Shinmoto, H., et al.: Small renal cell carcinoma: MRI with pathologic correlation. J. Magn. Reson. Imaging 8(3), 690–694 (1998)
10. Zhang, Z., Song, Y., Qi, H.: Age progression/regression by conditional adversarial autoencoder. In: The IEEE Conference on Computer Vision and Pattern Recognition (CVPR), vol. 2 (2017)

Accurate 3D Reconstruction of a Whole Pancreatic Cancer Tumor from Pathology Images with Different Stains

Mauricio Kugler[1]([⊠]), Yushi Goto[1], Naoki Kawamura[1], Hirokazu Kobayashi[1], Tatsuya Yokota[1], Chika Iwamoto[2], Kenoki Ohuchida[2], Makoto Hashizume[2], and Hidekata Hontani[1]

[1] Nagoya Institute of Technology, Gokiso-cho, Showa-ku, Nagoya, Japan
mauricio@kugler.com
[2] Kyushu University, 3-1-1 Maidaishi, Higashi-ku, Fukuoka, Japan

Abstract. When applied to 3D image reconstruction, conventional landmark-based registration methods tend to generate unnatural vertical structures due to inconsistencies between the employed model and the real tissue. This paper demonstrates a fully non-rigid image registration method for 3D image reconstruction which considers the spatial continuity and smoothness of each constituent part of the microstructures in the tissue. Corresponding landmarks are detected along the images, defining a set of trajectories, which are smoothed out in order to define a diffeomorphic mapping. The resulting reconstructed 3D image preserves the original tissue architecture, allowing the observation of fine details and structures.

1 Introduction

Histopathological image analysis refers to the use of microscopical images from histological sections in order to analyze, diagnose and prevent diseases. These images are often used to obtain a high-resolution three-dimensional (3D) reconstruction of the original tissue architecture. Even with recent advances in 3D medical imaging techniques, such as Magnetic Resonance (MR) and Computed Tomography (CT), histology imaging still presents superior resolution and it remains the main source of information for several kinds of diseases, including most types of cancer [6,8].

Several studies tackle the problem of reconstructing 3D structures from a given set of microscopic images of histological sections obtained from a single target tissue [7,10]. These reconstructed images can be used for the study and visualization of the anatomical structures themselves, or for registration between the microscopic images and a corresponding MR macro image for multiscale analysis [4].

Given a series of images $I_i(\cdot)$ $(i = 1, 2, \ldots, N)$ scanned from stained thin sections of a chemically fixed tissue, a 3D reconstruction can be obtained by stacking

© Springer Nature Switzerland AG 2018
D. Stoyanov et al. (Eds.): COMPAY 2018/OMIA 2018, LNCS 11039, pp. 35–43, 2018.
https://doi.org/10.1007/978-3-030-00949-6_5

up non-rigidly registered versions of such images. The registration is required due to independent translation, rotation and deformation of the histological images introduced by the process of sectioning the tissue and mouthing the sections into glass slides. The registered images $J_i(\mathbf{y}) = I_i(\psi^{-1} \circ \mathbf{y})$, obtained from the original images and the mapping ψ_i, are combined to create the full 3D reconstruction as follows:

$$R(y_1, y_2, y_3) = J_{y_3}(y_1, y_2), \quad (y_3 = 1, 2, \ldots, N). \tag{1}$$

Several registration methods have been proposed for 3D image reconstruction, which can be roughly classified into two categories: iconic (intensity-based) and geometric (landmark-based) methods [8]. In the first, images are registered by maximizing the similarities in intensity between corresponding pixels [1,3], while in the second, images are registered by minimizing the distance between corresponding points (landmarks) on the images. This research focus on the later category due to its computational efficiency.

Let the coordinates of a landmark P_i^j detected from $I_i(u_1, u_2)$ be denoted by $\mathbf{u}_i^j$, where $j = 1, 2, \ldots M$. Given two images, $I_i(\mathbf{u})$ and $I_{i+1}(\mathbf{u})$, many landmark-based methods compute the mapping ψ_{i+1} using a criterion for evaluating the degree of match between corresponding landmark locations, e.g. $\left\| \mathbf{u}_i^j - \psi_{i+1} \circ \mathbf{u}_{i+1}^j \right\|^2$ [2,11]. In other words, these methods prefer 3D images in which the corresponding landmarks are vertically aligned parallel to the y_3 axis. However, corresponding landmarks are often detected along an anatomical structure which is not necessarily all vertical. The criterion employed by these methods is inconsistent with 3D microscope images, and thus often results in unnatural large deformation. In contrast, a few registration methods use a criterion that evaluates not the location matching, but the smoothness of landmark trajectories [5], which is more consistent with real tissue architecture and hence is the criterion adopted by the proposed method.

The proposed method detects corresponding landmarks using template matching, and rejects unreliable landmarks based on its confidence. As a result, the trajectories of landmarks will be automatically terminated, for instance, at blurred or folded portions, which should not contain landmarks. Once the corresponding landmarks are detected from all given images, the non-rigid mapping of each image is simultaneously determined based on the smoothed trajectories. This strategy for automatically handling damaged image portions and the capability of processing large sets of images with multiple stains, together with the results of the KPC mouse pancreas reconstruction, are the key contributions of this research.

2   Proposed Method

The reconstruction is performed over a series of N microscopic images $I_i(\mathbf{u})$ acquired from a single tissue, where $i = 1, 2, \ldots N$ and $\mathbf{u} = (u_1, u_2)^T$ corresponds to coordinates on the original images. The process starts by roughly aligning the images by rigid registration [7], creating a new set $\tilde{I}_i(\mathbf{x}) = I_i(\rho^{-1} \circ \mathbf{u})$, where ρ_i is the rigid transformation and $\mathbf{x} = \rho_i(\mathbf{u})$.

The method then detects a set of corresponding landmark points P_i^j along the series of images $\tilde{I}_i(\mathbf{x})$. Another set of destination coordinates for the landmarks must then be calculated in order to correctly deform the images. Assuming that a set of corresponding points are usually detected along several cross-sections of the same anatomical structure, and that these structures are spatially smooth and continuous, the objective is to deform the image in order that each set of corresponding points is located along a smooth curve on the reconstructed image.

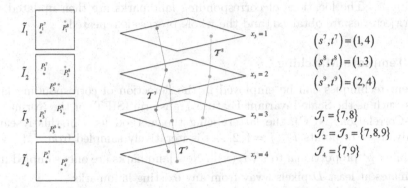

Fig. 1. Corresponding landmarks P_i^j and trajectories T^j (center), detected from four input images $\tilde{I}_i(\cdot)$ (left). Each trajectory T^j spans between s^j and t^j, and the set of indexes of trajectories intersecting each plane $x_3 = i$ is given by $\mathcal{J}_i$ (right).

The process is illustrated in Fig. 1. Let $\mathcal{P}^j = \left\{ P_i^j \,|\, i \in [s^j, t^j] \right\}$ denote the jth set of corresponding landmarks detected from a subset of N images, where $1 \leqslant s^j < t^j \leqslant N$, i.e. a set of corresponding points might be detected along only some of the N available images. Let $\tilde{R}(x_1, x_2, x_3) = \tilde{I}_i(x_1, x_2)$ denote a starting reconstruction obtained by stacking $\tilde{I}_i(\mathbf{x})$, where $x_3 = i$. The corresponding landmarks in $\mathcal{P}^j$ will then form a polygonal trajectory $T^j = P_{s^j}^j P_{s^j+1}^j \ldots P_{t^j}^j$ in $\tilde{R}(x_1, x_2, x_3)$ that extends from $x_3 = s^j$ to $x_3 = t^j$. Due to the independent non-rigid deformations of the histological images, these trajectories usually present a non-smooth jagged shape.

The destination coordinates of each landmark point are obtained by smoothing each trajectory T^j. The corresponding smoothed trajectory is denoted by $T_Q^j = Q_{s^j}^j Q_{s^j+1}^j \ldots Q_{t^j}^j$, where Q_i^j is the intersection between the smoothed trajectory with the plane $x_3 = i$. The set $\left\{ Q_i^j \,|\, j \in \mathcal{J}_i \right\}$ of all intersections crossing this plane, where $\mathcal{J}_i$ is the set of indexes of trajectories crossing the plane $x_3 = i$, is used to define a diffeomorphic mapping ϕ_i of $\tilde{I}_i(\cdot)$. The image $\tilde{I}_i(\mathbf{x})$ is then non-rigidly deformed with the obtained mapping in order to transport P_i^j to Q_i^j ($j \in \mathcal{J}_i$). The proposed method then reconstructs a 3D image $R(y_1, y_2, y_3) = J_{y_3}(y_1, y_2) = \tilde{I}_i(\phi_i^{-1} \circ \mathbf{y})$, where $y_3 = i$, $\mathbf{y} = (y_1, y_2)^T$, $\mathbf{y}_i^j = \phi_i \circ \mathbf{x}_i^j$ and $\mathbf{x}_i^j$ and $\mathbf{y}_i^j$ are, respectively, the coordinates of P_i^j and Q_i^j in the image $\tilde{I}_i(\mathbf{x})$.

Thus, the coordinates of the new points Q_i^j are obtained from $t^j - s^j + 1$ images $\tilde{I}_i$ ($i = s^j, \ldots, t^j$) simultaneously. It is worth mentioning that similar smooth trajectories for the landmarks cannot always be obtained by simply introducing rigidity regularization to the image deformation.

In order to obtain an accurate and stable detection of corresponding points, a coarse-to-fine approach is employed. The template size D is reduced as $D \leftarrow \gamma D$ with $0 < \gamma < 1$ after all the images $\tilde{I}_i$ ($i = 1, 2, \ldots, N$) are deformed by the mapping ϕ_i. The locations of corresponding landmarks are then updated and new trajectories are obtained, and the whole process is repeated.

2.1   Template Matching

Different techniques can be employed in the detection of corresponding landmarks, such as the Scale-Invariant Feature Transform (SIFT) or the Normalized Cross-Correlation (NCC), the latter being the method used in this research. Initially, a set of points P_i^j ($j = 1, 2, \ldots$) is iteratively sampled from $\tilde{I}_i(\cdot)$ with probability p_i^j proportional to $\left\| \nabla \tilde{I}_i(\mathbf{x}) \right\|$. New landmarks are only sampled from coordinates at least D pixels away from any existing landmarks.

False matchings are suppressed based on the confidence of the template matching given by the NCC and by applying backward template matching [13]. First, any landmark with an NCC value lower than a threshold is eliminated. After, backward template matching is applied in order to detect a landmark $\hat{P}_i^j$ corresponding to $\tilde{P}_{i+1}^j$. The landmark P_{i+1}^j is discarded if the distance between P_i^j and $\hat{P}_i^j$ is larger than a threshold.

2.2   Image Warping

The proposed method smooths the trajectories $\mathcal{T}^j$ in order to obtain the destination points Q_i^j. The coordinates $\mathbf{y}_i^j$ of the destination points are calculated by minimizing the square of the total variation of each trajectory and their square errors with a tradeoff parameter λ as follows:

$$\left\{ \mathbf{y}_{s^j}^j, \ldots, \mathbf{y}_{t^j}^j \right\} = \underset{\tilde{\mathbf{y}}_{s^j}^j, \ldots, \tilde{\mathbf{y}}_{t^j}^j}{\arg\min} \left(\sum_{i=s_i}^{t_i-1} \left\| \tilde{\mathbf{y}}_i^j - \tilde{\mathbf{y}}_{i+1}^j \right\|^2 + \lambda \left\| \tilde{\mathbf{y}}_i^j - \mathbf{x}_{i+1}^j \right\|^2 \right) . \quad (2)$$

This function is convex and has a unique minimizer, which can be analytically calculated. With the obtained destination coordinates $\mathbf{y}_i^j$ for each landmark $\mathbf{x}_i^j$, the diffeomorphic mapping ϕ_i can be computed using the well-known B-spline deformation method [9], which allows a diffeomorphic mapping with hard constrains $\mathbf{y}_i^j = \phi_i\left(\mathbf{x}_i^j\right)$.

3 Experimental Results

The reconstruction method was evaluated using a dataset of histological sections from the pancreas of a KPC mouse [12]. The extracted tumorous pancreas was fixed with formaldehyde, split into five 5 mm blocks, and each was sliced in a series of 4 μm-thick sections. After damaged samples were removed, the dataset contains around 2500 images, the majority (85%) stained with Hematoxylin & Eosin (HE) stain, but some with Antigen KI-67 (Ki67), Masson's Trichrome (MT) and Cytokeratin-19 (CK19) stains, which are interposed between HE stained images. The size of the original images is 100k × 60k pixels; however, a downsampled version of 15k × 10k pixels was used in some of the experiments.

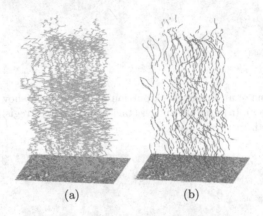

(a)   (b)

Fig. 2. Example of calculated trajectories: (a) original set of trajectories T^j and (b) corresponding smoothed set of trajectories T_Q^j. Some trajectories are fragmented as a result of the suppression of falsely matched landmarks.

Figure 2 shows actual calculated trajectories from a portion of the images. The original trajectories T^j are shown in Fig. 2(a), while the smoothed trajectories T_Q^j are shown in Fig. 2(b). Figure 3 shows cross-sections of $\tilde{R}(\cdot)$ and $R(\cdot)$, i.e. before and after smoothing. In Fig. 3(a), even the major structures are difficult to discern, while small structures, such as blood vessels and ducts, are virtually indistinguishable. The smoothed version shown in Fig. 3(b) enables the observation of fine details and small structures, which can then be labeled for further analysis. The hyperparameters were determined experimentally and are not critical for the overall performance of the method.

It is then possible to automatically label and reconstruct the structure of the central necrosed portion of the tumor and the pancreatic ducts, as shown in Fig. 4. This reconstruction reveals a high concentration of ducts around the necrosis.

Different microstructures require different stains in order to be accurately identified and labeled. Figure 5 shows registration results for neighbor sections of different stains, confirming the robustness of the proposed registration method.

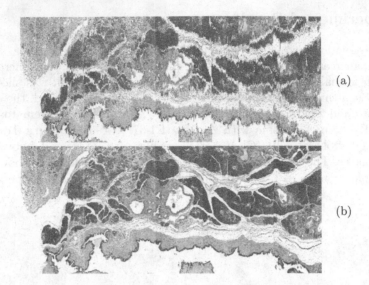

Fig. 3. Cross-sections of an image block portion of 810 images (shown horizontally for visualization purposes): (a) original images stacked-up, and (b) registered images using the proposed method.

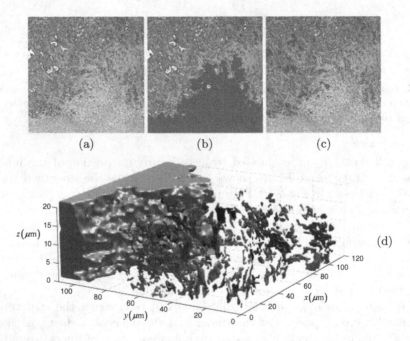

Fig. 4. Visualization of the internal microstructures from a portion of the full 3D reconstruction of a block of images: (a) registered HE stained image portion, (b) labeled necrosed portion of the tumor, (c) labeled pancreatic ducts, and (d) combined 3D reconstruction of ducts (blue and green) and necrosis (red). (Color figure online)

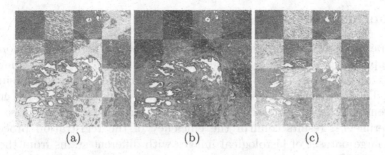

(a)   (b)   (c)

Fig. 5. Registration between adjacent images with different stains: (a) HE & CK19, (b) HE & MT and (c) HE & Ki67.

The Ki67 stain can indicate high cell proliferation. After reconstruction, sections stained with Ki67 were automatic labeled based on hue and the density of brown spots was calculated using Kernel Density Estimation (KDE), as shown in Fig. 6. By linearly interpolating along all sections and combining this results with the anatomical reconstruction of the tumor block, it is possible to observe the high proliferation regions on the external portion of the tumor.

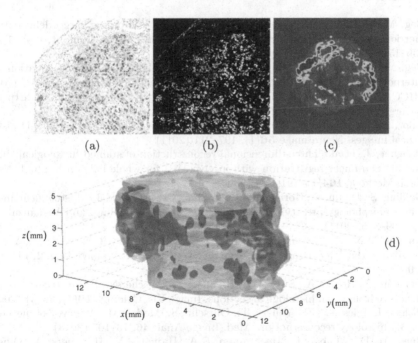

Fig. 6. Labeling, detection and reconstruction of active regions within the tumor from Ki67 stained images: (a) Ki67 stained image portion, (b) automatic labeling of active cells (brown spots), (c) high density (red) regions of active cells (overall image), and (d) reconstruction of active areas and superposition with full block 3D reconstruction, which contains 810 images of 15k × 10k pixels, or around 121.5×10^9 voxels. (Color figure online)

4   Conclusions

This paper describes a non-rigid registration method for 3D reconstruction from microscopic images of histological sections. The method explicitly constructs smooth trajectories in order to determine the deformation of all images simultaneously. This approach suppresses the unnatural vertical structures often created by conventional landmark-based methods, which process images pairwise.

Experimental results confirm the efficiency of the registration procedure, using a large dataset of histological images with different stains from the pancreas of a KPC mouse. The obtained reconstructions present continuous and smooth anatomical micro-structures, which could be successfully labeled and visualized.

Future work will include the registration of a new full-body KPC mouse image dataset, as well as a quantitative analysis of the registration results.

Acknowledgments. This research was supported by JSPS KAKENHI Grant Number 26108003.

References

1. Beg, M.F., Miller, M.I., Trouvé, A., Younes, L.: Computing large deformation metric mappings via geodesic flows of diffeomorphisms. Int. J. Comput. Vis. **61**(2), 139–157 (2005)
2. Beg, M.F., Miller, M., Trouvé, A., Younes, L.: The euler-lagrange equation for interpolating sequence of landmark datasets. In: Ellis, R.E., Peters, T.M. (eds.) MICCAI 2003. LNCS, vol. 2879, pp. 918–925. Springer, Heidelberg (2003). https://doi.org/10.1007/978-3-540-39903-2_112
3. Cifor, A., Bai, L., Pitiot, A.: Smoothness-guided 3-D reconstruction of 2-D histological images. NeuroImage **56**(1), 197–211 (2011)
4. Dauguet, J., et al.: Three-dimensional reconstruction of stained histological slices and 3D non-linear registration with in-vivo MRI for whole baboon brain. J. Neurosci. Methods **164**(1), 191–204 (2007)
5. Gaffling, S., Daum, V., Hornegger, J.: Landmark-constrained 3-D histological imaging: a morphology-preserving approach. In: Vision, Modeling, and Visualization, pp. 309–316 (2011)
6. Gurcan, M.N., Boucheron, L.E., Can, A., Madabhushi, A., Rajpoot, N.M., Yener, B.: Histopathological image analysis: a review. IEEE Rev. Biomed. Eng. **2**, 147–171 (2009)
7. Ourselin, S., Roche, A., Subsol, G., Pennec, X., Ayache, N.: Reconstructing a 3D structure from serial histological sections. Image Vis. Comput. **19**(1), 25–31 (2001)
8. Pichat, J., Iglesias, J.E., Yousry, T., Ourselin, S., Modat, M.: A survey of methods for 3D histology reconstruction. Med. Image Anal. **46**, 73–105 (2018)
9. Rueckert, D., Aljabar, P., Heckemann, R.A., Hajnal, J.V., Hammers, A.: Diffeomorphic registration using B-splines. In: Larsen, R., Nielsen, M., Sporring, J. (eds.) MICCAI 2006. LNCS, vol. 4191, pp. 702–709. Springer, Heidelberg (2006). https://doi.org/10.1007/11866763_86
10. Saalfeld, S., Cardona, A., Hartenstein, V., Tomančák, P.: As-rigid-as-possible mosaicking and serial section registration of large ssTEM datasets. Bioinformatics **26**(12), i57–i63 (2010)

11. Sotiras, A., Davatzikos, C., Paragios, N.: Deformable medical image registration: a survey. IEEE Trans. Med. Imaging **32**(7), 1153–1190 (2013)
12. Westphalen, C.B., Olive, K.P.: Genetically engineered mouse models of pancreatic cancer. Cancer J. **18**(6), 502–510 (2012)
13. Yin, Z., Collins, R.: Moving object localization in thermal imagery by forward-backward MHI. In: Proceedings of the Computer Vision and Pattern Recognition Workshop, pp. 133–133. IEEE (2006)

Role of Task Complexity and Training
in Crowdsourced Image Annotation

Nadine S. Schaadt[1], Anne Grote[1], Germain Forestier[2], Cédric Wemmert[3],
and Friedrich Feuerhake[1,4(✉)]

[1] Institute for Pathology, Hannover Medical School, Hannover, Germany
`feuerhake.friedrich@mh-hannover.de`
[2] IRIMAS, University of Haute Alsace, Mulhouse, France
[3] ICube, University of Strasbourg, Illkirch, France
[4] Institute for Neuropathology, University Clinic Freiburg,
Freiburg im Breisgau, Germany

Abstract. Accurate annotation of anatomical structures or pathological changes in microscopic images is an important task in computational pathology. Crowdsourcing holds promise to address this demand,
but so far feasibility has only be shown for simple tasks and not for
high-quality annotation of complex structures which is often limited by
shortage of experts. Third-year medical students participated in solving
two complex tasks, labeling of images and delineation of relevant image
objects in breast cancer and kidney tissue. We evaluated their performance and addressed the requirements of task complexity and training phases. Our results show feasibility and a high agreement between
students and experts. The training phase improved accuracy of image
labeling.

Keywords: Crowdsourcing · Human decision making
Image classification · Image delineation · Digital pathology
Annotation

1 Introduction

Crowdsourcing (CS) in digital pathology has been largely limited to less complex tasks such as identification of cancer cells [6,11], scoring of cell nuclei based
on immunohistochemistry (IHC) [2,4,6,11], malaria diagnostics [9], and creation
of training sets for convolutional neural networks [1,5]. In general, intrinsically
motivated contributors in voluntary CS perform better compared to paid "crowd-
workers" [10]. Quality of CS depends on training and adaptation of task design
to contributors' background knowledge [3,7].

This work was performed in the framework of SYSIMIT (FKZ:01ZX1308A), ILUMI-
NATE (FKZ:031 B0006C), and SYSMIFTA (FKZ:031L0085A) funded by BMBF.
Nadine S. Schaadt and Anne Grote contributed equally to this work.

D. Stoyanov et al. (Eds.): COMPAY 2018/OMIA 2018, LNCS 11039, pp. 44–51, 2018.
https://doi.org/10.1007/978-3-030-00949-6_6

In this paper, we investigate annotation of complex structures by medical students without pathology expertise but with profound understanding of anatomy and disease mechanism and a need to learn pathology as a strong incentive to recapitulate anatomy. We show that medical students can acquire skills to label images and delineate image objects in kidney and breast pathology, and discuss the influence of task complexity and training on CS approaches to produce high-quality annotations for machine learning.

2 Materials and Methods

2.1 Setting

We studied performance of a crowd of "educated" contributors: 142 third-year medical students, who were entering the curricular pathology course and thus had basic knowledge about microscopic anatomy but no expertise in pathology nor experience in annotating histological images.

We considered four independent experiments, each with 1–3 sessions on different days (Table 1). Each experiment started in a room equipped with computers with a short teaching session on relevant anatomical structures and pathological conditions, and explanations of the tools. The latter evolved from face-to-face lessons into a video tutorial, ready for use in experiment 4. The crowd were asked to work on two different tasks:

1. Labeling of regions of interest (ROIs)
   select one of several proposed categories for each of a set of images
   Used tools: software developed for the project that displays the current image, a progress line, and radio buttons for each class
2. Delineation of ROIs
   draw the outlines of all objects of some well-defined classes and mark the class names in an image showing a tissue region
   Used tools: Aperio ImageScope by Leica Microsystems (experiment 1), Cytomine [8] running on an own server (experiment 2–4)

Table 1. Overview over the crowds participating in independent experiments.

Crowd	Size	Session 1		Session 2		Session 3	
		Labeling	Delineation	Labeling	Delineation	Labeling	Delineation
Experiment 1	36	9	10	4	9	4	0
Experiment 2	14	4	12	0	6	0	0
Experiment 3	26	23	23	12	11	0	0
Experiment 4	66	41	27	28	22	0	0

The labeling task included an obligatory training phase in the beginning in which the correct solution was immediately shown to the participants and a test phase without feedback. Training for ROI delineation was introduced in experiment 4 as optional work on images with the possibility to switch on/off GT. Students received detailed feedback on both tasks after each session.

Images source were whole slide images (WSIs) from sections stained for H&E or IHC markers (ethical approval review board of Hannover Medical School).

2.2   Answer Aggregation and Evaluation

As final annotations, we aggregated individual statements as majority vote (MV: relative majority) or weighted vote (WV: weights calculated by training phase results of individuals). Equal votes result in unclassified objects and were counted as false negatives.

Two experts (one for each tissue type) provided annotations, such that there is a ground truth (GT) for each image to measure the performance of the crowd. To evaluate ROI labeling, we measured the accuracy averaged over each class,

$$ACC = \frac{1}{|C|} \sum_{i \in C} \frac{TP_i + TN_i}{TP_i + TN_i + FP_i + FN_i} \tag{1}$$

where C is the set of classes, TP, TN, FP, and FN are the numbers of true positives, true negatives, false positives, and false negatives, respectively. This is compared to the expected value of random labeling, estimated as $1/|C|$.

In ROI delineation, due to comparably large tissue areas without occurrence of any considered class, we calculated the F_1 score averaged over each class:

$$F_1 = 2 \cdot \frac{PPV \cdot TPR}{PPV + TPR} \tag{2}$$

with the precision (PPV) and the recall (TPR)

$$PPV = \frac{1}{|C|} \sum_{i \in C} \frac{TP_i}{TP_i + FP_i} \quad \text{and} \quad TPR = \frac{1}{|C|} \sum_{i \in C} \frac{TP_i}{TP_i + FN_i} \tag{3}$$

both computed per class and averaged over all classes. This was calculated for each pixel that is part of the tissue.

3   Results and Discussion

3.1   Feasibility and Role of Task Complexity

We compared the crowd results with expert annotations for high-level structures in breast and kidney tissue in independent experiments confirming feasibility even for complex ROIs representing pathologically relevant tissue conditions.

ROI Labeling. Our results suggested that the class complexity has a stronger effect on crowd performance than the tissue type as mistakes occurred predominantly in the distinction of classes defined by complex object features (Table 2).

In experiment 1, automatically detected ROIs intended to show epithelial structures in normal breast tissue was categorized into: "lobule", "duct", "FP" (session 1), and additionally "lobule with extralobular ducts" (session 2 and 3).

In experiment 2, the crowd classified images from breast cancer cases, distinguishing between (1) "technical artefact", (2) "invasive cancer", (3) "intraepithelial neoplasia", (4) "glandular epithelium", and (5) "other anatomical structure". As a single image could include normal and neoplastic structures, a more complex class definition was required. We used a hierarchical order such that the images should be classified by occurrence of highest order class. For example, if an image contained mainly glandular epithelium and some invasive tumor, then the image should be classified as "invasive tumor". Accuracy of WV (weighted by training phase precision) was 0.976. Even the lowest accuracy for individuals was detectably higher than the corresponding probability by chance.

In experiment 3, kidney structures from biopsies were labeled into four types ("normal", "pathologically changed", "sclerotic", "no" glomerulum). In both sessions, WV (session 1: 0.940, session 2: 0.832) was clearly higher than average but, some individuals outperformed the best combinations (Supp. Mat., Fig. 1). A particular challenge for the crowd was the class of "pathologically changed glomerula", most likely because the class definition included semi-quantitative criteria such as hypercellularity, thickened Bowman's capsule, mesangial sclerosis, collapse or retraction of the capillary tuft.

Experiment 4 considered four categories ("normal", "partially sclerotic", "sclerotic", "no" glomerulum). Highest accuracy was achieved by WV (session 1: 0.973, session 2: 0.942). In the case of "partially sclerotic glomerula", the precision was quite low for MV. Combining both classes "partially sclerotic glomerulum" and "sclerotic glomerulum" clearly increased the accuracy and precision.

Table 2. Overall accuracies (ROI labeling), displayed are expected value $(1/|C|)$, minimum (min), maximum (max), average (avg), and majority vote (MV).

Crowd	$1/|C|$	min	max	avg	MV
ex_1, se_1	0.333	0.789	0.923	0.878	0.937
ex_1, se_2	0.250	0.733	0.808	0.765	0.810
ex_1, se_3	0.250	0.813	0.869	0.847	0.879
ex_2, se_1	0.200	0.847	0.948	0.911	0.951

Crowd	$1/|C|$	min	max	avg	MV
ex_3, se_1	0.250	0.820	0.940	0.885	0.910
ex_3, se_2	0.250	0.730	0.840	0.791	0.818
ex_4, se_1	0.250	0.880	0.975	0.934	0.973
ex_4, se_2	0.250	0.838	0.928	0.891	0.942

ROI Delineation. To present the results, images are referred to as I, with experiment, session, and an id. For example, $I_{ex3,se2,1}$ denotes the first image of the second session of the third experiment. If necessary, participant groups are referred to in the image name as G with group number. Figure 1 shows the difference of the MV to the reference for one image from each experiment. Table 3 shows the overall F_1 scores for all experiments (further measures in Supp. Mat.

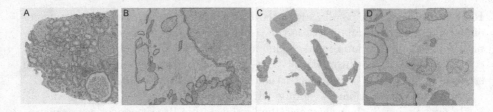

Fig. 1. Difference of majority vote to ground truth (GT) in examples of kidney (A: $I_{ex_1,se_1,1}$, C: $I_{ex_3,se_1,G_1,2}$) and breast (B: $I_{ex_2,se_1,2}$, D: $I_{ex_4,se_1,1}$) tissue. Green: agreement with GT, red: difference to GT. B: illustrates problems at tumor border. D: illustrates confused structures. (Color figure online)

Table 3. Overview overall F_1 Scores (ROI delineation), where n is the number of participants, MV the majority vote, and avg the average.

Image	n	MV	avg
$I_{ex_1,se_1,1}$	10	0.902	0.865
$I_{ex_1,se_2,1}$	9	0.645	0.606
$I_{ex_3,se_1,1}$	22	0.879	0.785
$I_{ex_3,se_1,G_1,2}$	8	0.789	0.740
$I_{ex_3,se_1,G_2,2}$	10	0.713	0.672
$I_{ex_3,se_2,1}$	11	0.884	0.744
$I_{ex_3,se_2,G_1,2}$	5	0.797	0.649
$I_{ex_3,se_2,G_1,3}$	5	0.940	0.848
$I_{ex_3,se_2,G_2,2}$	6	0.815	0.764
$I_{ex_3,se_2,G_2,3}$	6	0.726	0.621

Image	n	MV	avg
$I_{ex_2,se_1,1}$	9	0.616	0.551
$I_{ex_2,se_1,2}$	12	0.775	0.592
$I_{ex_2,se_2,1}$	6	0.694	0.598
$I_{ex_2,se_2,2}$	5	0.565	0.605
$I_{ex_4,se_1,1}$	27	0.716	0.661
$I_{ex_4,se_1,2}$	22	0.710	0.626
$I_{ex_4,se_2,1}$	22	0.535	0.530
$I_{ex_4,se_2,2}$	21	0.585	0.553

Table 2–5) indicating general feasibility. The quality decreases with increasing complexity, number of classes, and image size. For structures with well-defined borders, such as glomerula (kidney) or lobules (breast), the borders of the objects have been drawn quite accurately in contrast to fractal-like outlines of tumor.

Experiment 1 tested the crowd's delineation performance on two image subsets representing renal tissue, using a duplex staining for immune cells (session 1) or immune cells and vascular endothelium (session 2). For delineating "glomerulum", "artery", and "tubulus", the overall scores were distinctly lower for the second session than for the first session. Precision for the "artery" class in session 2 was markedly lower due to mislabeling of other blood vessels such as veins and smaller arterioles. To check how participants would be influenced by the provided classes, we used classes in session 2 that could potentially occur in kidney tissue but were not included in the specifically provided image. Several participants mistook narrow peritubular interstitial tissue for such a class ("collageneous tissue/septae"). We assume that in small, single images there is a tendency to annotate more objects in contrast to batches of larger images.

Experiment 2 tested a more complex setting for breast cancer and surrounding tissue. Classes e.g. included "invasive tumor", "duct", "lobule", and "large blood vessel". In most cases, the F_1 scores of MV were better than the F_1 scores on average. For the class "large blood vessel" in $I_{ex_2,se_1,2}$, for example, the recall value was on average 0.472 and for the MV 0.826, without loss of precision. Some objects in this complex setting, however, were challenging. For example, blood vessels in $I_{ex_2,se_2,1}$ and $I_{ex_2,se_2,2}$ were missed by two thirds of the crowd. Common differences between MV and GT occurred in (1) individual variations in the object border delineation, most pronounced at the tumor border and (2) confusions between the visually similar structures (epithelial/epitheloid) "lobule", "duct", and "invasive tumor".

Experiment 3 used eight WSIs of kidney tissue and focused on "glomerulum", "artery", and occasionally included "muscle". We split the crowd into roughly equally sized groups G_1 and G_2. In each session, both groups worked on a common image ($I_{ex_3,se_1,1}$ or $I_{ex_3,se_2,1}$, stained for H&E) and additionally annotated one further image(s) stained for a macrophage marker. The class "glomerulum" had the highest scores. In five images, its MV precision was higher than 0.990, with virtually no FPs, and the outlines of the glomerula were close to GT.

In experiment 4, four images of breast cancer were used, with similar complexity to experiment 2, but with more participants. Classes were "duct", "intraepithelial neoplasia", "tumor", "lobule", and "necrosis". The MV results were in the same range as for experiment 2. The results of experiments 2 and 4 suggested that most objects could be found reliably already with a small crowd while some difficult objects could not be identified by most participants.

Overall, there seemed to be a role for certain pathological changes mimicking or hiding ROIs: In the renal images (experiment 1 and 3), sclerotic glomerula and arteries were sometimes confused and arteries were also frequently completely missed. In two images (experiment 2 and 4), lobules with heavy immune infiltration were missed by all participants.

3.2   Role of Training Phase

ROI Labeling. For experiment 3, we compared the accuracy during the training phase, in which the correct label was shown to the participants immediately after their decision, with the test accuracy (Fig. 2A). Most students performed better during the test phase in both sessions, especially high-performer (based on test accuracy). Nevertheless, several results of the training phase were close to the test phase in session 1 (Spearman's correlation coefficient: 0.41). To investigate a suitable size of the training phase, we varied them in experiment 4 (three student groups in each session: 20, 40, or 60 images). For this, we kept the same images in the same order. The number of correctly labeled images was similar with a trend to increase with increasing size of trainings phase (Fig. 2 (B)). Students that participated in both session 1 and 2 showed higher correctness in the second training phase compared to students first time participating. Figure 2 (C) shows that the second training phase did also not increase their accuracy for the most difficult class of partially sclerotic glomerula. We conclude out of this, that the training phase covering a broad variability of representatives for each class was helpful to increase the performance of the crowd.

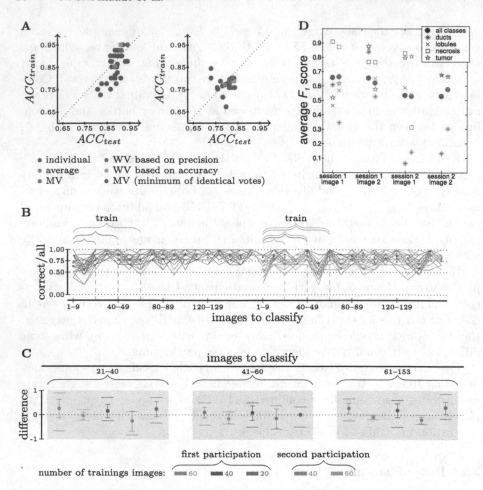

Fig. 2. Training phase effects in ROI labeling (A–C) and ROI delineation (D) A: Correlation between training and test accuracy for individuals (blue) and aggregations in experiment 3, session 1 (left) and session 2 (right). B: Changes of individual (lines) performance during experiment 4. C: Role of training phase length for difficult class "partially sclerotic glomerulum". Shown is the difference between accuracy of the first 20 images and of images 21–40, 41–60, and 61–153. D: Influence of size of optional training image (blue: 80% of test image size, red: 40% of test image size) on F_1 score. (Color figure online)

ROI Delineation. The participants could annotate a "training image" with the option to see the GT in experiment 4. To measure the training effect, we compared two group of individuals that either received a small training image (40% of test image size, not all classes represented) or a large training image (80% of test size). No clear effect of the size on F_1 score could be seen (Fig. 2D).

4   Conclusion

Our study shows general feasibility of CS for the annotation of complex histological images by participants with medical background, but without specific expert knowledge. To ensure annotation quality, it is necessary to design the tasks with well-defined objects and to include a sufficient training phase. Our approach can be adapted to individual project requirements and shows the importance of finding an adequate match between level of task complexity and previous knowledge of the crowd. Future work should focus on the comparison of "educated" contributors and nonexperts, and the usefulness of this type of noisy training data for machine learning.

Acknowledgements. We thank all students for contribution; M. Temerinac-Ott, Icube; R. Schönmeyer, C. Vanegas, Definiens for help in data selection; G. Stiller, M. Behrends, Peter L. Reichertz Institute for Medical Informatics; and A.-K. Rieke for the video.

References

1. Albarqouni, S., Baur, C., Achilles, F., Belagiannis, V., Demirci, S., Navab, N.: Aggnet: deep learning from crowds for mitosis detection in breast cancer histology images. IEEE Trans. Med. Imag. **35**, 1313–1321 (2016)
2. Della Mea, V., Maddalena, E., Mizzaro, S., Machin, P., Beltrami, C.A.: Preliminary results from a crowdsourcing experiment in immunohistochemistry. Diagn. Pathol. **9**, S6 (2014)
3. Hoßfeld, T., et al.: Best practices and recommendations for crowdsourced qoe-lessons learned from the qualinet task force crowdsourcing. In: QUALINET (2014)
4. Irshad, H., et al.: Crowdsourcing scoring of immunohistochemistry images: evaluating performance of the crowd and an automated computational method. Sci. Rep. **7** (2017)
5. Kim, E., Mente, S., Keenan, A., Gehlot, V.: Digital pathology annotation data for improved deep neural network classification. In: SPIE Medical Imaging, p. 101380D (2017)
6. Lawson, J., et al.: Crowdsourcing for translational research: analysis of biomarker expression using cancer microarrays. Br. J. Cancer **116**, 237–245 (2017)
7. Liu, S., Xia, F., Zhang, J., Wang, L., Wang, L.: How crowdsourcing risks affect performance: an exploratory model. Manag. Decis. **54**, 2235–2255 (2016)
8. Marée, R.: Collaborative analysis of multi-gigapixel imaging data using cytomine. Bioinformatics **32**, 1395–1401 (2016)
9. Mavandadi, S., et al.: Distributed medical image analysis and diagnosis through crowd-sourced games: a malaria case study. PloS One **7**, e37245 (2012)
10. Redi, J., Povoa, I.: Crowdsourcing for rating image aesthetic appeal: better a paid or a volunteer crowd? In: Proceedings of 2014 International ACM Workshop Crowdsourcing Multimedia, pp. 25–30. ACM (2014)
11. dos Reis, F.J.C., et al.: Crowdsourcing the general public for large scale molecular pathology studies in cancer. EBioMedicine **2**, 681–689 (2015)

Capturing Global Spatial Context for Accurate Cell Classification in Skin Cancer Histology

Konstantinos Zormpas-Petridis[1]([✉]), Henrik Failmezger[2],
Ioannis Roxanis[3], Matthew Blackledge[1], Yann Jamin[1],
and Yinyin Yuan[2]([✉])

[1] Division of Radiotherapy and Imaging,
The Institute of Cancer Research, London, UK
[2] Division of Molecular Pathology,
The Institute of Cancer Research, London, UK
[3] Royal Free London, NHS, London, UK
{konstantinos.zormpas-petridis,yinyin.yuan}@icr.ac.uk

Abstract. The spectacular response observed in clinical trials of immunotherapy in patients with previously uncurable Melanoma, a highly aggressive form of skin cancer, calls for a better understanding of the cancer-immune interface. Computational pathology provides a unique opportunity to spatially dissect such interface on digitised pathological slides. Accurate cellular classification is a key to ensure meaningful results, but is often challenging even with state-of-art machine learning and deep learning methods.

We propose a hierarchical framework, which mirrors the way pathologists perceive tumour architecture and define tumour heterogeneity to improve cell classification methods that rely solely on cell nuclei morphology. The SLIC superpixel algorithm was used to segment and classify tumour regions in low resolution H&E-stained histological images of melanoma skin cancer to provide a global context. Classification of superpixels into tumour, stroma, epidermis and lumen/white space, yielded a 97.7% training set accuracy and 95.7% testing set accuracy in 58 whole-tumour images of the TCGA melanoma dataset. The superpixel classification was projected down to high resolution images to enhance the performance of a single cell classifier, based on cell nuclear morphological features, and resulted in increasing its accuracy from 86.4% to 91.6%. Furthermore, a voting scheme was proposed to use global context as biological a priori knowledge, pushing the accuracy further to 92.8%.

This study demonstrates how using the global spatial context can accurately characterise the tumour microenvironment and allow us to extend significantly beyond single-cell morphological classification.

Keywords: Histology image processing · Hierarchical model
Cell classification

K. Zormpas-Petridis and H. Failmezger have contributed equally.

© Springer Nature Switzerland AG 2018
D. Stoyanov et al. (Eds.): COMPAY 2018/OMIA 2018, LNCS 11039, pp. 52–60, 2018.
https://doi.org/10.1007/978-3-030-00949-6_7

1   Introduction

Cell classification is an essential task in the histopathological characterisation of the tumour microenvironment. Differences in cell type abundance, regional distributions and spatial interactions can inform about the nature of disease, and provide robust markers of disease prognosis for risk-stratification [1]. In the new area of digital pathology, advanced image analysis can objectively, consistently and quantitatively characterise different components of the tumour and assist in tumour grading [2]. Accurate detection and classification algorithms are critical to assess the spatial distribution of all cell types.

Machine learning, and more recently, deep learning approaches have shown great promise in cell classification yielding high quality results [3–5]. However, even state-of-art algorithms tend to underperform in certain cases, as cell types often appear morphologically similar to each other or they overlap/touch. The existing pathological image analysis tools usually focus on individual cells' nuclei morphology with limited local context features, neglecting the tumour's global context.

Contextual information can be the key to further improve cell classification. Tumours are inherently heterogeneous, consisting of a mixture of tumour nests, lymphoid aggregates, stroma and other normal cell structures. Pathologists overcome the aforementioned issues by incorporating this context information and use tissue architecture, together with cell morphological features to accurately classify cells.

Currently, there is a lack of methods to effectively define the boundaries of tumour components and propagate contextual features down, to aid cell classification. Proposed methods to separate tumour tissue into regions of interest (ROIs) include classification of image patches into diagnostically relevant ROIs using a visual bag-of-word model [6] and deep learning approaches [7]. Another popular method to capture similar local regions is the segmentation of image using superpixels [8]. *Bejnordi et al.* used superpixels to identify regions (stroma, background, epithelial nuclei) in whole-slide images (WSIs) at different magnifications [9]. Also, *Beck et al.* developed a framework that classifies superpixels as epithelium or stroma and used this framework in order to uncover stromal features that are associated with survival [10]. However, melanoma histology is highly heterogeneous, posing a number of challenges to machine learning such as class imbalance, intra-class diversity, and ambiguous tumour component boundaries (Fig. 1A).

In this paper, we aim to overcome these challenges by effectively including a global tumour spatial context into single-cell classification. We propose a multi-resolution hierarchical framework, which captures the spatial global context at low magnification, by classifying superpixels into biologically meaningful regions (tumour area, normal stromal, normal epidermis and lumen/white space, Fig. 1B–C) and combining them with cell nuclei morphological features at high resolution to improve single cell classification (Fig. 2). We applied our algorithm on WSIs hematoxylin and eosin (H&E)-stained slides of melanoma skin cancer.

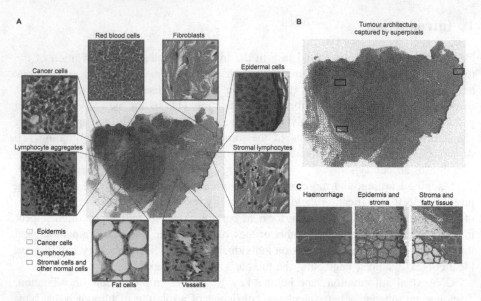

Fig. 1. The complex nature of melanoma architecture. A. Heterogeneous tumour stroma makes accurate cell classification a difficult task. B. Superpixels captures tumour global architecture by delineating the boundaries among heterogeneous tumour components, including haemorrhage area, fatty tissues, stromal regions, epidermis and cancer nests. C. Current classification scheme assigned these components accurately to their respective superpixel classes. Top: example images; Bottom: segmentation and classification using SLIC superpixels.

2   Methodology and Results

2.1   Data

58 full-face, H&E-stained section images from formalin-fixed, paraffin-embedded diagnostic blocks of melanoma skin cancer from the Cancer Genome Atlas were used. We scaled all digitized histology images to 20x magnification with a pixel resolution of 0.504 μm using Bio-Formats (https://www.openmicroscopy.org/bio-formats/). To set the ground truth for regional classification, an expert pathologist provided annotations on the slides for 4 different regions: tumour area, normal stroma, normal epidermis and lumen/white space. We randomly selected 21 images for training and reserved the remaining 37 images as an independent test set.

For single cell classification, 7 WSIs (representative size: 30000 × 30000 pixels) were split into subimages (tiles) of 2000 × 2000 pixels each. 3 WSIs were used for training and 4 for testing. Based on pathologist's input, we used 3863 cell nuclei (1320 cancer cells, 1100 epidermal cells, 751 lymphocytes, 692 stromal cells) from 82 subimages for training and 2405 cell nuclei (876 cancer cells, 602 epidermal cells, 417 lymphocytes, 510 stromal cells) from 224 subimages as an independent test set (Fig. 3A).

2.2 Superpixel Classification

First, whole-slide full resolution images were downscaled to 1.25x magnification to retain overall tumour structures, while reducing the noise. Reinhard stain normalisation [11] was applied to account for stain variabilities that could affect the classification [12].

Subsequently, the images were segmented using the simple linear iterative clustering (SLIC) superpixels algorithm [8], which is designed to provide roughly uniform superpixels. Choosing the optimal number of superpixels is important to ensure that the superpixels capture homogeneous areas and adhere to image boundaries. With the pathologist's input, we visually identified a size of superpixels that met these criteria and chose the number of superpixels automatically based on each image's size (Eq. 1).

$$N_i = ceiling\left(\frac{S_i}{U}\right) \tag{1}$$

where N_i is the number of superpixels in the ith image, S_i is the size of image i in pixels, and U (here U = 1250) is a constant held across all images that defined a desired size of the superpixels. This means, on average, a superpixel occupies an area of approximately 35pixel-by-35pixel, equivalent to 280 × 280 micron. The SLIC superpixels algorithm was proven to be computationally efficient, requiring only 3 s to segment a single downscaled image of 2500 × 2500 pixels using a 2.9 GHz Intel core i7 processor.

We identified 15477 superpixels belonging in tumour areas, 6989 in stroma areas, 141 in epidermis and 691 in lumen/white space for training by determining whether their central points fell within the regions annotated by the pathologist.

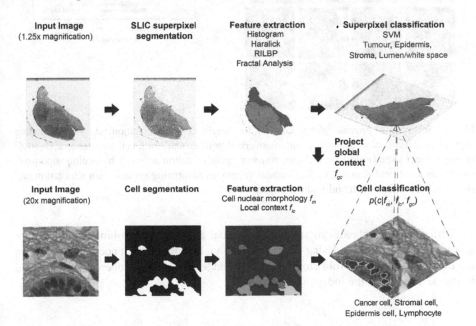

Fig. 2. Proposed hierarchical framework to project tumour global context onto single cell classification by integrating superpixel segmentation and classification.

Next, we extracted 4 types of features, totalling 85, from each superpixel, including 7 histogram features (mean value of hue, saturation and brightness, sum of intensities, contrast, standard deviation and entropy), and texture features (12 Haralick features [13], 59 rotation-invariant local binary patterns (RILBP), 7 segmentation-based fractal texture analysis (SFTA) features [14]). Features were standardized into z-scores. The mean values and SD of the features from the training set were used for the normalization of the test set. A support vector machine (SVM) with a radial basis function (RBF, γ = 1/number_of_features) was trained with these features to classify superpixels into 4 different categories: cancer, stroma, epidermis and lumen/white space. To solve the class imbalance problem for training, we randomly selected a subset of 5000 cancer and stroma superpixels and increased the penalty in the cost function for the epidermis and lumen/white space classes by a factor of 10.

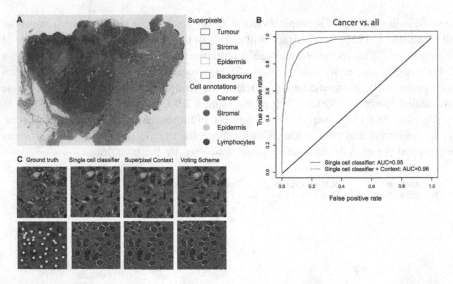

Fig. 3. Superpixels provide global context for single cell classification. A. Representative superpixel classification of tumour regions overlaid with ground-truth cell annotations. B. ROC curves (cancer vs. all) illustrate the improvement in classification accuracy by adding superpixel context as additional features. C. Representative images comparing ground truth annotation and single cell classifiers with and without superpixels.

Performance of classification using individual and various combinations of feature sets was tested (Table 1). Using all 85 features, yielded the highest accuracy (97.7% in the training set using 10-fold cross validation and 95.7% in 2997 superpixels annotated in the 37 images of the independent test set).

Table 1. Accuracy matrix of the superpixels' classification for single sets of features (left) and various combinations (right).

Features	Accuracy (%)	Feature combinations	Accuracy
Hist.	95.9%	Hist. + Haralick	97.3%
Haralick	91.4%	Hist. + RILBP	96.3%
RILBP	88.8%	Hist. + SFTA	96.9%
SFTA	85.2%	Hist. + Haralick + RILBP	97.1%
		Hist. + Haralick + SFTA	97.1%
		Hist. + Haralick + RILBP + SFTA	**97.7%**

2.3 Cell Classification Based on Nuclear Features

Image processing was carried out using the Bioconductor package EBImage [15]. Cell nuclei were extracted by Otsu thresholding; morphological opening was used to delete noisy structures in the image and clustered nuclei were separated by the Watershed algorithm. For every nucleus, 91 morphological features (f_m) were extracted [16]. Three local features f_{lc} were added: the number of nuclei neighbours in a distance of 25 μm, the density at the particular cell position, and the size of the surrounding cytoplasm, calculated by thresholding the image's red channel after excluding the nuclei pixels.

For single cell classification, a Support Vector Machine (SVM) with a RBF ($\gamma = 1/\text{number\_of\_features}$) kernel was trained with these features and achieved 86.4% accuracy (Table 3) in the test dataset. As expected, the classifier underperformed in distinguishing between epidermal and cancer cells (Table 2, Fig. 3C), due to their similar morphology and related local context (both exist in crowded environments).

Table 2. Confusion matrix of the Single Cell Classifiers. C: cancer, E: epidermis, L: lymphocyte, S: stromal. Red text highlight the confusion between cancer cells and epidermis cells.

		Morphology				Morphology + global context				Morphology + Voting scheme			
		C	E	L	S	C	E	L	S	C	E	L	S
	cancer	715	127	20	14	786	27	23	40	844	0	30	2
Classes	epidermis	96	496	5	5	44	550	2	6	48	515	17	22
	lymphocyte	7	12	397	1	12	5	399	1	14	1	401	1
	stromal	18	4	17	471	11	12	18	469	10	4	23	473

2.4 Cell Classification with Context

Two different schemes were used in order to integrate regional classification with cell classification. First, the type of area a single cell belonged to, provided by the super-pixel classification, was added to the morphological feature set as the global context feature (f_{gc}). This reduced the misclassification between epidermis and cancer cells in large degree (Table 2) and led to a much higher accuracy (91.6%, Table 3, Fig. 3BC) compared to the cell-morphology based classifier.

Secondly, global context given by superpixels served as biological *apriori* knowledge to correct single cell classifications. E.g. stromal cells seldom exist in non-stromal regions, while, cancer cells should only exist in tumour regions and epidermal cells should be found only in epidermal regions. Lymphocytes, however, can infiltrate into both tumour regions and stroma, but are rarely found in epidermis [17]. The regional context of a cell is thus of great importance for its annotation. We subsequently implemented an iterative voting scheme for cells in stromal and tumour regions:

$$c_i = \begin{cases} epidermis & if & s = epidermis \\ t & if\ else & t = s \lor t = lymphocyte \\ k & else & k = s \lor k = lymphocyte \end{cases} \quad (2)$$

Where c_i is the cell at position i, $t \in \{cancer, stromal, epidermis, lymphocyte\}$ is the most probable annotation of the cell in the SVM, $s \in \{cancer, stromal, epidermis\}$ is the annotation of the cell's superpixel, $k \in \{cancer, stromal, epidermis, lymphocyte\}$ is the annotation with the next highest unchecked probability in the SVM. This voting scheme (Eq. 2) was applied for all cell nuclei and resulted in 92.8% accuracy in the test dataset (Fig. 3C, Table 3).

Table 3. Accuracy of the classifiers.

Method	Accuracy	Precision	Recall
Single cell	86.4%	87.4%	87.9%
Single cell + context	91.6%	91.5%	92.2%
Voting scheme	92.8%	92.8%	92.7%

3 Discussion

Accurate characterisation of the tumour microenvironment is crucial for understanding cancer as a highly complex, non-autonomous disease. In this study, we built a hierarchical framework to mirror the way pathologists perceive tumour architecture. Despite recent advances in image analysis, there are limitations associated with cell classification strictly based on their nuclear morphology or local contextual features, due to phenotypical similarities; this can be overcome by incorporating the global spatial context. Multi-resolution or superpixel-based methods have been successfully proposed for identifying ROIs [9] and unsupervised segmentation [18, 19]. Here, our methodology enables the classification of cancer and diverse microenvironmental cells with high accuracy through the implementation of contextual features and a voting scheme.

We demonstrated that the addition of global context feature from superpixels improved cell classification based only on nuclei morphology from 86.4% to 92.8%. After the validation of our method on a larger dataset, we intend to combine it with more sophisticated cell segmentation and classification deep learning algorithms [5] and pave the way towards automatic scoring to assist in the stratification of melanoma patients. Such system can provide a better understanding of the cancer-immune cell interface, cell-stroma interactions and predictive biomarkers of response to novel therapies, including immunotherapy, which has radically changed melanoma patient survival. Also, the proposed framework can be easily adapted and used to study other cancer types.

4 Conclusion

We have demonstrated that our multi-resolution approach, incorporating spatial tissue context improves the accuracy of automated cell classification from digital histopathology compared to our conventional single cell classification methodology based solely on nuclear morphology in clinical melanoma samples.

References

1. Gurcan, M.N., et al.: Histopathological image analysis: a review. IEEE Rev. Biomed. Eng. 2, 147–171 (2009)
2. Kothari, S., et al.: Pathology imaging informatics for quantitative analysis of whole-slide images. J. Am. Med. Inform. Assoc. 20(6), 1099–1108 (2013)
3. Chen, C.L., et al.: Deep learning in label-free cell classification. Sci. Rep. 6, 21471 (2016)
4. Khoshdeli, M., et al.: Detection of nuclei in H&E stained sections using convolutional neural networks. In: 2017 IEEE EMBS International Conference Biomedical & Health Informatics (BHI) (2017)
5. Sirinukunwattana, K., et al.: Locality sensitive deep learning for detection and classification of nuclei in routine colon cancer histology images. IEEE Trans. Med. Imaging 35(5), 1196–1206 (2016)
6. Mercan, E., et al.: Localization of diagnostically relevant regions of interest in whole slide images: a comparative study. J. Digit. Imaging 29(4), 496–506 (2016)
7. Araújo, T., et al.: Classification of breast cancer histology images using convolutional neural networks. PLoS One 12(6), e0177544 (2017)
8. Achanta, R., et al.: SLIC superpixels compared to state-of-the-art superpixel methods. IEEE Trans. Pattern Anal. Mach. Intell. 34(11), 2274–2282 (2012)
9. Bejnordi, B.E., et al.: A multi-scale superpixel classification approach to the detection of regions of interest in whole slide histopathology images. Presented at the 19 March 2015
10. Beck, A.H., et al.: Systematic analysis of breast cancer morphology uncovers stromal features associated with survival. Sci. Transl. Med. 3(108), 108ra113 (2011)
11. Reinhard, E., et al.: Color transfer between images. IEEE Comput. Graph. Appl. 21(5), 34–41 (2001)
12. Khan, A.M., et al.: A nonlinear mapping approach to stain normalization in digital histopathology images using image-specific color deconvolution. IEEE Trans. Biomed. Eng. 61(6), 1729–1738 (2014)
13. Haralick, R.M., et al.: Textural features for image classification. IEEE Trans. Syst. Man. Cybern. 3(6), 610–621 (1973)

14. Imran, M., et al.: Segmentation-based fractal texture analysis and color layout descriptor for content based image retrieval. In: 2014 14th International Conference on Intelligent Systems Design and Applications, pp. 30–33. IEEE (2014)

15. Pau, G., et al.: EBImage–an R package for image processing with applications to cellular phenotypes. Bioinformatics **26**(7), 979–981 (2010)

16. Yuan, Y., et al.: Quantitative image analysis of cellular heterogeneity in breast tumors complements genomic profiling. Sci. Transl. Med. **4**, 157 (2012)

17. Spetz, A.-L., et al.: T cell subsets in normal human epidermis. Am. J. Pathol. **149**, 2 (1996)

18. Bankhead, P., et al.: QuPath: open source software for digital pathology image analysis. Sci. Rep. **7**(1), 16878 (2017)

19. Wright, A.I., et al.: Incorporating local and global context for better automated analysis of colorectal cancer on digital pathology slides. Procedia Comput. Sci. MIUA **90**(2016), 125–131 (2016)

Exploiting Multiple Color Representations to Improve Colon Cancer Detection in Whole Slide H&E Stains

Alex Skovsbo Jørgensen[1]([✉]), Jonas Emborg[2], Rasmus Røge[3,4],
and Lasse Riis Østergaard[1]

[1] Department of Health Science and Technology, Aalborg University,
Aalborg, Denmark
asj@hst.aau.dk
[2] Diagnostics and Genomics Group, Dako Denmark A/S, an Agilent
Technologies Company, Glostrup, Denmark
[3] Institute of Pathology, Aalborg University Hospital, Aalborg, Denmark
[4] Department of Clinical Medicine, Aalborg University, Aalborg, Denmark

Abstract. Currently, colon cancer diagnosis is based on manual assessment of tissue samples stained with hematoxylin and eosin (H&E). This is a high volume, time consuming, and subjective task which could be aided by automatic cancer detection. We propose an algorithm for automatic cancer detection within WSI H&E stains using a multi class colon tissue classifier based on features extracted from 5 different color representations. Approx. 32000 tissue patches were extracted for the classifier from manual annotations of 9 representative colon tissue types from 74 WSI H&E stains. Colon tissue classifiers based on gray level or color features were trained using leave-one-out forward selection. The best colon tissue classifier was based on color texture features obtaining an average tissue precision-recall (PR) area under the curve (AUC) of 0.886 and a cancer PR-AUC of 0.950 on 20 validation WSI H&E stains.

Keywords: Classification · H&E stain · Colon · Cancer
Machine learning

1 Introduction

Colorectal cancer is the third most common form of cancer worldwide and the fourth most common cause of death from cancer. [1] Currently, the gold standard colon cancer diagnosis is manual histopathological assessment of tissue structures within biopsies or tissue samples. Whole slide image (WSI) hematoxylin and eosin (H&E) stains allow pathologists to assess the tissue structures to detect cancer tissue and either perform a diagnosis or determine if special stains has to be used for further diagnosis. This is a high volume and subjective task and subject to inter- and intra-observer variation in the diagnosis which may lead to suboptimal treatment of the patient. [2] Digitization of WSI H&E stains has

© Springer Nature Switzerland AG 2018
D. Stoyanov et al. (Eds.): COMPAY 2018/OMIA 2018, LNCS 11039, pp. 61–68, 2018.
https://doi.org/10.1007/978-3-030-00949-6_8

led to development of algorithms to aid pathologists in reducing workload and inter- and intra-observer variation in the diagnostic work flow. Automatic cancer detection within WSI H&E stains can be used as first step for an automated analysis of WSI H&E stains e.g. for prescreening of slides to detect regions with cancer tissue and discard slides with obvious benign tissue for further analysis. However, WSI H&E stains can contain many different benign and malignant tissue structures with different appearances, which can complicate development of automated cancer detection algorithms. Some benign tissue structures appear very dissimilar to cancer tissue and can be easy to classify while others more closely resembles cancer tissue e.g. mucosa tissue in colon. This may be illustrated by extracting features from manual tissue annotations of representative colon tissue structures and be used to take into consideration during algorithm development. Additionally, accurate tissue classification between representative tissue structures may be used e.g. to determine degree of tumor infiltration and tumor tissue composition within the WSI H&E stains as cancer tissue can consist of different sub-tissue structures e.g. necrosis.

Currently, patch based methods are popular to characterize the appearance of tissue structures and has been used to extract both gray level [6] and color based texture and intensity features for tissue classification. [3–9] Kather et al. [6] proposed a patch based multi class classification framework to discriminate between eight colon tissue types to determine tumor composition in colorectal cancer using texture features. The study showed promising results reporting a 87.4% classification accuracy between the eight tissue structures. However, no color information was exploited in the study and it was only applied on a limited data set of 10 independent H&E stains. However, many studies has proven that color features are useful for tissue classification in H&E stains. [3–9] However, the studies has only extracted features from one or two color representations even though many color transformations exists each which may provide unique information useful for tissue classification. Therefore, it would be interesting to assess classification performance when utilizing all the color information available in the images.

To the best of our knowledge, no previous studies has explored using features from a broad range of color representations for multi class colon tissue classification to detect colon cancer in WSI H&E stains. Therefore, the purpose of this study was to develop a patch based framework to detect colon cancer tissue within WSI H&E stains using a multi class colon tissue classifier trained using texture and intensity features from five color representations from patches extracted from manual tissue annotations of representative colon tissue structures.

2   Methods

A patch based multi class colon tissue classification framework was developed to detect cancer tissue within WSI H&E stains. Tissue patches were extracted from manual annotations of nine representative colon tissue structures to obtain color based texture and intensity features from the tissue structures. The features were used for training and validation of colon tissue classifiers designed to classify extracted tissue patches within a WSI H&E stain and obtain tissue probabilities (P_{tissue}) of each colon tissue structure. Two multi class colon tissue classifiers were trained based on two initial feature sets (gray level vs. color based features) to assess feasibility in using features from many color representations for classification between multiple tissue structures.

2.1 Image Data

A total of 94 colon WSI H&E stains from independent subjects were available for the study (46 containing only non-neoplastic tissue and 48 adenocarcinoma). The tissue samples were fixed in formaline prior to embedding in paraffin. The tissue blocks were cut in 4 μm sections on a microtome at room temperature. Specimens were stained with H&E on a Dako CoverStainer with Dako Ready-to-Use reagents using the manufacturer's validated protocol. The stained slides were scanned on Phillips IntelliSite Ultra-Fast Scanner (40x; 0.25 μm^2/pixel) and were processed on the 20x resolution level. The WSI H&E stains were randomly divided into a training set (38 adenocarcinoma and 36 non-neoplastic) and validation set (10 adenocarcinoma and 10 non-neoplastic). A pathologist annotated each specimen with ROI's indicating gross regions containing cancer. Additionally, manual annotations of 9 representative colon tissue types (mucosa, muscle, fat, inflammation, red blood cells, necrosis, mucous, connective tissue, and cancer) were obtained. The manual annotations in the validation slides were all assessed and either approved or discarded by an experienced pathologist.

Patches with a size of 128 × 128 pixels were extracted from the manual annotations to obtain training and validation data from each colon tissue type. For the patch extraction, background pixels were first identified to discard patches without tissue information. Background pixels were defined as pixels with an intensity below 0.05 in the S-channel from the HSV color space, which was determined empirically. After the thresholding, background regions with an area smaller than a cell nuclei (700 pixels selected based on manual segmentations of representative cell nuclei within the data) was included as tissue pixels to ensure pixels within hypochromatic cells nuclei were included as tissue information. Finally, patches with more that 90% overlap with a manual tissue annotation and containing at least 30% tissue pixels were selected for training and validation data for each tissue structure. The criteria caused fat tissue to be excluded as a tissue class as patches overlapping with fat annotations consistently contained less than 30% tissue pixels. Each patch were labeled according to the tissue label of the overlapping manual segmentation. The final training set consisted of 4628 Mucosa, 4015 muscle, 5368 fat, 1435 inflammation, 4691 blood cells, 4396 necrosis, 609 mucous, 4300 connective tissue, and 4606 cancer patches and the validation set of 804 Mucosa, 722 muscle, 1189 fat, 305 inflammation, 282 blood cells, 932 necrosis, 6 mucous, 670 connective tissue, and 1006 cancer patches.

2.2 Feature Extraction

The feature extraction obtains features from the tissue patches for training and validation of the colon tissue classifier. Classic texture and intensity features were selected to assess contribution when using many color representations for the classification. Two initial feature sets were extracted for the study for comparative analysis: (1) Gray level texture features consisting of the 80 features proposed by Kather et al. [6] (2) Color texture features consisting of 7 intensity histogram and 18 gray level co-occurrence matrix (GLCM) texture features extracted from each color channel from RBG, HSV, CIELab, CMYK, and H&E color deconvolution as well as mean RGB and RGB gradient feature images (18 feature images in total). The algorithm proposed by Ruifrok et al. [10] was used for the color deconvolution. The intensity histogram features consisted of mean, standard deviation, coefficient of variation, skewness, kurtosis, 3rd moment, and entropy. The GLCM features were extracted using five distances (1, 3, 7, 15, and 20), four angles (0, 45, 90, and 135°), and 32 gray levels. For each distance the GCLM's were averaged to obtain rotation invariant features. The GLCM features proposed by Haralick et al. [11] were extracted: autocorrelation, contrast, correlation, cluster prominence, cluster shade, dissimilarity, energy, entropy, homogeneity, maximum probability, variance, sum average, sum variance, sum entropy, difference entropy, information measure of correlation, inverse difference, and inverse difference moment. In total, the color feature set contained 1746 features for each tissue patch. The feature values were z-score normalized.

2.3 Colon Tissue Classifier

The extracted features were used to train 7 colon tissue classifiers using 7 initial feature sets: One only containing gray level texture, 5 only containing features from each separate color representation, and one containing all color based features. During a feature analysis it was observed that the feature distributions of each tissue class approx. could be modeled with multivariate Gaussian distributions. Therefore, for simplicity of the study, a Bayes classifier [12] was used for classification and to obtain P_{tissue} for each tissue structure. Each classifier was trained using a leave-one-out forward selection procedure for feature reduction to obtain the most compact and discriminative feature set for each classifier and prevent over-fitting. In each leave-one-out, all tissue patches extracted from one WSI H&E stain was used as validation data and all the other tissue patches extracted from the training data slides were used for training. Precision-recall area under the curve (PR-AUC) for each tissue class and average tissue PR-AUC (PR-AUC$_{avg}$) were used as performance metrics. PR-AUC for each tissue class were obtained by defining tissue patches from one tissue class as true positives (TP) class and all other tissue patches as true negatives (TN). In each iteration in the feature selection, the feature with the highest cancer PR-AUC (PR-AUC$_{canc}$) that also improved PR-AUC$_{avg}$ was selected to improve cancer classification while also improving the overall tissue classification. The feature selection was stopped when the improvement in PR-AUC$_{canc}$ or PR-AUC$_{avg}$

was less than 0.005 between iterations. For validation, the classification performance of the trained models were assessed on tissue patches extracted from the 20 validation WSI H&E stains using the same TP and TN definitions and performance metrics as in the model training. Additionally, one-vs-all PR-AUC's for each separate tissue type as well as one-vs-one PR-AUC between cancer and each separate tissue type were obtained for the best colon tissue classifier to highlight classification performance between different tissue types.

3   Results

3.1   Colon Tissue Classifier

PR-AUC$_{canc}$ and PR-AUC$_{avg}$ obtained on the training and validation data in each colon tissue classifier can be seen in Table 1 (PR-AUC's obtained during training are given in the brackets). The classifier obtained using all color features obtained superior PR-AUC$_{canc}$ compared to the other classifiers with a PR-AUC$_{cancer}$ of 0.930 and 0.950 in the training and validation data, respectively. Additionally, the all color feature classifier obtained high PR-AUC$_{avg}$'s of 0.886 and 0.836 on the training and validation data, respectively. Therefore, the all color feature classifier were selected as the best colon tissue classifier. PR-AUC's for each tissue type in the selected colon tissue classifier can be seen in Table 2. The tissue types with the worst classification performance were mucosa, inflammation, and mucous with PR-AUC's below 0.9. Additionally, results of the one-vs-one classification between patches from cancer and each separate tissue type can be seen in Table 3. A PR-AUC above 0.95 were obtained between cancer and each tissue types but the lowest classification performance were obtained between cancer and mucosa, inflammation, and necrosis patches, respectively.

Table 1. PR-AUC's obtained for each of the 7 colon tissue classifiers. PR-AUC's are given as validation PR-AUC [training data PR-AUC].

	Gray level texture	RGB	HSV	H& E color deconvolution	CIELab	CMYK	All color features
PR-AUC$_{avg}$	0.699 [0.677]	0.867 [0.855]	0.838 [0.823]	0.885 [0.846]	0.843 [0.874]	0.837 [0.895]	0.886 [0.836]
PR-AUC$_{canc}$	0.886 [0.836]	0.906 [0.861]	0.885 [0.916]	0.870 [0.877]	0.871 [0.836]	0.900 [0.875]	0.950 [0.930]

The selected features for the final colon tissue classifier consisted of two intensity features (coefficient of variation (Y-channel) and skewness (Eosin-channel)) and 7 texture features (information measure of correlation (C-channel, distance 3), information measure of correlation (V-channel, distance 3), correlation (L-channel, distance 3), information measure of correlation (RGB gradient, distance 1), correlation (Chromaticity A-channel, distance 15), difference entropy

(H-channel, distance 3), sum variance (S-channel, distance 3)). The selected features were obtained from 9 different color channels and channels from each of the five color representations were represented in the selected features. This indicate that using multiple color representations can provide additional information for the tissue classification. The selected texture features were mainly obtained with a GCLM distance of 1 or 3, indicating that local texture patters contained the most significant information in discriminating between tissue structures.

Table 2. PR-AUC's obtained for each tissue class in the selected colon tissue classifier. PR-AUC's are given as validation PR-AUC [training data PR-AUC].

	Cancer	Mucosa	Muscle	Blood cells	Inflammation	Necrosis	Mucous	Connective tissue
PR-AUC	0.950	0.881	0.942	0.900	0.891	0.924	0.626	0.974
	[0.930]	[0.865]	[0.936]	[0.835]	[0.951]	[0.805]	[0.410]	[0.957]

Application of the selected colon tissue classifier on three representative subimages of the H&E stains from the study can be seen in Fig. 1. First column show classification of benign colon tissue where only a few local patches are misclassified as cancer. It can also be seen that the colon tissue classifier can discriminate between different benign tissue structures such as red blood cells (green), muscle (cyan), connective tissue (brown), and mucousa (magenta). The two other columns shows subimages containing cancer tissue in the bottom to right corner in the images. Most of the patches located in cancer tissue are correctly classified as cancer (yellow patches) and most of the benign tissue are not classified as cancer. However, small misclassification problems between cancer tissue and mucosa can sometimes be observed (top of Fig. 1c). Additionally, the classifier could detect necrosis (white) within the tumor tissue in Fig. 1f.

4   Discussion

We have presented a framework for detecting cancer in WSI H&E stains of colon tissue using a multi class colon tissue classifier based on color intensity and texture features. The colon tissue classifier trained based on all color features obtained the best performance with a $PR-AUC_{avg}$ of 0.886 and $PR-AUC_{cancer}$ 0.950, respectively compared to the other classifiers based only on the gray level features proposed by Kather et al. [6] or features from separate color representations. The final feature set consisted of 9 intensity and texture features obtained from feature images from five different color representations. The study indicate that using information from multiple color representations can improve tissue classification within WSI H&E stains.

Table 3. One-vs-one PR-AUC's obtained on the validation data between cancer and each tissue class in the final colon tissue classifier based on all color features

	Mucosa	Muscle	Blood cells	Inflammation	Necrosis	Mucous	Connective tissue
PR-AUC	0.959	0.997	0.999	0.975	0.979	0.999	0.998

The worst one-vs-one PR-AUC's were obtained between cancer and mucosa, inflammation, and necrosis, respectively. This may be explained as tumor tissue may contain inflammatory cells and necrosis which may appear within the cancer annotations and the fact that colon adenocarcinoma originates from mucosa making cancer appearance more similar to mucosa than the other tissues. This confirm our hypothesis that some benign tissue structures makes accurate cancer classification more difficult within WSI H&E stains of colon tissue compared to

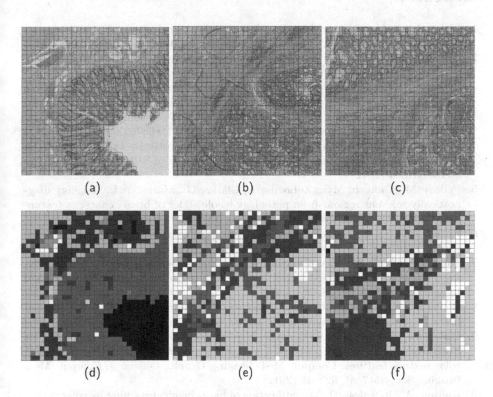

Fig. 1. Application of the colon tissue classifier in three representative H&E stain images (columns). First row show the location of the extracted patches in the original image. Second row show a color coded image of the tissue class with the highest P_{tissue} obtained within the patch (yellow = cancer, magenta = mucosa, cyan = muscle, green = blood cells, blue = inflammation, white = necrosis, red = mucous, and brown = connective tissue). (Color figure online)

others. This may have to be taken into consideration to improve classification accuracy when designing algorithms for automated cancer detection.

Stain normalization was not applied during this study as the framework was developed on WSI H&E stains stained with a standard staining protocol which minimized stain variation between slides. Therefore, the current framework is only expected to be robust to small variations in stain intensity. Stain normalization should therefore be applied as a preprocessing step to use the framework on WSI H&E stains stained with other staining protocols to ensure the generalized performance. Still, the results indicate that a good overall classification accuracy can be obtained between benign tissue and cancer using data obtained from manual tissue annotations of representative tissue structures. The framework can also be used to indicate the location of cancer tissue within WSI H&E stains which may be used for further analysis e.g. for cancer grading.

References

1. Ferlay, J., et al.: Cancer incidence and mortality worldwide: sources, methods and major patterns in GLOBOCAN 2012. Int. J. Cancer **386**(136), E359–E386 (2015)
2. Ismail, S.M., et al.: Observer variation in histopathological diagnosis and grading of cervical intraepithelial neoplasia. BMJ **298**(March), 707–710 (1989)
3. Doyle, S., Madabhushi, A., Feldman, M., Tomaszeweski, J.: A boosting cascade for automated detection of prostate cancer from digitized histology. In: Larsen, R., Nielsen, M., Sporring, J. (eds.) MICCAI 2006. LNCS, vol. 4191, pp. 504–511. Springer, Heidelberg (2006). https://doi.org/10.1007/11866763_62
4. Bahlmann, C., et al.: Automated detection of diagnostically relevant regions in H&E stained digital pathology slides. In: Proceedings of SPIE, Medical Imaging **8315**, 831504 (2012)
5. Peikari, M., Gangeh, M.J., Zubovits, J., Clarke, G., Martel, A.L.: Triaging diagnostically relevant regions from pathology whole slides of breast cancer: a texture based approach. IEEE Trans. Med. Imaging **35**(1), 307–315 (2016)
6. Kather, J.N., et al.: Multi-class texture analysis in colorectal cancer histology. Sci. Rep. **6**(1), 27988 (2016)
7. Tabesh, A., et al.: Automated prostate cancer diagnosis and Gleason grading of tissue microarrays. In: Proceedings of the SPIE International Symposium on Medical Imaging, vol. 5747, pp. 58–70 (2005)
8. Bejnordi, B.E., et al.: Automated detection of DCIS in whole-slide H & E stained breast histopathology images. IEEE Trans. Med. Imaging **35**, 1–10 (2016)
9. DiFranco, M.D., O'Hurley, G., Kay, E.W., Watson, R.W.G., Cunningham, P.: Ensemble based system for whole-slide prostate cancer probability mapping using color texture features. Comput. Med. Imaging Graph.: Official J. Comput. Med. Imaging Soc. **35**(7–8), 629–45 (2011)
10. Ruifrok, A., Johnston, D.: Quantification of histochemical staining by color deconvolution. Anal. Quant. Cytol. Histol. **23**, 291–299 (2001)
11. Haralick, R.M., Shanmugam, K., Dinstein, I.: Textural features for image classification (1973)
12. Duda, R.O., Hart, P.E., Stork, D.G.: Pattern Classification, 2nd edn. Wiley, Hoboken (2012)

Leveraging Unlabeled
Whole-Slide-Images for Mitosis Detection

Saad Ullah Akram[1,2](✉), Talha Qaiser[2], Simon Graham[2], Juho Kannala[3],
Janne Heikkilä[1], and Nasir Rajpoot[2,4]

[1] Center for Machine Vision and Signal Analysis (CMVS),
University of Oulu, Oulu, Finland
sakram@ee.oulu.fi
[2] Tissue Image Analytics (TIA) Lab, University of Warwick, Coventry, UK
[3] Department of Computer Science, Aalto University, Espoo, Finland
[4] The Alan Turing Institute, London, UK

Abstract. Mitosis count is an important biomarker for prognosis of various cancers. At present, pathologists typically perform manual counting on a few selected regions of interest in breast whole-slide-images (WSIs) of patient biopsies. This task is very time-consuming, tedious and subjective. Automated mitosis detection methods have made great advances in recent years. However, these methods require exhaustive labeling of a large number of selected regions of interest. This task is very expensive because expert pathologists are needed for reliable and accurate annotations. In this paper, we present a semi-supervised mitosis detection method which is designed to leverage a large number of unlabeled breast cancer WSIs. As a result, our method capitalizes on the growing number of digitized histology images, without relying on exhaustive annotations, subsequently improving mitosis detection. Our method first learns a mitosis detector from labeled data, uses this detector to mine additional mitosis samples from unlabeled WSIs, and then trains the final model using this larger and diverse set of mitosis samples. The use of unlabeled data improves F1-score by ∼5% compared to our best performing fully-supervised model on the TUPAC validation set. Our submission (single model) to TUPAC challenge ranks highly on the leaderboard with an F1-score of 0.64.

Keywords: Mitosis detection · Computational pathology
Breast cancer · Self-supervised learning · Semi-supervised learning

1 Introduction

Precise quantification of mitotic figures in hematoxylin and eosin (H&E) stained slides is of high clinical significance in understanding the proliferation activity of cells within tumor regions. In breast cancer, mitosis count is a significant prognostic biomarker and one of the most important criteria for cancer grading. In routine clinical practice, a pathologist visually examines H&E stained slides

© Springer Nature Switzerland AG 2018
D. Stoyanov et al. (Eds.): COMPAY 2018/OMIA 2018, LNCS 11039, pp. 69–77, 2018.
https://doi.org/10.1007/978-3-030-00949-6_9

under the microscope. This conventional way of mitosis counting is extremely time-consuming and tedious, as a pathologist has to perform this task for several high-power-fields (HPF) in multiple whole-slide-images (WSI) for a single patient. In addition, this process is extremely subjective with noticeable disagreement between different pathologists [17], mainly due to the inherent ambiguity and difficulty of the task. Automated methods have the potential to overcome the inter- and intra-observer variability by producing more consistent results and freeing up valuable pathologist's time which can be better spent on understanding the aggressiveness of disease and stratified medicine.

Mitosis detection is considered a challenging task even for experienced pathologists due to the variations in morphological appearance of mitotic cells (see Fig. 2). These variations are caused by various factors including the mitotic phase, staining variability and tissue damage during the slide preparation. Mitotic cells typically look like hyper-chromatic objects without nuclear membrane, which have hairy extensions of nuclear material [17]. However, there are many instances in which these characteristic features are difficult to spot and subjective decisions have to be made. These subjective decisions can be challenging due to the presence of other cells, such as apoptotic (programmed cell death) cells, which have very similar appearances. In addition, mitotic cells are significantly less in number as compared to other malignant and healthy cells. Hence, the detection of mitotic cells naturally suffers from a huge class imbalance.

In recent years, several mitosis detection challenges have been organized: MITOS12 [14]; AMIDA13 [18]; MITOS14 [1]; and TUPAC [2]. These challenges have tackled increasingly more difficult scenarios and have greatly helped to advance mitosis detection research. Regardless, one of the shortcomings of these contest datasets is the limited number of mitosis samples and pathology centers from which the data is acquired, limiting the generalization ability of trained mitosis detectors. State-of-the-art deep learning based methods require a large number of annotated samples for training. One approach for increasing the number of annotations is semi-supervised learning, which can make good use of readily available unlabeled data in histopathology.

With regards to the previously mentioned challenges, we propose a deep learning based self-supervised algorithm that makes use of both labeled HPFs and unlabeled WSIs to train the mitosis detector. We show that mitosis detection can benefit from larger datasets and in the absence of a large amount of labeled data, leveraging an unlabeled dataset can lead to a significant improvement in the performance of current mitosis detection methods.

2   Related Research

Automated methods have shown great promise in detecting mitotic cells [5, 7, 11] within histological images in recent years. Mitosis detection was among the first few problems where deep learning showed its potential, when a convolutional neural network (CNN) [5] out-performed competing methods by a large margin in MITOS12 challenge [14]. This method detected mitosis by scanning an image in a sliding window manner. This was followed by a fully convolutional version, which significantly reduced the test inference time and won the AMIDA13 challenge by a convincing margin [18].

Since then, there have been many advances to improve the performance of these methods. Some of the most recent methods have made use of deeper [11], wider [19] and cascades of networks [4]. Recently, the use of crowd-sourced annotations [3] has been explored for this task and despite these annotations being noisy, the proposed aggregation layer can lead to good performance. PHH3 labeling has also been used to create larger mitosis detection datasets [16].

In the last couple of years, semi- and weakly-supervised methods have shown their potential in improving performance beyond the limits of state-of-the-art fully-supervised methods. Sun et al. [15] showed the importance of dataset size and made use of noisy web-scale dataset in addition to ImageNet to achieve state-of-the-art performance on various important vision tasks including image classification, object detection and segmentation. Radosavovic et al. [12] made use of self-supervision and a large unlabeled dataset to improve performance on object and human keypoint detection tasks. These methods are able to utilize large unannotated datasets, in addition to annotated datasets, and improve performance over fully-supervised techniques.

3   Method

Our method consists of 3 stages shown in Fig. 1. First, we train a mitosis detector on exhaustively labeled high-power-fields (HPFs) (Sect. 3.1); second, we mine unannotated whole-slide-images (WSIs) for additional mitosis patches (Sect. 3.2); and finally, we re-train the mitosis detector using patches from both HPFs and WSIs (Sect. 3.1).

3.1   Mitosis Detector

We use a 12-layer ResNet [6], shown in Fig. 1b, for mitosis detection.

Training: This network is trained using patches of 128×128, which have the label of either background or foreground (mitosis). These patches are sampled from the following four sets depending on the training stage: *BG-Rand*, *BG-Hard*, *FG-Lab* and *FG-WSI* (see Fig. 1a).

Set *BG-Rand* contains patches, which are sampled randomly with uniform probability from HPFs such that their distance is greater than a fixed threshold from previously selected background patches and all mitotic cells in that HPF.

BG-Hard consists of hard-negative mined patches, i.e. the false positive patches extracted by applying the model (@ 30k iteration) to training HPFs. The detections for each HPF are post-processed using non-maximum suppression (NMS), with a distance of 50 pixels, to remove duplicate.

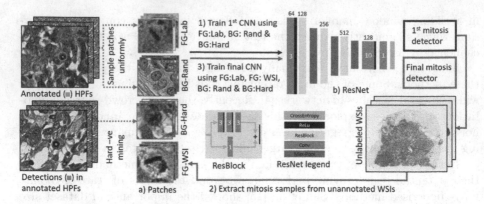

Fig. 1. Overview of the method. First, a mitosis detector is trained using labeled HPFs. Then, additional mitosis are mined from unannotated WSIs and finally, the mitosis detector is re-trained using patches from both labeled HPFs and unlabeled WSIs.

Set *FG-Lab* contains patches centered at the annotated mitosis positions and set *FG-WSI* includes mitosis patches extracted from the unlabeled whole-slide-images (WSIs), details of which are provided in Sect. 3.2.

The network weights are optimized using ADAM optimizer [9], which minimizes the cross-entropy loss. We use a learning rate of 0.0001 for the first 20k iterations, which is lowered to 0.00001 for the next 30k iterations and for the final 10k iterations, it is dropped to 0.000001. We use a batch size of 64 with a fixed ratio of mitotic:background patches (2:62).

For the initial model, which makes use of only the labeled data, we train it using patches from sets *BG-Rand* and *FG-Lab* for the first 30k iterations. Then, hard-negative mining is used to obtain *BG-Hard* patches from the training HPF. The next 30k iterations sample background patches from *BG-Hard*.

For the final model, which utilizes both the labeled and unlabeled data, the foreground (mitosis) patches are sampled from *FG-Lab* and *FG-WSI* sets with a fixed ratio and the background patches are sampled from *BG-Rand* for first 30k iterations, then hard-negative mining is used to obtain background patches *BG-Hard* and the model is trained for another 30k iterations.

Data Augmentation: Contrast variation is one of the biggest challenges in automated analysis of histopathology images. Slides prepared using different staining protocols and/or imaged using different scanners can have very different color distributions (see Fig. 2). We use contrast transfer [13] to make our network robust to these variations. During training, with a fixed probability, we change the mean and standard deviation (in LAB color-space) of the HPF from which a patch is sampled to the values from another randomly selected HPF. We also use flipping, rotation and jitter to augment training samples.

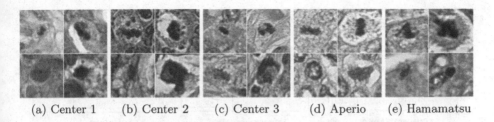

(a) Center 1   (b) Center 2   (c) Center 3   (d) Aperio   (e) Hamamatsu

Fig. 2. Top row: mitotic samples. Bottom row: hard-negative samples. (a–c) are from TUPAC dataset and were collected from 3 different pathology centers. (d–e) are from MITOS14 dataset and they were scanned using two different scanners.

Testing: At test time, we apply our method in a fully-convolutional manner. Since the test images can be too large to fit in the current GPU memory, we process them in windows of size 512×512. We pad (64 pixels) images by mirroring and use overlapping (120 pixels) windows to handle zero-padding in convolutional layers.

Once we have the probability maps of all windows predicted by our network, we stitch and re-size them to the original image size. This is followed by NMS with radius of 30 pixels to discard duplicate detections. Finally, we remove all detections with score below 0.5.

3.2 Mining Mitosis from Unlabeled WSIs

Current mitosis detectors typically have a high false negative rate at low values of false positive rate (see Fig. 3c), so we use unlabeled data only for mining mitosis samples. In order to ensure that the training does not derail, it is important that the additional mitosis samples have very few false positives (FPs). We achieve this by using test-time data augmentations (flipping and rotations), which removes many FPs with high score for only few of these transformations.

We apply our mitosis detector to whole-slide-images (WSIs) in a sliding window manner. Processing WSIs can be computationally expensive and applying our mitosis detector on a single WSI can takes hours. In order to reduce the computational demands, we skip windows which have mean RGB value greater than a fixed threshold. Similarly, if the maximum score in a window is below a threshold for the first transformation, then additional transformations are skipped.

4   Experiments

Dataset: We evaluate our method on two public mitosis datasets: TUPAC [2] and MITOS14 [1]. We split these datasets into 80% train/20% val sets at case level. Figure 2 shows some mitosis and hard-negative patches from these datasets.

TUPAC (**T**) training set contains 1,552 mitotic cells annotated in 656 HPFs, which are selected from 73 cases acquired from 3 different pathology labs. This dataset also includes 500 WSIs from 500 cases in the training set which are

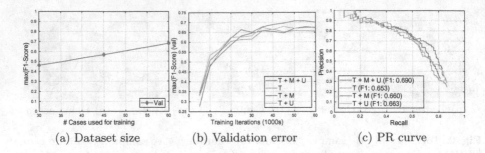

(a) Dataset size   (b) Validation error   (c) PR curve

Fig. 3. (a) Performance improves almost linearly with the TUPAC training data size. (b) TUPAC validation error during training for various models. (c) Precision-recall curve (TUPAC-val). F1-Scores @ threshold = 0.5 are listed in the legend.

Table 1. Performance on both TUPAC/MITOS14-val sets for various combinations of training sets: TUPAC (T), MITOS14 (M) and unlabeled WSIs (U).

Train set	T	M	Train set	T	M
	max (F1)	max (F1)		max (F1)	max (F1)
T	0.655	0.552	T + U	0.673	0.579
M	0.378	0.678	M + U	0.466	0.674
T + M	0.653	0.660	T + M + U	0.684	0.655

not annotated for mitosis. We use these WSIs as unlabeled data (**U**) for our experiments. MITOS14 (**M**) training set consists of 1,502 mitotic cells marked in 1,200 HPFs from 11 cases, each scanned using two different scanners: Aperio Scanscope XT and Hamamatsu Nanozoomer 2.0-HT.

Metrics: We use the same criteria as TUPAC challenge [2] for evaluating mitosis detection performance. A detection is considered as true positive (TP) if it is within a radius of 30 pixels from any unmatched ground truth (GT). All other detections are considered false positives (FP) and the GT annotations without any detection in 30 pixel radius are counted as false negative (FN). Then, recall (R), precision (P) and F1-Score (F1) are computed. When comparing different models, we use maximum F1-Score at any threshold value; this mitigates the problem of high sensitivity of F1-score to the selected threshold.

Dataset Size: We select 13 cases as validation set and then train three models with the same hyper-parameters but with 30, 45 and 60 cases in the training set. The results are plotted in Fig. 3a, which shows that the performance improves almost linearly as we increase the training set size.

Cross-Dataset Generalization: Table 1 presents performance òn TUPAC (T) and MITOS14 (M) validation sets for models trained using various combinations of training sets. The performance on TUPAC-val drops sharply when only MITOS14 and unlabeled data (**M** and **M + U** in Table 1) is used for training, which can be explained by the fact that TUPAC data is collected from multiple pathology centers and as a result it has much larger variation than MITOS14. The performance on MITOS14 is much more consistent irrespective of which training set is used for training, indicating that at least some cases in TUPAC data are representative of this dataset.

Impact of Unlabeled WSIs: The experiments with unlabeled data (Table 1) show that the performance on TUPAC-val set improves considerably for any combination of TUPAC and MITOS14 dataset, indicating that mitosis samples mined from unlabeled WSIs provide additional non-trivial information which is helpful for learning a better model. For MITOS14-val on the other hand, the use of unlabeled data only helps when MITOS14-train data is not used during training. This is due to the unlabeled data having much larger contrast variation, which makes the model generalize better (as performance on TUPAC always improves) but does not provide much additional help when training set already includes data from MITOS14.

Figure 3b shows the performance on TUPAC-val during training for few models trained using different combinations of **T**, **M** and **U** sets and Fig. 3c shows the precision-recall curve for these models. Hard-negative mining is performed at 30k iterations and it consistently improves the F1-score by 5–10%.

Case-Level Performance: Mean and standard deviation of F1-Score at case level for models trained using only **T** (0.622 ± 0.236), **T + M** (0.619 ± 0.227) and **T + M+ U** (0.636 ± 0.191) also indicate that using only MITOS14 data in addition to TUPAC has little overall benefit as this data is from a different source. However, when unlabeled data from TUPAC is used, it leads to much greater improvement in performance and the variation across cases reduces as well, indicating that this automatically-labeled data is better for training than the manually labeled data which comes from a different source.

Comparison with State-of-the-Art: Table 2 compares the performance of our method with top ranked methods on both TUPAC and MITOS14 datasets. On MITOS14, our method has much higher F1-score on validation set (0.642) compared to top ranked method on test set (0.482). On TUPAC test set, our method ranks 3rd[1] with F1-score of 0.64. We would like to point out that our submission consists of a single model (@threshold = 0.5) with minimal post-processing (i.e. only NMS). Techniques such as test-time augmentation and ensembling can potentially improve performance considerably. With ensembling alone leading to ~10% improvement in F1-score of some methods [4, 16]) on this task.

---

[1] Out of 20 methods listed at http://tupac.tue-image.nl/node/62.

Table 2. Performance on TUPAC/MITOS14. The results in first two rows are on validation set, while the rest are on test sets.

	TUPAC				MITOS14		
	F1	Recall	Precision		F1	Recall	Precision
Ours (T + M + U)	0.690	0.661	0.722	Ours (T + M + U)	0.642	0.605	0.683
Ours (T)	0.653	0.621	0.688	Ours (M)	0.620	0.496	0.828
Lunit [11]	0.652	–	–	CasNN [4]	0.482	0.507	0.460
IBM [19]	0.648	–	–	DeepMitosis [10]	0.437	0.443	0.431
Ours (T + M + U)	0.640	0.671	0.613	–	–	–	–

5   Discussion

In this paper, we have presented a semi-supervised mitosis detection method, which benefits from a large amount of unlabeled data, that is readily available in histopathology. This method has the potential to benefit other detection and segmentation tasks in many medical and biological research areas, where labeled data is scarce and unlabeled data is abundant. It is clear from current methods [8,12] and our own results that once a method reaches a sufficient performance level, it can be used to generate pseudo-labels for unlabeled data, which can then be used to re-train it, improving performance. However, it is worth investigating, where that sufficient performance level lies for various tasks.

An important future direction is to explore the upper limit of performance that is attainable with current annotation procedures. This upper-limit exists, due to the unavoidable noise in pathologist mitosis annotation. State-of-the-art methods already obtain similar performance to that of pathologists, but these methods have poor generalization when applied to data from different sources [16]. Using larger and more diverse datasets can resolve this issue and semi-supervised techniques can be highly beneficial in these scenarios. In addition, more objective annotation techniques (e.g. using PHH3 labeling [18]) have the potential of improving performance considerably, which may result in future automated methods outperforming experienced pathologists.

References

1. MITOS-ATYPIA-14 (2014). https://mitos-atypia-14.grand-challenge.org/home/
2. Tumor Proliferation Assessment Challenge (2016). http://tupac.tue-image.nl/
3. Albarqouni, S., et al.: AggNet: deep learning from crowds for mitosis detection in breast cancer histology images. TMI **35**, 1313–1321 (2016)
4. Chen, H., et al.: Mitosis detection in breast cancer histology images via deep cascaded networks. In: AAAI (2016)
5. Cireşan, D.C., Giusti, A., Gambardella, L.M., Schmidhuber, J.: Mitosis detection in breast cancer histology images with deep neural networks. In: Mori, K., Sakuma, I., Sato, Y., Barillot, C., Navab, N. (eds.) MICCAI 2013. LNCS, vol. 8150, pp. 411–418. Springer, Heidelberg (2013). https://doi.org/10.1007/978-3-642-40763-5_51

6. He, K., Zhang, X., Ren, S., Sun, J.: Deep residual learning for image recognition. In: CVPR (2016)
7. Khan, A.M., Eldaly, H., Rajpoot, N.M.: A gamma-gaussian mixture model for detection of mitotic cells in breast cancer histopathology images. J. Pathol. Inform. **4**, 11 (2013)
8. Khoreva, A., et al.: Simple does it: weakly supervised instance and semantic segmentation. In: CVPR (2017)
9. Kingma, D.P., Ba, J.: Adam: a method for stochastic optimization. In: ICLR (2015)
10. Li, C., Wang, X., Liu, W., Latecki, L.J.: DeepMitosis: mitosis detection via deep detection, verification and segmentation networks. Med. Image Anal. **45**, 121–133 (2018)
11. Paeng, K., Hwang, S., Park, S., Kim, M.: A unified framework for tumor proliferation score prediction in breast histopathology. In: Cardoso, M. (ed.) DLMIA/ML-CDS 2017. LNCS, vol. 10553, pp. 231–239. Springer, Cham (2017). https://doi.org/10.1007/978-3-319-67558-9_27
12. Radosavovic, I., et al.: Data distillation: towards omni-supervised learning. In: CVPR (2018)
13. Reinhard, E., Adhikhmin, M., Gooch, B., Shirley, P.: Color transfer between images. IEEE Comput. Graph. Appl. **21**, 34–41 (2001)
14. Roux, L., et al.: Mitosis detection in breast cancer histological images an ICPR 2012 contest. J. Pathol. Inform. **4**, 8 (2013)
15. Sun, C., Shrivastava, A., Singh, S., Gupta, A.: Revisiting unreasonable effectiveness of data in deep learning era. In: ICCV (2017)
16. Tellez, D., et al.: Whole-slide mitosis detection in H&E breast histology using PHH3 as a reference to train distilled stain-invariant convolutional networks. TMI **37**(9), 2126–2136 (2018). https://doi.org/10.1109/TMI.2018.2820199
17. Veta, M., et al.: Mitosis counting in breast cancer: object-level interobserver agreement and comparison to an automatic method. PLoS One **11**, e0161286 (2016)
18. Veta, M., et al.: Assessment of algorithms for mitosis detection in breast cancer histopathology images. Med. Image Anal. **20**, 237–248 (2015)
19. Zerhouni, E., Lanyi, D., Viana, M., Gabrani, M.: Wide residual networks for mitosis detection. In: ISBI (2017)

Evaluating Out-of-the-Box Methods for the Classification of Hematopoietic Cells in Images of Stained Bone Marrow

Philipp Gräbel[1][(✉)], Martina Crysandt[2], Reinhild Herwartz[2],
Melanie Hoffmann[2], Barbara M. Klinkhammer[3], Peter Boor[3],
and Tim H. Brümmendorf[2], and Dorit Merhof[1]

[1] Institute of Imaging and Computer Vision, RWTH Aachen University,
Kopernikusstraße 16, 52074 Aachen, Germany
{philipp.graebel,dorit.merhof}@lfb.rwth-aachen.de
[2] Department of Hematology, Oncology,
Hemostaseology and Stem Cell Transplantation,
University Hospital RWTH Aachen University,
Pauwelsstraße 30, 52074 Aachen, Germany
[3] Institute of Pathology, University Hospital RWTH Aachen University,
Pauwelsstraße 30, 52074 Aachen, Germany

Abstract. Compared to the analysis of blood cells in microscope images of peripheral blood, bone marrow images are much more challenging for automated cell classification: not only are the cells more densely distributed, there are also significantly more types of hematopoietic cells. So far, several attempts have been made using custom image features and prior knowledge in form of cytoplasm and nuclei segmentations or a restricted number of cell types in peripheral blood. Instead of hand-crafting features and classification methods for bone marrow images, we compare several well-known methods on our more challenging dataset and we show that while generic classical machine learning approaches cannot compete with specialized algorithms, even out-of-the-box deep learning methods already yield valuable results. Our findings indicate that automated analysis of bone marrow images becomes possible with the advent of convolutional neural networks.

Keywords: Classification · Bone marrow · White blood cells
Hematopoietic cells · Residual network

1 Introduction

Automated classification of blood cells made significant progress in recent years: novel methods and applications provide clinical experts with tools that yield quantifiable results in short time [1,2]. Part of this is the analysis of white blood cells (leukocytes) found in peripheral blood. Since these are relatively sparse and show great contrast to background and red blood cells (erythrocytes),

© Springer Nature Switzerland AG 2018
D. Stoyanov et al. (Eds.): COMPAY 2018/OMIA 2018, LNCS 11039, pp. 78–85, 2018.
https://doi.org/10.1007/978-3-030-00949-6_10

segmentation is straight forward. However, as only mature cells are present in healthy peripheral blood, classification of those is less challenging compared to bone marrow, where more cell types in various degrees of maturation are present. This makes distinguishing between those cell types a challenging task even for humans.

As the distribution of cell-classes reflects the hematopoietic process of bone marrow and is severely affected by diseases such as leukemia, a tool for reliable, automated classification and counting would be a great support for medical experts. Such an application could provide impartial and reproducible results as a basis for diagnostic decisions. Even today, the tedious task of counting cells is still performed manually by medical experts.

Previous research on the classification of hematopoietic cells [1] suggested a pipeline for the classification of white blood cells in blood smear images. The authors made use of morphological features, which they can extract based on a segmentation of cell cytoplasm and nucleus. Due to the higher density of cells, this approach is not feasible with bone marrow images, as morphological operations for segmentation require a better separation between individual cells. Other specialized features based on automatically segmented regions described by shape, color and texture yielded a good separation of hematopoietic cells [3].

In a more recent work, convolutional neural networks have been proposed for the classification of hematopoietic cells [2]. They used images obtained from blood smears, which make it possible to conveniently detect and segment the white cell lineage. Based on texture features, three classes of cells were separated with a Support Vector Machine, followed by a classification of three other cell types using features from a neural network and a Random Forest Classifier. This method achieves good results in blood smear images.

As opposed to research on the analysis of white blood cells in blood smear images, research on the classification of hematopoietic cells in bone marrow is rare and often only possible under certain constraints or simplifications: for example, manually segmented images of cropped cells from different maturity stages of the granulopoiesis are given as prior knowledge [4]. Based on the segmentation of nucleus and cytoplasm, it is possible to craft specialized features that make classification possible.

To the best of our knowledge, classification has not been researched in detail for a wider range of hematopoietic cells in bone marrow images without prior knowledge such as manual segmentations. In this paper, we compare and evaluate classical machine learning pipelines with feature extraction and classification for their suitability to perform this challenging task. Classification is performed on images with a centered hematopoietic cell each – without providing anyfurther details such as segmentation of cytoplasm and nucleus or removal of neighbouring cells. Next to classical machine learning, we also consider modern classification methods based on convolutional neural networks.

2   Dataset

Bone marrow samples are acquired and processed by the Department of Hematology, Oncology, Hemostaseology and Stem Cell Transplantation of the University Hospital, RWTH Aachen University. By applying Pappenheim staining, cell nuclei and different types of cytoplasm can be more easily distinguished. The samples were scanned at the Institute of Pathology of the University Hospital, immersed in oil with 63× magnification. Medical experts of the Department of Hematology annotated representative excerpts and assigned a label to every cell and cell-like artifact.

For the following evaluations, we used patches of size 200 × 200 pixels containing bone marrow hematopoietic cells. While most cells fit completely into a patch of this size, some cell types – e.g., megakaryocytes – are slightly cropped. Neighboring cells are often present as well.

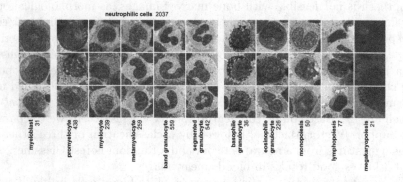

Fig. 1. Three examples of each cell class as well as their sample size in the dataset.

Our dataset comprises 2478 patches with leukocytes separated into the classes denoted in Fig. 1. Additionally, we have 3331 cell-like artifacts and non-leukocytes.

3   Methods

This paper provides a thorough analysis of various algorithms to classify hematopoietic cells in microscope images of stained bone marrow, which are briefly presented in the following paragraphs.

3.1   Classical Machine Learning Algorithms

Classical approaches comprise a feature extraction and a classification stage. The feature extractor provides an abstract representation of the image that the classifier uses to predict its class.

Feature Extraction. In a preliminary evaluation on a smaller subset, we compared several broadly applicable and openly available feature extractors, namely Dense SIFT, HOG, LBP and Daisy. Since we perform classification purely on image data without prior knowledge of segmentations of nucleus and cytoplasm, shape and contour features were not applicable. To have a manageable number of tests, we focused on Histogram of Oriented Gradients (HOG) [5] and Local Binary Pattern (LBP) [6] as representative, well-performing features for the final evaluation presented in this work. The Histogram of Oriented Gradients feature is computed for every color channel with 50 pixels per cell and 4 cells per block. We also applied the same descriptor to an image transformed into polar coordinates (SHOG). Local Binary Pattern is computed on every color channel as well, using the uniform variant, which is rotation and grayscale invariant.

In an attempt to combine classical and deep machine learning approaches, it is also possible to use the output of a convolutional neural network as a feature vector [7]. To this end, we extracted features from the output of the convolutional part of a ResNet18 and a ResNet152 (cf. Sect. 3.2) pretrained on ImageNet. Every feature descriptor was normalized to zero-mean and unit-variance prior to training.

Classification. Based on feature descriptors, a classifier is able to separate two or more classes. Each of the features presented in Sect. 3.1 is fitted using a Linear Support Vector Machine (SVM), [8] a Radial Basis Function Support Vector Machine (RBF) [9], a Random Forest Classifier (RFC) [10], and the AdaBoost Classifier (ABC) [11]. In addition to annotated training data, those algorithms require the user to define several hyper-parameters such as a kernel and a penalty parameter for SVMs or the number of decision trees in a Random Forest.

3.2 Deep Neural Networks

Neural networks with a large number of parameters have become a powerful tool for classification tasks for all kinds of medical image processing and analysis tasks [7]. A deep neural network in its simplest form has several convolutional layers (the output of which can be used as a feature vector as mentioned above) followed by a few fully connected layers (serving as the classification part of the network). A popular model for image classification tasks is the *ResNet*, a deep residual network [12]. It is common practice to initialize the weights of these networks based on a pre-training on a larger image database such as *ImageNet*. This not only drastically reduces the amount of training data necessary for successful network training, it also leads to quicker convergence.

For our evaluation, we chose two variants of ResNet: ResNet18 and ResNet152. As with the training of classical classifiers, a set of hyper-parameters needs to be fine-tuned to achieve high precision. These include, for example, learning rate and data batch size.

We used an Adam optimizer for training these networks, as it showed superior performance compared to Stochastic Gradient Descent (SGD) in preliminary tests. Furthermore, we turn one crossvalidation fold of the training set into a validation set to use the validation error as an early stopping criterion.

4   Experimental Setup

Each classifier is evaluated in several experiments: binary classification between white blood cells and other cells (experiment named *leukocytes*), classification of the grade of maturity of neutrophilic cells (ignoring other cell types, experiment named *neutrophilic cells*), and classification between all mentioned cell types (ignoring non-leukocytes, experiment named *all*).

For each test, we performed a three-fold crossvalidation to obtain the overall results. In every fold, we performed hyper-parameter optimization with six-fold crossvalidation as an inner loop. We also performed very basic data augmentation by using random rotation and a slight translation of up to 10 pixels.

Hyper-parameter Optimization. Every classifier was optimized in terms of its hyper-parameters by using random search over an appropriate range of parameter values. While being able to test 150 hyper-parameter sets with classical models, time constraints only allowed the evaluation of 10 sets with deep learning approaches.

We optimized the linear SVM with respect to the penalty term and the class weights and the radial SVM additionally with respect to the gamma term. For Random Forest we considered the number of estimators, class weights, and maximum depth, and for AdaBoost the number of estimators and learning rate. All of these hyper-parameters were randomly chosen and not algorithmically optimized. We selected these particular sets of parameters for optimization due to their significant impact on training process and classifier quality.

m-score. A significant class-imbalance renders the commonly used accuracy-score infeasible, as smaller classes can be completely misclassified while still having a high overall accuracy. To mitigate this, we used the sum of the logarithmic per-class F1-scores, called *m-score* in the remainder of this paper, as a measure for the selection of the best performing hyper-parameter set: $m = \sum_n^N \log(\mathrm{f1}_n)$ with N classes and $\mathrm{f1}_n$ the F1-score of class n. To prevent negative infinity if at least one class has an F1-score of zero, we capped the individual F1-scores at 0.01.

Implementation. The project is implemented in *Python*, using *Scikit-Learn* for the classical pipeline and *Pytorch* for the deep neural networks. Both pre-trained models were obtained from the *Torchvision* package. Model training was performed on a single *GeForce GTX 1080 Ti* GPU each.

5   Results

The results are shown in Fig. 2: at a first glance it is obvious that all features, classifiers and networks achieve the best results in the binary classification task of distinguishing between leukocytes and cell-like artifacts. Particularly the commonly used image features HOG and LBP reach acceptable scores. Using outputs of the ResNet's convolutional parts did not perform equally well. Training a deep neural network directly, however, yielded excellent results.

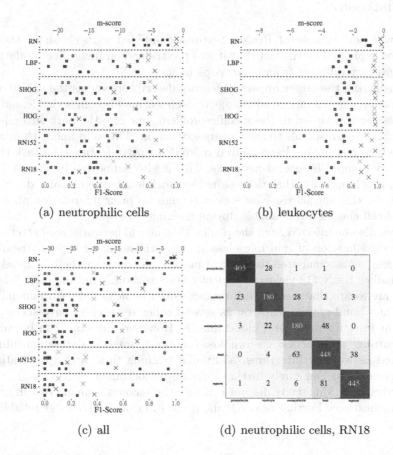

Fig. 2. (a)–(c): results of all three experiments. F1-Scores are marked with a square (bottom horizontal axis), the m-score is marked with an 'x' (top horizontal axis). Grouped by feature descriptor, from bottom to top there are the classifiers SVM (blue), RBF (red), ABC (green), RFC (orange). On the top there are the ResNet18 (lower) and the ResNet152 (upper). (d): confusion matrix for ResNet18 in the *neutrophilic cells* experiment. (Color figure online)

Likewise, neural networks performed significantly better than classical methods in both the *neutrophilic cells* and the *all* experiment. They did not only achieve higher scores, the individual class scores were also more closely together.

Both experiments are demanding for general purpose descriptors and standard classification algorithms: in almost every single test, at least one class – usually one of the underrepresented classes – is completely misclassified. Neural networks, however, still provide valuable results.

The confusion matrix corresponding to classification of neutrophilic cells with a ResNet18 shows that most of the errors are just one class off when considering the ongoing maturation process.

6   Discussion

The lower performance of ResNet features combined with classical classifiers might be explained by the fact that it was pre-trained on ImageNet – a database of images very different from microscope images.

Considering the larger impact of feature descriptors versus the choice of the classification algorithm, one advantage of deep learning becomes visible: not only the classifier is adapted to the specific problem, but also the part of the neural network that corresponds to the feature extraction. This is in line with previous work, where researchers hand-crafted more descriptive features – a task that is inherently covered by neural networks. This is also supported by the confusion matrix, which indicates that the residual network learned significant, descriptive features for this special use-case – even though no prior information about the order of cell classes was available during training.

It can also be inferred from the results that neural networks are better suited for the classification of multiple classes: they still achieved superior results in these cases while compared to classical machine learning methods. Already the more shallow ResNet18 yielded sufficient results without requiring further layers.

We are aware that better performance could be achieved with manual fine-tuning and hand-crafted features, as several other researchers have successfully shown at least for peripheral blood images. However, the same can be said for deep learning approaches: custom loss-functions, smarter data-augmentation, dedicated network architectures, additional training time, and optimisation of hyper-parameters can surely further increase performance.

Furthermore, this study did not evaluate the amount human error in ground truth annotations. Further research about the inter- and intra-rater reliability is needed.

7   Conclusion and Outlook

In this paper, we provide a comparison of several common approaches for the classification of hematopoietic cells. We considered a wide range of white cells in different stages of maturity in the bone marrow, which is a very challenging scenario. We showed that while most classical machine learning algorithms perform poorly, deep neural networks yield promising results.

Particularly the fact that they perform better in multi-class classification is important for the automated analysis of bone marrow smear images, which show an increased number of cell types. It stands to reason that more specialized architectures will be able to yield results accurate enough for medical analysis based on images of bone marrow instead of peripheral blood smears. This paper provides a baseline evaluation for yet to come improvements to neural networks.

Acknowledgement. This work was supported by the German Research Foundation (DFG) under grant no. ME3737/3-1.

References

1. Piuri, V., Scotti, F.: Morphological classification of blood leucocytes by microscope images. In: IEEE International Conference on Computational Intelligence for Measurement Systems and Applications, CIMSA 2004, pp. 103–108. IEEE (2004)
2. Zhao, J., Zhang, M., Zhou, Z., Chu, J., Cao, F.: Automatic detection and classification of leukocytes using convolutional neural networks. Med. Biol. Eng. Comput. **55**(8), 1287–1301 (2017)
3. Putzu, L., Caocci, G., Di Ruberto, C.: Leucocyte classification for leukaemia detection using image processing techniques. Artif. Intell. Med. **62**(3), 179–191 (2014)
4. Theera-Umpon, N., Dhompongsa, S.: Morphological granulometric features of nucleus in automatic bone marrow white blood cell classification. IEEE Trans. Inf. Technol. Biomed. **11**(3), 353–359 (2007)
5. Dalal, N., Triggs, B.: Histograms of oriented gradients for human detection. IEEE Comput. Soc. Conf. Comput. Vis. Pattern Recogn. **1**, 886–893 (2005)
6. Ojala, T., Pietikainen, M., Harwood, D.: Performance evaluation of texture measures with classification based on Kullback discrimination of distributions. In: Proceedings of the 12th IAPR International Conference on Pattern Recognition, vol. 1, pp. 582–585. IEEE (1994)
7. Litjens, G., et al.: A survey on deep learning in medical image analysis. Med. Image Anal. **42**, 60–88 (2017)
8. Chang, C.C., Lin, C.J.: LIBSVM: a library for support vector machines. Trans. Intell. Syst. Technol. (TIST) **2**(3), 27 (2011)
9. Fan, R.E., Chang, K.W., Hsieh, C.J., Wang, X.R., Lin, C.J.: LIBLINEAR: a library for large linear classification. J. Mach. Learn. Res. **9**(Aug), 1871–1874 (2008)
10. Breiman, L.: Random forests. Mach. Learn. **45**(1), 5–32 (2001)
11. Hastie, T., Rosset, S., Zhu, J., Zou, H.: Multi-class AdaBoost. Stat. Interface **2**(3), 349–360 (2009)
12. He, K., Zhang, X., Ren, S., Sun, J.: Deep residual learning for image recognition. In: Proceedings of the IEEE Conference on Computer Vision and Pattern Recognition, pp. 770–778 (2016)

DeepCerv: Deep Neural Network for Segmentation Free Robust Cervical Cell Classification

O. U. Nirmal Jith[1(✉)], K. K. Harinarayanan[1], Srishti Gautam[2],
Arnav Bhavsar[2], and Anil K. Sao[2]

[1] Aindra Systems, Bangalore, India
{nirmal,hari}@aindra.in
[2] School of Computing and Electrical Engineering,
Indian Instiute of Technology, Mandi, India
srishti_gautam@students.iitmandi.ac.in
{arnav,anil}@iitmandi.ac.in

Abstract. Automated classification of cervical cancer cells has the potential to reduce high mortality rates due to cervical cancer in developing countries. However traditional algorithms for the same depend on accurate segmentation of cells, which in itself is an open problem. Often the algorithms are also not evaluated by considering the huge inter-observer variability in ground truth labels. We propose a new deep learning algorithm that does not depend on accurate segmentation by directly classifying image patches with cells. We evaluate the proposed algorithm on the popular Herlev dataset and show that it achieves state of the art accuracy while being extremely fast. The experimental results are also demonstrated using AIndra dataset collected by us, which also captures the inter observer variability.

Keywords: Cervical cancer · Cell classification · PAP-test
Deep learning · Dataset · Inter-observer reliability · Cohen's kappa

1 Introduction

Cervical cancer is the second most common cancer among women, with more than half a million new cases reported every year [1]. However systematic screening of cervical cancer using Papanicolaou test (PAP-test) can reduce mortality rate by 70% or more [6]. PAP-test consists of a cytologist scanning a slide of vaginal smear, typically at 400x magnification. At this magnification, cytologist has to look at thousands of field of views raising the possibility of fatigue, thereby restricting the number of samples observed to 70 per day [5]. In light of these challenges automation of cervical cancer screening has the potential to significantly improve healthcare.

O. U. Nirmal Jith and K. K. Harinarayanan—Equal contribution.

© Springer Nature Switzerland AG 2018
D. Stoyanov et al. (Eds.): COMPAY 2018/OMIA 2018, LNCS 11039, pp. 86–94, 2018.
https://doi.org/10.1007/978-3-030-00949-6_11

In this work we propose a new deep learning algorithm for classification of cervical cancer cells. The algorithm is intended to be usable in a health centre with limited computing resources. Hence it is designed to be extremely fast and lightweight. The algorithm surpasses state of the art performance in Herlev dataset while being robust to segmentation errors. We also conducted experimets using our AIndra dataset. This new cervical cell dataset contains annotations for nuclear boundary and labels by multiple expert annotators. This dataset enables novel analysis and interpretation of classification, segmentation and detection algorithms.

The combination of algorithm and dataset enables unique evaluation strategy. By taking into account the inter-observer variability, we are able to unearth interesting insights into the data which is not evident on the surface during evaluation. To best of our knowledge, no work has been reported, which demostrate the effect of inter-observability for PAP semear images.

2   Related Works

2.1   Performance Measures

Common measures of performance like accuracy, precision and recall presuppose the existence of unique ground truth. This assumes that the disagreement between two observers on a classification label is quite small. In many medical problems, this assumption does not hold true. As an example, [17] found that only 35% of their PAP-test samples have unanimous agreement between pathologists. Hence, it is essential to include inter-observer variability in analysing algorithms.

A common strategy to deal with inter observer variance is to remove samples where observers disagree, but this willreduce the difficulty of problem by removing ambiguous samples. Another approach is to take a majority vote with odd number of observers. This method forces a label on samples which are fundamentally ambiguous. Consequently training/evaluation with this data penalises algorithm on samples where pathologists were indecisive. The consensus in medical community to deal with this challenge is to use measures like percentage agreement between the observers [10] or Cohen's kappa coefficient (κ) [4,10] which discount observer agreement due to random chance. We refer to percentage agreement between the observers using the symbol Θ in the following sections.

2.2   Datasets

The most popular dataset for evaluating cervical cancer cell classification is the Herlev dataset [12]. It consists of 917 high quality images of single cells in seven classes. During data collection, cells were labelled by two cyto-technicians. Cells that were labelled differently were discarded. Consequently the dataset has artificially reduced difficulty as discussed in Sect. 2.1. Though there are other datasets like HEMLBC [20] none of them provide annotations from multiple pathologists, ruling out inter observer variability analysis.

2.3   Algorithms for Cervical Cancer Cell Classification

During the past decade, extensive research has been devoted towards accurate classification of cells for automating PAP test. Most of the methods looked at classification of single cells into various stages of carcinoma [3,13]. These methods in turn relied on accurate segmentation of cell or nucleus. However state of the art segmentation algorithms [3,8,16,18] do not provide the needed segmentation accuracy. As an illustrative example, the best segmentation algorithm achieve a ZSI score of 0.92 on Herlev dataset [19]. When this performance is taken into account classification accuracy drops [20].

An interesting approach that does not rely on accurate segmentation is classification of image patches [11,14]. These patches consists of a cell and its immediate neighbourhood. The recent work [20] used this patch based method. However their evaluation does not involve challenges like clumping, staining variation, overlapping cell etc. The algorithm proposed by [20] is also computationally expensive, taking around 3.5 seconds per input. Given that there are typically upto 300,000 cells in a single slide [12], the algorithm will not be usable on a clinical device.

3   Our Contributions

In this work, we propose a new deep learning algorithm for classification of cervical cancer cells. We also illustrate a unique evaluation strategy.

Salient Aspects of the Proposed Algorithm

- Not reliant on accurate segmentation of nucleus or cytoplasm
- Extremely fast and lightweight
- Surpasses state of the art performance on Herlev dataset

Salient Aspects of Evaluation Method

- Accounts for inter-observer variability using our AIndra dataset
- Includes common evaluation strategy as a subset
- Brings out latent information in the dataset

3.1   AIndra Dataset

This dataset consists, 140 images of conventional PAP smear with sizes varying from 640×480 to 1280×720 pixels with a total of 1201 cells. Each image consists of multiple epithelial cells along with granulocytes like neutrophils. The images also exhibit clumping, defocus blur, staining variation etc. Few sample images from the dataset are shown in Fig. 1. Each epithelial cell in the dataset has its nuclear boundary marked and is classified by a cyto-patholgist (annotator-1) and a cyto-technician (annotator-2) according to Bethesda system [15]. Unlike other datasets, we retain both labels. Among the annotated nucleus, we have an Θ of 76.55% and κ of 0.61.

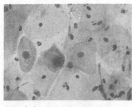 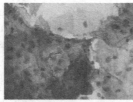 

(a) Good quality   (b) Clumping   (c) Defocus blur

Fig. 1. Sample images containing epithelial cells that exemplify challenges in datasets with respect to image quality, cell distribution and blurring

This dataset is the first cervical cancer cell dataset with multiple expert annotations enabling inter observer variability analysis. Since the dataset contains images with multiple cells, it can be used for benchmarking detection and segmentation of nuclei. The dataset also enables the use of features that are external to epithelial cells, like presence of neutrophils.[1]

3.2 DeepCerv: Network architecture

Our Network design is guided by the twin goals of accuracy and speed. Hence the network takes in raw RGB pixel values without any data prepossessing. Design of the network follows the observation that neural networks for medical image analysis, display adequate performance with low depth. This observation is validated by popular networks in literature like in [21]. The network constitutes the initial three layers of AlexNet [7], batch normalization layer to reduce overfitting, followed by a fully connected layer as depicted in the Table 1

Table 1. Network architecture

	Input	conv1	maxpool1	BN1	Conv2	maxpool2	BN2	conv3	dense1
Filter size		11×11	3×3		5×5	3×3		3×3	
Channels	3	96		96	256		256	384	2
Stride		4			5			1	

The network is designed to process image patches of size (99,99,3) because, at 40X resolution complete cell information will be captured in 99×99 field of view. It classifies cells into two classes; normal and abnormal. The abnormal class consists of the abnormal classes in the Bethesda system and normal class captures the rest.

Model Size and Inference Time:- Our implementation of the network in tensorflow is only 6 MB in size without any optimizations like weight quantization. The network is extremely fast, processing an input in 1.7 ms on a Nvidia

---

[1] Presence of neutrophils signifies inflammation in that region.

Geforce GTX930 M GPU with only 5–10% utilisation. With an estimated count of 300,000 cells per slide [12], our network takes 8.5 min per slide as compared to 12 days for [20]. This performance is without any discount for the low end gpu we use in comparison to the TITAN Z used in [20]. The small size of network coupled with extremely fast performance enable future applications for cervical cancer screening on mobile devices.

3.3   Nucleus Detection for Generating DeepCerv Input

DeepCerv expects input to be single cell images. But all the images in AIndra dataset contains multiple cells, even overlapping cells. Hence we pass these images through a cell detection algorithm to detect cell regions. The algorithm will do a contrast adaptive histogram equalisation (CLAHE) followed by thresholding on the AIndra dataset images to get a binary image. Connected components from this binary image is analysed and those with less than 20% overlap with any groundtruth epithelial cells are rejected. Remaining connected components are considered to be potential nuclei. Though the algorithm is simple, it achieves reasonable performance as given in Table 2. We denote this method by the label SEED.

Table 2. Localization performance of detection algorithm for SEED

Ground truth nucleus	Localization on single nucleus	Localization on multiple nuclei	Missed nucleus
1201	753	49	34(2%)

We also use the annotated ground truth to generate input patches. These patches would be free of segmentation errors and hence would serve as a benchmark of DeepCerv performance. This is similar to the strategies employed in other segmentation free algorithms like [20]. To generate the data we crop a patch of fixed size around the centroid of ground truth. We use the label GND to refer this strategy in following sections.

4   Experiments and Results

DeepCerv is evaluated on Herlev dataset and in the AIndra dataset. In line with other work in literature we report accuracy on Herlev dataset. The reason for better performance of features, estimated using first few layers of the networks, could be that they capture low-level features of cell image, such as texture and smoothness of the cell boundary, hence become decisive features for abnormal cell and supported by Bethseda system also. On the proposed dataset we use inter observer agreement and Cohen's kappa as per the discussion in Sect. 2.1.

Experiment Setup:- We have used Stochastic gradient descent(SGD) optimiser for training the network described in Sect. 3.2, with parameter setup:- *learning rate*= 0.0001, *decay*= $1e^{-6}$, *momentum* = 0.9. The network performance is improved by using the data augmentation methods like *image rotation, width/height shift* and *horizontal flip*

4.1 Experiments on Herlev Dataset

We did a 5 fold cross-validation on this dataset and the results are in Table 3. It is to be noted that no prepossessing of any sort is involved apart from resizing. The seven classes in the dataset were converted to two classes by combining all abnormal classes in Herlev into one and all the normal classes into another. The table clearly shows that we are able to achieve state of the art performance on Herlev dataset.

Table 3. Performance on Herlev dataset

Algorithm	Ensemble [2]	GEN-1nn [9]	DeepPap [20]	ANN [3]	**DeepCerv**
Accuracy	96.5%	96.8%	98.3%	99.3%	99.6%

4.2 Experiments on the AIndra Dataset

Experiments on Data Where Annotators Agree. The discussion in Sect. 2.1 brought forward the impact of various strategies to deal with inter-observer variance. Since the strategy of discarding samples where annotators do not agree is prevalent in literature, we explore the impact of such a strategy. Hence in this experiment we use the cells on which annotators agree, hereafter referred as common data. We perform 5-fold crossvalidation on common data using DeepCerv. From the results given in Table 4 we can see that the percentage agreement and κ between algorithm and common data (80.53%, 0.57) is close to that between two annotators (76.55%, 0.61). When the same experiment is repeated on SEED, the results do not show significant change thereby validating robustness of DeepCerv to segmentation errors.

Table 4. Performance on data where annotators agree

Data	Annotator	Algorithm agreement with annotator	
		Θ	κ
GND	Common	80.53%	0.57
SEED	Common	78.83%	0.54

Performance on Data from Individual Annotators. The strategy of training and testing on common data illustrates the performance of DeepCerv on comparatively error free data. However as this strategy reduces the problem complexity, the results do not reflect the performance on a practical problem. A better estimate would be to see how DeepCerv performs when the data contains cells that are ambiguous and labels have randomness associated with them. Consequently, in this experiment we generate data by using all cells annotated by individual annotators. Similar to the earlier section, we perform 5-fold cross-validation on this data using DeepCerv. The result of the experiment is given in Table 5.

Table 5. Performance on data from individual annotators

Data	Annotator	Algorithm agreement with annotator		Between annotator agreement	
		Θ	κ	Θ	κ
GND	Annotator-1	82.34%	0.62	76.55%	0.61
	Annotator-2	76.04%	0.45	76.55%	0.61
SEED	Annotator-1	79.03%	0.56	NA	NA
	Annotator-2	73.75%	0.43	NA	NA

The results in Table 5 shows observations in line to that of common data. While the algorithm is able to achieve high percentage agreement on both annotators, the performance on annotator-1 exceeds that on annotator-2. Surprisingly algorithm is able to achieve better agreement with annotator-1 than with common data in the previous section. These observations may indicate low annotation consistency of annotator-2 in comparison to annotator-1. Interestingly this aligns with the annotator profiles given in Sect. 3.1 and acts as a validation to the algorithm.

5   Conclusions

In this work, we proposed a new deep learning algorithm for classification of cervical cells. The algorithm is able to surpass state of the art performance in Herlev dataset while being extremely fast in comparison to similar work on cervical cancer cell classification in literature. The algorithm in virtue of high accuracy and speed has the potential to enable automated cervical cancer screening on low power devices while the AIndra dataset allows novel analysis that is much closer to real world applications. Through the combination of algorithm and dataset, we are able to provide novel analysis that brings forward the importance of considering inter observer reliability in context of medical problems and the insights it can provide on data.

References

1. Bengtsson, E., Malm, P.: Screening for Cervical Cancer Using Automated Analysis of PAP-Smears. Computational and Mathematical Methods in Medicine **2014**, 1–12 (2014)
2. Bora, K., Chowdhury, M., Mahanta, L.B., Kundu, M.K., Das, A.K.: Automated classification of Pap smear images to detect cervical dysplasia. Computer Methods and Programs in Biomedicine **138**, 31–47 (2017)
3. Chankong, T., Theera-Umpon, N., Auephanwiriyakul, S.: Automatic cervical cell segmentation and classification in Pap smears. Computer Methods and Programs in Biomedicine **113**(2), 539–556 (2014)
4. Cohen, J.: A Coefficient of Agreement for Nominal Scales. Educational and Psychological Measurement **20**(1), 37–46 (1960)
5. Elsheikh, T.M., Austin, R.M., Chhieng, D.F., Miller, F.S., Moriarty, A.T., Renshaw, A.A.: American Society of Cytopathology: American Society of Cytopathology workload recommendations for automated Pap test screening: Developed by the productivity and quality assurance in the era of automated screening task force. Diagnostic Cytopathology **41**(2), 174–178 (2013)
6. Kitchener, H.C., Castle, P.E., Cox, J.T.: Chapter 7: achievements and limitations of cervical cytology screening. Vaccine **24**, S63–S70 (2006)
7. Krizhevsky, A., Sutskever, I., Hinton, G.E.: Imagenet classification with deep convolutional neural networks. In: Advances in Neural Information Processing Systems. pp. 1097–1105 (2012)
8. Lu, Z., Carneiro, G., Bradley, A.P., Ushizima, D., Nosrati, M.S., Bianchi, A.G.C., Carneiro, C.M., Hamarneh, G.: Evaluation of Three Algorithms for the Segmentation of Overlapping Cervical Cells. IEEE Journal of Biomedical and Health Informatics **21**(2), 441–450 (2017)
9. Marinakis, Y., Dounias, G., Jantzen, J.: Pap smear diagnosis using a hybrid intelligent scheme focusing on genetic algorithm based feature selection and nearest neighbor classification. Computers in Biology and Medicine **39**(1), 69–78 (2009)
10. McHugh, M.L.: Interrater reliability: The kappa statistic. Biochemia Medica **22**(3), 276–282 (2012)
11. Nanni, L., Lumini, A., Brahnam, S.: Local binary patterns variants as texture descriptors for medical image analysis. Artificial Intelligence in Medicine **49**(2), 117–125 (2010)
12. Norup, J.: Classification of Pap-Smear Data by Tranduction Neuro-Fuzzy Methods. masterThesis, Technical University of Denmark, DTU, DK-2800 Kgs. Lyngby, Denmark (2005)
13. Phoulady, H.A., Zhou, M., Goldgof, D.B., Hall, L.O., Mouton, P.R.: Automatic quantification and classification of cervical cancer via adaptive nucleus shape modeling. In: Image Processing (ICIP), 2016 IEEE International Conference On. pp. 2658–2662. IEEE (2016)
14. Sokouti, B., Haghipour, S., Tabrizi, A.D.: A framework for diagnosing cervical cancer disease based on feedforward MLP neural network and ThinPrep histopathological cell image features. Neural Computing and Applications **24**(1), 221–232 (2014)
15. Solomon, D.: The 2001 Bethesda SystemTerminology for Reporting Results of Cervical Cytology. JAMA **287**(16), 2114 (2002)

16. Song, Y., Zhang, L., Chen, S., Ni, D., Lei, B., Wang, T.: Accurate Segmentation of Cervical Cytoplasm and Nuclei Based on Multiscale Convolutional Network and Graph Partitioning. IEEE transactions on bio-medical engineering **62**(10), 2421–2433 (2015)
17. Young, N.A., Naryshkin, S., Atkinson, B.F., Ehya, H., Gupta, P.K., Kline, T.S., Luff, R.D.: Interobserver variability of cervical smears with squamous-cell abnormalities: A philadelphia study. Diagnostic cytopathology **11**(4), 352–357 (1994)
18. Zhang, L., Kong, H., Liu, S., Wang, T., Chen, S., Sonka, M.: Graph-based segmentation of abnormal nuclei in cervical cytology. Computerized Medical Imaging and Graphics **56**, 38–48 (2017)
19. Zhang, L., Kong, H., Ting Chin, C., Liu, S., Fan, X., Wang, T., Chen, S.: Automation-assisted cervical cancer screening in manual liquid-based cytology with hematoxylin and eosin staining. Cytometry. Part A: The Journal of the International Society for Analytical Cytology 85(3), 214–230 (Mar 2014)
20. Zhang, L., Lu, L., Nogues, I., Summers, R.M., Liu, S., Yao, J.: DeepPap: Deep Convolutional Networks for Cervical Cell Classification. IEEE Journal of Biomedical and Health Informatics **21**(6), 1633–1643 (2017)
21. Zhu, X., Yao, J., Zhu, F., Huang, J.: Wsisa: Making survival prediction from whole slide histopathological images. In: IEEE Conference on Computer Vision and Pattern Recognition. pp. 7234–7242 (2017)

Whole Slide Image Registration
for the Study of Tumor Heterogeneity

Leslie Solorzano[1]($\boxtimes$), Gabriela M. Almeida[2,3,4]($\boxtimes$), Bárbara Mesquita[2,3]($\boxtimes$),
Diana Martins[2,3]($\boxtimes$), Carla Oliveira[2,3,4]($\boxtimes$), and Carolina Wählby[1]($\boxtimes$)

[1] Center for Image Analysis, Uppsala University, Uppsala, Sweden
{leslie.solorzano,carolina.wahlby}@it.uu.se
[2] i3S, Instituto de Investigação e Inovação em Saúde,
Universidade do Porto, Porto, Portugal
[3] Ipatimup, Institute of Molecular Pathology and Immunology,
University of Porto, Porto, Portugal
{bmesquita,galmeida,dianam,carlaol}@ipatimup.pt
[4] Faculty of Medicine of the University of Porto, Porto, Portugal

Abstract. Consecutive thin sections of tissue samples make it possible
to study local variation in e.g. protein expression and tumor heterogeneity by staining for a new protein in each section. In order to compare
and correlate patterns of different proteins, the images have to be registered with high accuracy. The problem we want to solve is registration of gigapixel whole slide images (WSI). This presents 3 challenges:
(i) Images are very large; (ii) Thin sections result in artifacts that make
global affine registration prone to very large local errors; (iii) Local affine
registration is required to preserve correct tissue morphology (local size,
shape and texture). In our approach we compare WSI registration based
on automatic and manual feature selection on either the full image or
natural sub-regions (as opposed to square tiles). Working with natural
sub-regions, in an interactive tool makes it possible to exclude regions
containing scientifically irrelevant information. We also present a new
way to visualize local registration quality by a Registration Confidence
Map (RCM). With this method, intra-tumor heterogeneity and characteristics of the tumor microenvironment can be observed and quantified.

Keywords: Whole slide image · Registration · Digital pathology

1 Introduction

In digital histopathology, whole slide imaging (WSI), denotes the scanning of
standard glass slides containing tissue to produce digital images that can be
stored, analyzed, annotated and shared remotely and locally [1,2]. With WSI
comes a set of challenges, comparable to geographical information systems and

C. Wählby—European Research council for funding via ERC Consolidator grant
682810 to C. Wählby.

© Springer Nature Switzerland AG 2018
D. Stoyanov et al. (Eds.): COMPAY 2018/OMIA 2018, LNCS 11039, pp. 95–102, 2018.
https://doi.org/10.1007/978-3-030-00949-6_12

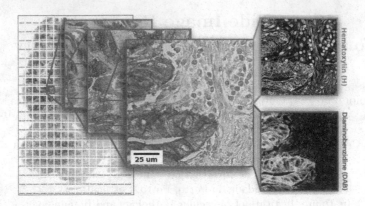

Fig. 1. (Left) 320 5000 pixels^2 tiles are represented. One tile is "zoomed in" up to a full resolution of 0.25 μm per pixel. (Right) Two colors for H and DAB are shown in separate tiles after color deconvolution.

astronomical images. WSI have sizes ranging from 35,000 to 200,000 pixels^2. These images require special formats, readers, standards and hardware to be read and processed. Processing of this type of images requires either computers with enough memory or smart strategies for partitioning the data without losing the spatial context.

Usually, a tissue section scanned with brightfield microscopy is stained with two or more stains. The tissue is "photographed" with visible light and the result is a microphotograph collected in 8-bit pixels in RGB space. Figure 1 shows an example of a WSI of gastric cancer tissue stained with Hematoxylin and Diaminobenzidine (H and DAB). H stains for sub-cellular compartments present in all the tissue such as the cellular membranes, cytoplasms and nuclei, allowing to delimit and resolve objects of interest. DAB on the other hand stains for a specific protein only present in certain cell types and sub-cellular compartments.

Conventional image processing libraries are not designed to process WSI. Most of them were conceived to index up to 2^{32} pixels and fail when naively prompted to open special headers and formats like those used for WSI. Images can be stored in a pyramidal fashion, where several resolutions of the image are saved in the same file (or in separate files) and can be accessed according to the desired resolution level. In order to create correct and efficient tools for pathologists, new and faster methods for image processing, such as convolution, filtering, registration, annotation and visualization have to be developed for this special type of images.

In the study of tumor microenvironments (TME) and intra-tumor heterogeneity, consecutive thin sections of tissue are stained so that a common structure is revealed [3,4], for instance with H, while each section is stained separately with DAB for a different protein. These sections are then scanned into a WSI that can be partitioned into new image channels by color deconvolution [5], making it possible to process general tissue information from H, and specific information from DAB, separately. After the jump from analog to digital pathology, new methods are needed to help automate the analysis of big amounts of WSI. But each WSI is not useful on its own, information has to be spatially aligned and there is a need for better methods of WSI registration and assessment of its quality.

Pixel perfect alignment in WSI is impossible, particularly because structures in consecutive tissue sections should appear in different locations and sizes. This is a common problem in 3D tissue reconstruction. A choice has to be made, either the image is deformed to try and achieve pixel perfect alignment or a different transformation is chosen. Image registration (image spatial alignment) is commonly expressed as an optimization problem which can be solved by iteratively searching for transformation parameters such that the distance between a pair of points (feature-based) or images (intensity-based) is minimized. The transformation can be rigid, affine or non-rigid (deformable, or non-linear). An affine transformation is done to preserve points, parallel lines and ratios between structures.

In research, many kinds of registration methods are continuously experimented with, but according to [6], lack of genericity and robustness has prevented the inclusion of deformable methods in commercial software for clinical diagnoses while globally rigid registration is the most frequently used approach. The reason is the ability to control and study the parameters for registration. With affine methods, fewer parameters have to be set as opposed to deformable approaches, and the distance between spatial landmarks can be minimized without compromising the structural integrity of the tissue. In affine transformations, the values of the affine matrix represent the amount of stretching and skewing which can be constrained. In multimodal imaging, entropy based measures are the most common distance measure in intensity based registration [7]. Within the feature based registration approaches, a spatial correlation between features is used to find the transformation. Depending on the desired resolution the alignment is expected to achieve, different registration schemes can be planned, where successive registration methods can be applied. There is no silver bullet and some methods may become irrelevant for different types of applications. It is very common to start with a rougher and less expensive global alignment in lower resolutions, followed by affine or non-rigid deformations. For instance, in histology images [4], uses a global rigid transformation followed by local non-rigid transformations of selected ROIs guided by intensity based information. In [8,9] there are: a feature based rigid transformation followed by feature based local refinements. Compared to previous works, our applications require cellular alignment, and a quantifiable confidence for it.

The main contributions of this paper are a piece-wise approach for image registration and visualization of registration confidence. The piece-wise approach lets us overcome the challenges of handling large WSI and tissue artifacts. We create a Registration Confidence Map (RCM) as well as a stain Combination Quantication Map (CQM) to visualize local structure and colocalization of proteins within the tissue sample. Automated and semi-automated registration methods require fast and efficient visualization of the images to immediately detect or correct mistakes. Even when handling big amounts of data, visualization is important before making any kind of prognostic assessment.

2   Materials and Methods

2.1   Image Data

We received a set of WSI of thin consecutive sections of gastric tumor, where each section spans the thickness of a single cell. All sections were exposed to the same H, but each DAB represents a unique protein per tissue section. Three proteins codenamed A, B and C are presented here. Pixel size is 0.25 μm, and the full images are in average 90000 pixels^2. Despite careful sample preparation, artifacts such as folds, rips and loss of tissue were observed in all images.

2.2   WSI Preprocessing

We first separate H which reveals common tissue structures from the specific DAB in all slides. In the end, two images of the same size as the original are saved, one representing H and one representing DAB. For this purpose, we calculated a stain matrix so that a pixel in RGB space can be projected to an H and DAB space. We acquired a new palette on each tissue section. We implemented color deconvolution based on the work of [5,10], the optical density (OD) matrix is calculated and then its inverse allows to send RGB points to H and DAB space.

After color deconvolution we contrast stretched all channels to the 99th percentile followed by a gamma correction with gamma 1.85. This is important since even the same stain can produce different results depending on how much it can actually penetrate a cell and parts of the tissue. It is important to note as well that in this case we are not quantifying the amount of a given stain but its presence. There has been much controversy as to decide if the pixel intensity represents the amount of stain and the nature of this representation [11].

As part of the WSI preprocessing we also developed a web-interface for manual selection of pairs of natural sub-regions and control points. The web interface is based on OpenSeaDragon [12] and enables quick zooming, panning and selection of regions. Image artifacts such as folds and tears are easy to spot visually, and we selected pairs of high-quality image regions from consecutive tissue sections. An example of sub regions is shown in Fig. 2. These sub-regions served as input for feature selection and registration as described below.

2.3   Feature Extraction and Image Registration

After regions are selected, common spatial features were found in each pair of corresponding regions and then these features were matched thus finding a transformation. Our innovation comes from the use of only the H channel, the channel that contains the common tissue structure and morphology.

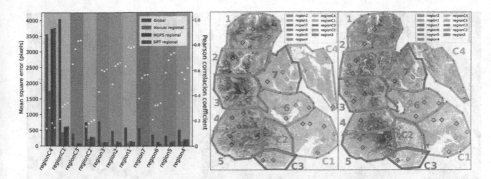

Fig. 2. Regions, MSE of landmarks after registration and PCC for each region. Bars show the MSE of distance between landmarks in pixels and the white dots show the PCC where 1 means complete colocalization.

To extract features we compare three methods, namely Scale Invariant Feature Transform (SIFT), Multi-Scale Oriented Patches (MOPS) and manually using a visualization and tool we developed for this purpose. Using Fiji [13], a list of possible pairs of feature sets are found and matched using RANSAC.

We thereafter found a transformation by matching features using the Blendenpik [14] least squares minimization as used by Numpy. This results in a transformation matrix that is applied either to a specific region or to the whole tissue. If regions are used, one must ensure that the regions have sufficient overlap to avoid introducing holes in the resulting image in case a transformation becomes too big. Nevertheless big transformations are not expected between adjacent regions.

2.4 Evaluation and Creation of RCM and CQM

We evaluated global registration results using the Pearson correlation coefficient (PCC) [15] with the two H as input, including all pixel pairs where at least one of the pixels was greater than zero. Next, we also evaluated the image registration result by defining colocalization as the percentage of pixel pairs where both intensity values are above the Costes threshold [16] which estimates the maximum threshold for each color below which pixels do not show any statistical correlation. Table 1 presents the PCC and the percentages of colocalized image pixels for each of the proposed approaches.

Table 1. PCC table for proteins A and B.

Images	PCC total	PCC coloc	%A Vol	%B Vol	%A >th	%B >th
Affine global	0.438	0.1437	63.32%	71.45%	66.03%	72.95%
Manual regional	0.697	0.6387	85.36%	86.81%	88.50%	89.35%
MOPS regional	0.696	0.6417	85.10%	86.60%	88.46%	88.98%
SIFT regional	0.697	0.6475	85.10%	86.58%	88.47%	88.97%

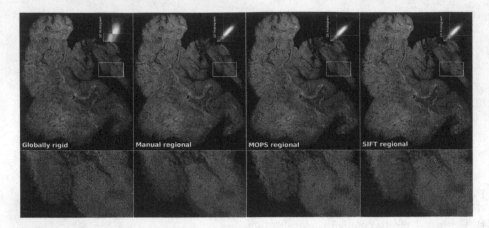

Fig. 3. RCM of stain A and B. Global affine, manual regional, MOPS regional and SIFT regional approaches to registration. Each has it's own 2D histogram color coded with the frequencies of intensity pairs. Thresholds are also shown. White means the pixels are colocated, while red and green shows regions where the H staining for slide A or B do not match. A closeup within region C4 is added to show a higher level of detail. (Color figure online)

Next we created the Registration Confidence Map (RCM) by subsampling the WSI by a factor 200. Using the Costes thresholds described above, output pixels were either colorcoded white (A >t and B >t), red (A >T, B ≤t), green (A ≤t, and B >T) or black (A ≤t, B ≤t).

Finally, a map of protein colocalizations was created, referred to as the Colocalization Quantification Map (CQM). Again, Costes thresholds were applied and each pixel color coded based on what combination of stains reached above the intensity threshold. An example is shown in in Fig. 4.

3 Results

In the presented results, three proteins codenamed A, B and C are registered and analyzed for colocalization in the tissue. Their CQM and RCM are shown in Figs. 3 and 4, respectively.

Image registration can be evaluated by finding distance between landmarks after transformation, using intensity based methods, or by comparison of transformation parameters As seen in [17], we consider landmark and contour based methods insufficient for our purpose. Distance between landmarks can be minimized but this does not guarantee alignment of the rest of the tissue. For this reason we consider RCM a good measure to find the locations within the tissue sections that can be trusted and to which degree. Nevertheless in Fig. 2 we offer a comparison of the mean square error (MSE) of landmark distances for each approach. It is worth noting that in region C4 points have been manually selected intentionally to force a fit of the tissue. For this reason the manual landmarks approach optimizes the distance between landmarks and shows an

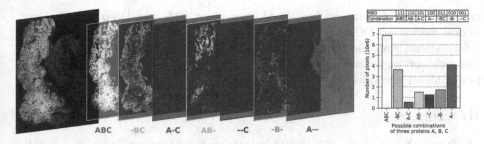

Fig. 4. CQM of three slides stained for different proteins A, B and C. Different combinations of proteins can be analyzed and their colocalization quantified per pixel. Given that three proteins are being studied, the results are shown in RGB color space. The plot shows the amount per protein combinations in the WSI. (Color figure online)

apparently good result while it can be visually assessed in Fig. 3 that region C4 is not properly aligned and that probably should be separated in several regions. Additionally, regions C2 and C3 each achieve a small distance between landmarks yet they have very different PCC. This confirms that landmark based evaluation should not be the only method to evaluate registration. In Table 1 several values are offered to quantify and interpret what is observed in the RCM.

– PCC total: PCC for all the non zero-zero pixels in the image.
– PCC coloc: PCC for pixels where both H levels are above their respective threshold (white box in the 2D histogram).
– %A vol (and %B vol): the number of pixels in the white box divided by the sum of white and green (or red for B) boxes.
– %A >t (and %B >t): the sum of the intensities of the pixels in the white box divided by the sum of intensities on both white box and green box (or red for B).

Figure 4 shows the overlap in expression of three proteins, all registered using our method. These CQM will be used in overlap quantification for tumor characterization studies.

4   Conclusions

We present a new way to do WSI registration and a new way to evaluate it based on the comparison of tissue morphology by using PCC as well as spatial maps of RCM and CQM. This general framework could be used for any staining protocol (HDAB, H&E, etc.) so long as the stain used for matching is comparably consistent across tissue sections. These maps do not only provide an excellent visual representation of spatial heterogeneity of tumor tissues, but can also serve as computer-generated input for training deep convolutional neural networks. Many learning approaches today rely on manual annotations, which are often highly variable between expert pathologists. Tumor niches and normal tissue can instead be automatically defined from our CQM, and Deep convolutional neural networks (DCNNs) can be trained to detect corresponding structural information (from the H-channel) in a clinical setting. This tool is expected to improve our understanding on ITH with a potential impact for definition of personalized therapies.

References

1. Sucaet, Y., Waelput, W.: Digital Pathology. SCS. Springer, Heidelberg (2014). https://doi.org/10.1007/978-3-319-08780-1
2. Ameisen, D.: Towards better digital pathology workflows: programming libraries for high-speed sharpness assessment of whole slide images. Diagn. Pathol. 9(Suppl. 1), S3 (2014). https://doi.org/10.1186/1746-1596-9-S1-S3
3. Spagnolo, D., et al.: Platform for quantitative evaluation of spatial intratumoral heterogeneity in multiplexed fluorescence images. Cancer Res. 77, e71–e74 (2017). American Association for Cancer Research
4. Moles Lopez, X., et al.: Registration of whole immunohistochemical slide images: an efficient way to characterize biomarker colocalization. J. Am. Med. Inform. Assoc. 22(1), 86–99 (2015)
5. Ruifrok, A.C.: Quantification of histochemical staining by color deconvolution. Anal. Quant. Cytol. Histol. 23, 291–299 (2002)
6. Viergever, M., et al.: A survey of medical image registration. Med. Image Anal. 33, 140–144 (2016)
7. Gurcan, M.N., et al.: Histopathological image analysis: a review. IEEE Rev. Biomed. Eng. 2, 147 (2009)
8. Cooper, L., et al.: Feature-based registration of histopathology images with different stains: an application for computerized follicular lymphoma prognosis. Comput. Methods Programs Biomed. 96(3), 182–192 (2009)
9. Trahearn, N., et al.: Hyper-stain inspector: a framework for robust registration and localised co-expression analysis of multiple whole-slide images of serial histology sections. Sci. Rep. 7, 5641 (2017)
10. Wemmert, C., et al.: Stain unmixing in brightfield multiplexed immunohistochemistry. In: 2013 IEEE International Conference on Image Processing (2013)
11. van Der Laak, J.A., et al.: Hue-saturation-density (HSD) model for stain recognition in digital images from transmitted light microscopy (2000)
12. An open-source, viewer for high-resolution zoomable images, in JavaScript. https://openseadragon.github.io. Accessed 13 May 2018
13. Image J Feature extraction. https://imagej.net/feature_extraction. Accessed 24 Sept 2015
14. Avron, H., et al.: Blendenpik: Supercharging LAPACK's least-squares solver. SIAM J. Sci. Comput. 32(3), 1217–1236 (2010)
15. Oheim, M., Li, D.: Quantitative colocalisation imaging: concepts, measurements, and pitfalls. In: Shorte, S.L., Frischknecht, F. (eds.) Imaging Cellular and Molecular Biological Functions. Principles and Practice, pp. 117–155. Springer, Heidelberg (2007). https://doi.org/10.1007/978-3-540-71331-9_5
16. Costes, S.V., et al.: Automatic and quantitative measurement of protein-protein colocalization in live cells. Biophys. J. 86, 3993–4003 (2004)
17. Rohlfing, T.: Image similarity and tissue overlaps as surrogates for image registration accuracy: widely used but unreliable. IEEE Trans. Med. Imaging 31(2), 153–163 (2012)

Modality Conversion from Pathological Image to Ultrasonic Image Using Convolutional Neural Network

Takashi Ohnishi[1]([⊠]) [iD], Shu Kashio[2], Takuya Ogawa[2], Kazuyo Ito[3],
Stanislav S. Makhanov[4], Tadashi Yamaguchi[1], Yasuo Iwadate[5],
and Hideaki Haneishi[1]

[1] Center for Frontier Medical Engineering, Chiba University,
Chiba 263-8522, Japan
t-ohnishi@chiba-u.jp
[2] Graduate School of Science and Engineering,
Chiba University, Chiba 263-8522, Japan
[3] Graduate School of Engineering, Chiba University, Chiba 263-8522, Japan
[4] Sirindhorn International Institute of Technology, Thammasat University,
Khlong Luang 12120, Pathum Thani, Thailand
[5] Graduate School of Medicine, Chiba University, Chiba 260-8670, Japan

Abstract. Relation analysis between physical properties and microstructure of the human tissue has been widely conducted. In particular, the relationships between acoustic parameters and the microstructure of the human brain fall within the scope of our research. In order to analyze the relationship between physical properties and microstructure of the human tissue, accurate image registration is required. To observe the microstructure of the tissue, pathological (PT) image, which is an optical image capturing a thinly sliced specimen has been generally used. However, spatial resolutions and image features of PT image are markedly different from those of other image modalities. This study proposes a modality conversion method from PT image to ultrasonic (US) image including downscale process using convolutional neural network (CNN). Namely, constructed conversion model estimates the US from patch image of PT image. The proposed method was applied to the PT images and we confirmed that the converted PT images were similar to the US images from visual assessment. Image registration was then performed with converted PT and US images measuring the consecutive pathological specimens. Successful registration results were obtained in every pair of the images.

Keywords: Modality conversion · Pathological image · Ultrasonic image
Convolutional neural network

1 Introduction

In recent years, the physical properties of human tissue such as mechanical, optical and acoustic properties have been widely measured. In addition, these properties have been compared with microstructure such as the distribution of the cell nuclei and the running

© Springer Nature Switzerland AG 2018
D. Stoyanov et al. (Eds.): COMPAY 2018/OMIA 2018, LNCS 11039, pp. 103–111, 2018.
https://doi.org/10.1007/978-3-030-00949-6_13

direction of nerve fiber [1–3]. The microstructure of tissue can be acquired as pathological (PT) images which are optical images of thinly sliced specimens. Methodologies of multi-modal analysis using such PT images and other modal images have been widely developed. We also have been analyzing a relationship between acoustic parameters and microstructures of the human brain using PT images and microscopic ultrasonic (US) images.

To compare the physical properties and the microstructure at the same location using multi-modal images, accurate image registration is required. Previous studies employed landmark-based or semi-automatic methods [4–7]. However, a correction of local differences was too difficult because tissue characteristics in the PT image are not taken into consideration in these methods, which makes detection of the corresponding landmarks difficult. In this case, an intensity-based registration may be more promising.

When the PT image is used in the intensity-based registration, a spatial resolution of the PT image can be a hurdle because it is extremely higher than that of other image modalities such as computed tomography, magnetic resonance imaging and US image. For example, the spatial resolution of the PT image is approximately 230×230 nm^2 while the spatial resolution of the US image measured by an US microscopic system is approximately 8×8 μm^2 at most. Therefore, when a pixel is selected on the US image during the registration process, the corresponding pixel value is calculated from 35×35 pixels on the PT image. In this situation, the spatial resolution of the PT image is generally adjusted to almost the same as those of another image using an averaging and downsampling technique before the image registration. However, such a simple downscaling processing eliminates microscopic pattern that each organ inherently possesses and leads to a decline in registration accuracy.

To enhance each structural component in the PT image and achieve the highly accurate image registration, we introduced an image feature conversion method combining with the downscale process. This study tuned up a conversion method assuming the image registration between PT and US images. As a preliminary experiment, simple affine registration was conducted.

2 Methods

The proposed method consists of two steps as shown in Fig. 1. In the model construction step, the landmark-based registration with PT and US images was conducted. US image was moved to the coordinate system of PT image in this registration process. Because the original PT image was too large, a region of interest was set to the PT image. Rescaled PT image was generated using the simple average method and then binarized with discriminant analysis method. The landmarks were detected by AKAZE feature detector [8] from the binarized PT image. Outliers of the landmarks were removed by random sampling consensus [9]. This registration results must be visually confirmed by the operator. The conversion model was constructed with the original PT image and the registered US image using convolutional neural network (CNN) [10]. Figure 2a shows the flow of the conversion model construction. Some patch images were extracted from the original PT image. The conversion model estimates an US signal from each small patch image. Estimated US signals p_i were compared with actual US signals l_i. CNN was optimized until the cost function becomes minimum. The cost function which is also called a loss function was defined as:

$$\text{Cost}(\mathbf{P}, \mathbf{L}) = \sum_{k=1}^{N} |p_k - l_k|. \tag{1}$$

Here, k and N describe the index of patch image and the total number of patch image input into the CNN, respectively. These processes were repeated until iteration number reached to a predefined limit. The framework of the CNN is shown in Fig. 2b. There were two convolution layers and two pooling layers followed by dropout and fully connected layers. CNN construction has to be conducted once before the actual registration.

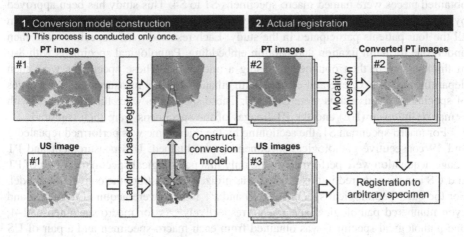

Fig. 1. Outline of the proposed method.

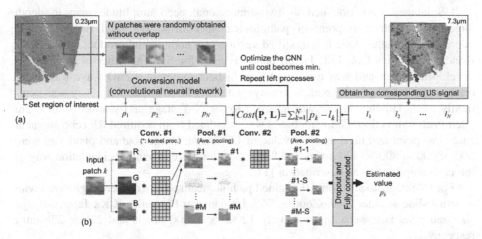

Fig. 2. (a) Flow of conversion model construction, (b) Framework of convolutional neural network.

In the second step, the obtained PT images for the image registration were converted by the constructed model. Affine registration including shift, rotation and scaling operations were then conducted. Normalized cross correlation (NCC) and Powell-Brent method were utilized as a similarity measure between converted PT and US images and an optimization method, respectively.

3 Experiments

3.1 Data

Brain tumors have been resected from four patients as a normal clinical procedure. After the surgery, the resected tumors were further resected into some pieces. These obtained pieces were named macro-specimens S1 to S4. This study has been approved by the Ethical Review Board of our University and we obtained informed consent from all the four patients participated in the study. Each resected macro-specimen was then undergone formalin fixation and paraffin embedding. Pathological specimens with 8-μm thickness were then obtained by using a microtome. These specimens were then deparafinized with xylene and cleaned with ethanol. For US measurement, the images of specimens in this status were captured. These specimens were further stained with hematoxylin-eosin (HE) and the PT images of the specimens were then captured.

For macro-specimen S1, the sectioning by the microtome was performed repeatedly and 19 consecutive pathological specimens were obtained. US measurement and PT image acquisition were performed on the only first pathological specimen. A pair of PT and US images acquired in this process was utilized to construct the conversion model. For the other pathological specimens, US and PT images were acquired from odd and even numbered pathological specimens, respectively. As for macro-specimens S2–4, one pathological specimen was obtained from each macro-specimen and a pair of US and PT images was acquired in each macro-specimen just as the pair of PT and US image of S1.

US images were obtained as two-dimensional echo amplitude map in depth direction at each scan point of pathological specimens. We used two ultrasonic microscopic systems. One is a modified version of a commercial product (AMS-50SI, Honda Electronics Co., Ltd, Japan) and was used for S1. The other is an in-house developed system and was used for S2–4. In both systems, a ZnO wave transducer (Fraunhofer IMBT, St. Ingbert, Germany) with 250 MHz center frequency was commonly used. This transducer was attached to the X-Y stage and scanned with 8-μm pitch in each direction. Echo amplitude was calculated from acquired RF echo signal at each scan point and used as pixel value of US image. Image size and pixel size were 300×300 to 800×800 pixels and 8.0×8.0 μm^2/pixel. Detailed calculation way of the echo amplitude was described in [11].

For PT image acquisition, HE-stained pathological specimens were digitalized with a virtual slide scanner (NanoZoomer S60, Hamamatsu Photonics K.K., Japan). Image size and pixel size were approximately 12000×12000 pixels and 228×228 nm^2, respectively.

3.2   Results and Discussions

In this study, two kinds of experiments were conducted. In the first experiment, applicability of the conversion model was evaluated with PT and US images obtained from the same macro-specimen S1. A conversion model was constructed with a pair of PT and US images and applied to other nine PT images. Image registration was then undergone. The first US image was used as reference image and the other images were registered into the first US image. To evaluate the versatility of the conversion model, another experiment was conducted with the images of S2–4. A conversion model was constructed with the images of S2 and applied to the PT images of S3 and S4. The patch size for the conversion model was set to 32 × 32 pixels. Namely, the pixel size after conversion were 7.30 × 7.30 μm². The number of iteration, the number of patch image (batch size), a learning rate and a dropout rate for CNN were set to 20,000, 100, 0.001 and 0.5, respectively.

Experiment 1: Study on the Consecutive Specimens Resected From a Patient
Figure 3 shows a result of conversion model construction. As yellow arrows in Fig. 3c indicate, black spots were clearly enhanced after conversion. From visual assessment, the converted PT image was similar to the US image compared with the simply downscaled PT image. The constructed conversion model was applied to other PT images. Conversion results are shown in Fig. 4. Because US images corresponding to PT images were not acquired, we could not evaluate the conversion effect quantitatively. However, we confirmed that image features of all converted images were similar to those of US image shown in Fig. 3d.

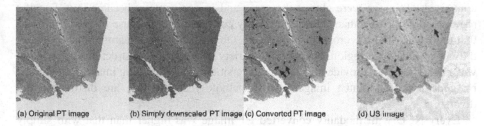

(a) Original PT image   (b) Simply downscaled PT image (c) Converted PT image   (d) US image

Fig. 3. Result of conversion model construction. (a) Original and (b) Simply downscaled PT image, (c) Converted PT image, (d) US image. Yellow arrows indicate enhanced regions by the conversion. It should be noted that image size of the original PT image was larger than that of other three images in practice. (Color figure online)

Registration results are shown in Fig. 5. All images including both US and PT images were successfully registered. Because image features of converted PT and US images were similar, NCC provided acceptable results.

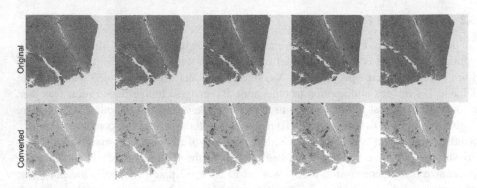

Fig. 4. Conversion results of S1. Top: Original PT images. Bottom: Converted PT images using the same model in Fig. 3.

Experiment 2: Study on the Specimens Resected From Different Patients

A conversion model was constructed with the image dataset of S2 and applied to the image datasets of S3 and S4. Resultant images are shown in Fig. 6. NCC between the US image and each downscaled PT image was calculated at the best match position. The best match positions for each image were manually decided by the operator. Calculated NCCs are shown in Table 1.

Some structures in the converted PT image of S2 were slightly enhanced as shown in enlarged region of Fig. 6. For image dataset of S3, tendency of conversion result was similar to S2. We confirmed the effect of the modality conversion, however it was less than that in the previous experiment. For the image dataset of S4, pixel values of the converted PT image were almost the same and image contrast became low. Specific region could not be enhanced after modality conversion. The cause of this result was that there were many necrosis regions on S4. On the other hand, learning image dataset did not include such region. Namely, the variety of pathological structures to be learnt was not sufficient to provide a versatile conversion model. Learning image dataset must be generated from patch images whose pathological structure are distinctive and diverse.

Every NCC with modality converted PT image was higher than that with simply downscaled PT image. Even though simple downscale method led to high NCC, the proposed method further improved it because the conversion model was optimized with S2. NCCs of the other two datasets were also improved and achieved more than 0.9 using the conversion model. Although we could not visually confirm the effectiveness of the proposed method, histogram or spatial distribution of pixel value was similar to that of the US image. From these results, we can expect that the proposed method will be able to produce the better registration than the simple downscale method.

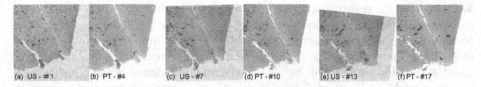

Fig. 5. Example of registration results. (a) Reference US image, (c)(e) Registered US images, (b)(d)(f) Registered PT images. The numbers describe pathological specimen number.

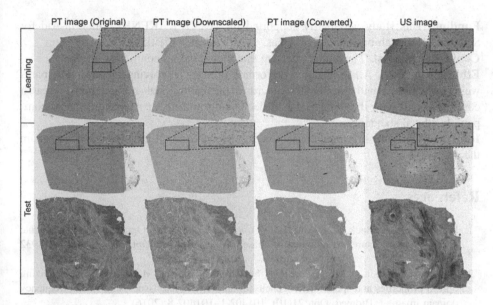

Fig. 6. Conversion results of S2–4. Top: Learning image dataset (S2), Middle and bottom: Test image dataset (S3 and S4).

Table 1. Normalized cross correlation between ultrasonic and each downscaled pathological image at best match position.

	S2	S3	S4
Simply downscaled	0.956	0.796	0.803
Modality converted	0.978	0.955	0.915

4 Conclusions

To conduct the image registration with ultrasonic image, we proposed a CNN-based modality conversion method for pathological image. From visual assessment, converted pathological images were similar to the ultrasonic image compared with the simply downscaled pathological image. Therefore, we will be able to obtain the highly accurate registration result without additional intelligent and/or complicated registration method.

We found that the registration results were successfully obtained using classical similarity measure. For future work, we would like to increase the various image datasets because the CNN can not estimate acceptable US signal from unseen image pattern. Quantitative assessment of the image registration and comparison with the landmark-based registration method will be also conducted. In addition, because pathological specimen might be locally deformed in the staining process, a non-rigid registration method would be introduced.

Compliance with Ethical Standards

Funding. This study was partly supported by MEXT KAKENHI, Grant-in-Aid for Scientific Research on Innovative Areas, Grant Number 17H05278.
Conflict of Interest. The authors declare that they have no conflict of interest.
Ethical Approval. All procedures performed in study involving human participants were in accordance with the ethical standards of the institutional and/or national research committee and with the 1964 Helsinki declaration and its later amendments or comparable ethical standards.
Informed Consent. Informed consent was obtained from all individual participants in the study.

References

1. Ban, S., Min, E., Baek, S., Kwon, H.M., Popescu, G., Jung, W.: Optical properties of acute kidney injury measured by quantitative phase imaging. Biomed. Opt. Express **9**(3), 921–932 (2018)
2. Nandy, S., Mostafa, A., Kumavor, P.D., Sanders, M., Brewer, M., Zhu, Q.: Characterizing optical properties and spatial heterogeneity of human ovarian tissue using spatial frequency domain imaging. Biomed. Opt. **21**(10), 101402-1–101402-8 (2016)
3. Rohrbach, D., Jakob, A., Lloyd, H.O., Tretbar, S.H., Silberman, R.H., Mamou, J.: A novel quantitative 500-MHz acoustic microscopy system for ophthalmologic tissues. IEEE Trans. Biomed. Eng. **64**(3), 715–724 (2017)
4. Choe, A.S., Gao, Y., Li, X., Compthon, K.B., Stepniewska, I., Anderson, A.W.: Accuracy of image registration between MRI and light microscopy in the ex vivo brain. Magn. Reson. Imaging **29**(5), 683–692 (2011)
5. Goubran, M., et al.: Registration of in-vivo to ex-vivo MRI of surgically resected specimens: a pipeline for histology to in-vivo registration. J. Neurosci. Method **241**, 53–65 (2015)
6. Elyas, E., et al.: Correlation of ultrasound shear wave elastography with pathological analysis in a xenografic tumour model. Sci. Rep. **7**(1), 165 (2017)
7. Schalk, S.G., et al.: 3D surface-based registration of ultrasound and histology in prostate cancer imaging. Comput. Med. Imaging Graph. **47**, 29–39 (2016)
8. Alcantarilla, P.F., Nuevo, J., Bartoli, A.: Fast explicit diffusion for accelerated features in nonlinear scale spaces. In: Proceedings of British Machine Vision Conference, pp. 13.1–13.11 (2013)
9. Fischler, M.A., Bolles, R.C.: Random sample consensus: a paradigm for model fitting with applications to image analysis and automated cartography. Commun. ACM **24**(6), 381–395 (1981)

10. Lawrence, S., Giles, C.L., Tsoi, A.C., Back, A.D.: Face recognition: a convolutional neural-network approach. IEEE Trans. Neural Netw. **8**(1), 98–113 (1997)
11. Kobayashi, K., Yoshida, S., Saijo, Y., Hozumi, N.: Acoustic impedance microscopy for biological tissue characterization. Ultrasonics **54**(7), 1922–1928 (2014)

Structure Instance Segmentation in Renal Tissue: A Case Study on Tubular Immune Cell Detection

T. de Bel[(✉)], M. Hermsen, G. Litjens, and J. van der Laak

Diagnostic Image Analysis Group, Department of Pathology,
Radboud University Medical Center, Nijmegen, The Netherlands
thomas.debel@radboudumc.nl

Abstract. In renal transplantation pathology, the Banff grading system is used for diagnosis. We perform a case study on the detection of immune cells in tubules, with the goal of automating part of this grading. We propose a two-step approach, in which we first perform a structure segmentation and subsequently an immune cell detection. We used a dataset of renal allograft biopsies from the Radboud University Medical Centre, Nijmegen, the Netherlands. Our modified U-net reached a Dice score of 0.85 on the structure segmentation task. The F1-score of the immune cell detection was 0.33.

1 Introduction

An important step in many automated medical image analysis applications is semantic segmentation. It describes the task of assigning a class to each pixel of an image. Since the advent of deep learning, great leaps in semantic segmentation performance have been achieved. One of the first applications of deep learning to semantic segmentation came from the Long et al. They used fully convolutional neural networks (F-CNN), omitting the fully connected layers, to perform end-to-end semantic segmentation, achieving state-of-the-art results on multiple datasets [7]. Recently, other fully convolutional networks emerged, achieving great results in semantic segmentation tasks. For instance, U-net applies deconvolutions and skip connections to achieve precise segmentation results [11]. Chen et al. used a contour-aware network to accurately segment gland instances, winning the MICCAI 2015 gland segmentation challenge [1].

The development of high resolution whole slide image (WSI) scanners paved the way for applying semantic segmentation techniques to histopathological slides [3]. Applications in pathology are seen in, for example, cancer diagnosis [6]. However, applications in the kidney are less common. Recent literature on renal tissue segmentation has focused mostly on the automatic segmentation and detection of glomeruli [9,12].

With thousands of people on the waiting list for kidney transplantation in Europe, correct diagnosis and treatment of rejection of transplanted kidneys is

© Springer Nature Switzerland AG 2018
D. Stoyanov et al. (Eds.): COMPAY 2018/OMIA 2018, LNCS 11039, pp. 112–119, 2018.
https://doi.org/10.1007/978-3-030-00949-6_14

of great importance to prevent the loss of transplanted organs. The Banff classification is an important guideline in such diagnosis [8]. The Banff classification is based on a set of gradings, one of which is the estimation of tubulitis: inflammation of the tubular structures in the kidney. This inflammation is graded by quantifying the amount of immune cells in the tubular structures. The assessment of tubulitis is known to be labor intensive and may possess limited accuracy [2]. As a case study, we build upon our structure segmentation results to quantify tubular inflammation. We attempted to identify and count immune cells within the tubular structures, taking a step towards automatic Banff grading.

Specifically, in this paper we present an approach for automatic multi-class instance segmentation of renal tissue. Instance segmentation extends semantic segmentation by requiring each object to be individually detected. We include a total of seven structures of the kidney anatomy in our segmentation task: proximal, distal and atrophic tubuli, normal and sclerotic glomeruli, arteries and the capsule. Furthermore, we explicitly define a 'background' class, which consists of all tissue that does not fall into one of the included structures. We propose an adapted U-net, which has shown excellent results on segmentation tasks, especially in microscopic imaging.

2   Methods

We require highly detailed segmentation of tubular structures, as we need to be able to locate and count the immune cells within. To this end, we opted for a multi-task setting in which we simultaneously perform a 8-class structure segmentation and a 2-class tubulus border segmentation. The border ground truth maps were created by extracting the contour from each tubulus annotation and dilating it with a disc filter with pixel radius 3.

We modified the original U-net by adding a second decoder pathway before the fourth max-pooling layer. The outputs of the structure and border segmentations for both decoders were separately summed to obtain the output of the network before the soft-max layer. The addition of the second decoder is schematically drawn in Fig. 2. The skip-connections were added to the second decoder in similar fashion as the original implementation. Utilizing max-pooling layers can result in loss of information, which is detrimental for precise segmentation. We hypothesize that adding the second decoder alleviates this problem. After only three max-pooling operations, the first decoder has retained more information. This is especially important at instance boundaries. Simultaneously, the second decoder has a greater receptive field, which is important for correctly classifying larger structures.

The structure segmentation results are used as a basis for the immune cell detection by isolating the tubular structures. We used the same adapted U-net for the immune cell detection task. To generate a ground truth, we converted immune cell annotations, which were dots roughly located at the center, to circular structures with a radius of ten pixels. An overview of all steps performed is shown in Fig. 1.

Fig. 1. Overview of the steps.

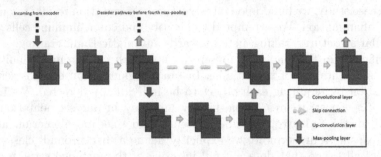

Fig. 2. Bottom part of the U-net with the extra decoder.

During training, batches were filled by uniformly sampling pixels across classes from the WSIs. We generated the patches from the WSI, centered around the sampled pixels. This resulted in an imbalance of class occurences, due to differences in size and occurrences of the classes. To account for this, we applied a loss weight to each pixel based on its class. Per batch we calculate the pixel-weight for each class as follows:

$$P_c = \frac{\sum_{c \in C}(N_c)/|C|}{N_c}, \tag{1}$$

with P_c as the weight for class C and N_c as the amount of pixels of class c in the batch.

We separately calculated the border segmentation loss and the structure segmentation loss, leading to the overall loss formula:

$$\mathcal{L}_{total}(X, W) = \lambda \cdot ||W||_2^2 + L_s(X, W) + L_b(X, W) + \omega_O \cdot \sum_{o \in O}(L_o(X, W)), \tag{2}$$

where L_b and L_b represent the structure and border loss terms, respectively. $||W||_2^2$ denotes the ℓ_2 regularization term on all weights W. The L_0 terms represent the border and structure segmentation results of the individual decoders. We add these terms to force both encoder branches to learn discriminative features. The term ω_O is halved after each epoch, until it becomes so small that it is essentially dropped from the loss function, leaving only the L_s, L_b and regularization terms. We use cross-entropy for all loss terms, incorporating the P_c class weight:

$$L(X, W) = -\frac{1}{N} \sum_{x \in X} P_{c_x} \log(p_x(x, W)), \tag{3}$$

with p_x as the classification output after applying a soft-max, and N as the number of pixels in X.

3 Dataset

3.1 Data Selection

We used a dataset of biopsies originating from a clinical trial, studying the effect of the medicine Rituximab on the incidence of biopsy proven acute rejection within 6 months after transplantation [13]. All biopsies were taken at the Radboud University Medical Centre, Nijmegen, The Netherlands between December 2007 and June 2012. The slides are stained with periodic acid-Schiff (PAS) and digitized using 3D Histech's Panoramic 250 Flash II scanner. We selected subsets of the dataset for the structure segmentation task and the immune cell detection task, which are described below.

3.2 Structure Segmentation

We selected a total of 24 WSIs from the dataset for the structure segmentation task. In each of the 24 slides, we randomly selected one or two rectangular regions, of approximately 4000 by 3000 pixels, in which all structures of interest were exhaustively annotated. After annotating all structures of interest, every unassigned pixel was placed in the 'background' class. Annotations were produced by a technician with expertise in renal biopsy histopathology and revised by a pathology resident, under consultation of an experienced nephropathologist. A total of 37 fully annotated rectangular areas were obtained.

3.3 Immune Cell Detection

Annotating immune cells is a difficult task, as it is hard to distinguish tubulus infiltrating lymphocytes from tubular epithelial cells [10]. We selected 5 slides to be fully annotated. There is no overlap between the slides for the structure segmentation and the immune cell detection. To obtain accurate data, we opted to make annotations in a PAS-stain and a CD3-stain of the same slide. First, lymphocytes were annotated in the PAS-stained section. This section was then re-stained with an immunohistochemical staining that shows presence of CD3, a surface marker only present on T-lymphocytes, which is the dominant immune cell population in tubular inflammation. After re-scanning of the slide, new annotations based on intratubular CD3-positivity were conducted. The annotation of both stains was performed by a technician with expertise in the field of renal biopsy histopathology. By intersecting the annotations from both stains, we made a third group of annotated immune cells, which were both visible in the PAS-stain and have been verified by the CD3-stain as truly being T-lymphocytes. Taking the intersection of the two sets resulted in a total amount of 891 immune cell annotations across the five slides. This third group was used for training of the immune cell detection.

4   Experiments and Results

4.1   Implementation Details

A set of parameters was shared across the experiments. For the weights we used He-initialization [4]. The initial learning rate was set at 0.0005, using Adam optimization [5]. The learning rate was halved when no improvement occurred in 5 epochs. Square input patches were sampled from the WSIs with a size of 412 pixels. The structure segmentation task was trained and tested on images at 10x magnification and the immune cell detection task at 40x magnification. We trained the structure segmentation and immune cell detection networks for 60 and 50 epochs, respectively. An epoch consisted of 50 training iterations with batch size of 6. Training was performed on an NVIDIA Titan X GPU.

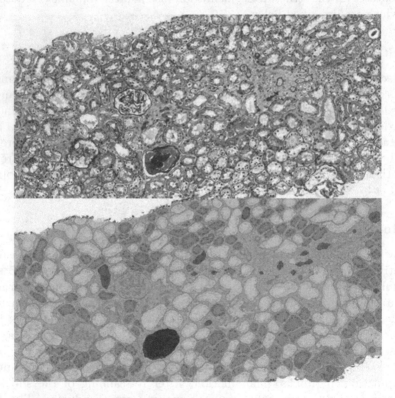

Fig. 3. Visual example of the segmentation results. In the bottom image, the segmentation has been laid over the original image. The colors correspond to the following classes: light-blue/background, green/glomeruli, purple/sclerotic glomeruli, yellow/proximal tubuli, orange/distal tubuli and pink/atrophic tubuli. (Color figure online)

WSI analysis

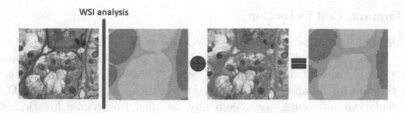

Fig. 4. Comparison of the structure segmentation with and without border segmentation. From left to right: original patch, structure segmentation, border segmentation and final result.

4.2 Structure Segmentation

We performed the segmentation task in 5-fold cross-validation, where we divided the dataset in nineteen slides for training and five slides for evaluation, for each fold. We assessed the performance of the structure segmentation by calculating the Dice score per class. Given the set of ground-truth pixels Y and the set of manually segmented pixels X, the Dice score was calculated for each class separately, as follows:

$$D(Y_c, X_c) = \frac{|Y_c \cap X_c|}{|Y_c| + |X_c|}, \tag{4}$$

where c denotes the class. The overall Dice score is calculated by taking the average of the classes, weighted by the pixel-contribution of each class. Scores per class can be seen in Table 1. An example of the segmentation results is shown in Fig. 3. We demonstrate the efficacy of separate structure and border segmentation in Fig. 4. We can observe from these examples that our method was able to accurately separate the boundaries of the instances.

Table 1. Dice-scores of the extended U-net for all classes. The overall score is calculated from the average of the classes, weighted by their pixel contribution.

Class	Dice-score
Background	0.88
Glomeruli	0.92
Proximal tubuli	0.89
Distal tubuli	0.81
Atrophic tubuli	0.51
Sclerotic glomeruli	0.59
Arteries	0.54
Capsule	0.60
Overall	0.85

4.3   Immune Cell Detection

We trained the immune cell detection network twice in 5-fold cross-validation on the dataset of five slides, using one slide for evaluation. After a round of training, we identified tubular epithelial cells of which the network produced a high likelihood for immune cells and annotated these as false positives. During the second training round, we specifically sampled from these locations with higher probability. Post-processing of the likelihood maps was used to convert the probabilistic segmentation to detections. The probabilistic segmentations where thresholded at a likelihood of 0.9. Subsequently, connected component analysis was performed to remove segmentation with a size smaller than 300 pixels. We randomly picked a slide for tuning post-processing, and left it out of the evaluation. We report the precision, recall and F_1-score for each individual slide in Table 2. The annotations in the CD3-stain were used as the ground truth. The detection performance of the network is compared with the performance of the technician based on PAS-stain annotations. Despite the low scores, it can be seen that the F1-score of our network is slightly higher than that of the technician.

Table 2. Immune cell detection performance of the technician and the network. For the technician, the PAS-annotations are compared with the CD-3 annotations. For the network, the CD3-stain was used as ground truth.

	TP	FP	FN	Precision	Recall	f1
Technician performance						
Slide 1	287	565	372	0.34	0.44	0.380
Slide 2	60	579	93	0.09	0.39	0.152
Slide 3	309	736	508	0.30	0.38	0.332
Slide 4	107	351	113	0.23	0.49	0.316
Total	763	2231	1086	0.25	0.41	0.315
Network performance						
Slide 1	187	188	472	0.50	0.28	0.362
Slide 2	37	453	116	0.08	0.24	0.115
Slide 3	252	261	565	0.49	0.31	0.379
Slide 4	92	181	128	0.34	0.42	0.373
Total	568	1083	1281	0.34	0.31	0.325

4.4   Discussion

In this paper, we presented a multi-task approach for accurately segmenting structures in renal tissue. We combined structure and border segmentation, which resulted in an accurate segmentation with separation of adjacent tubular instances. Our structure segmentation can be used as a basis for applications that can further help the pathologist in diagnostic decision making. As an example, detection and quantification of glomeruli can readily be implemented from our results. Some classes had few annotations, resulting in lower Dice score. We built upon the structure segmentation by using it as the basis for the detection of intratubular immune cells. This case study was severely limited by the small dataset of five WSIs. In future work, we want to improve our immune cell detection by increasing the size of and quality of our dataset. We plan to infer a tubular infiltration Banff grading from the immune cell detections, rendering us able to compare our approach with the pathologists' manual grading.

References

1. Chen, H., Qi, X., Yu, L., Heng, P.-A.: DCAN: deep contour-aware networks for accurate gland segmentation. In: Proceedings of the IEEE Conference on Computer Vision and Pattern Recognition, pp. 2487–2496 (2016)
2. Elshafie, M., Furness, P.N.: Identification of lesions indicating rejection in kidney transplant biopsies: tubulitis is severely under-detected by conventional microscopy. Nephrol. Dial. Transplant. **27**(3), 1252–1255 (2011)
3. Gurcan, M.N., Boucheron, L.E., Can, A., Madabhushi, A., Rajpoot, N.M., Yener, B.: Histopathological image analysis: a review. IEEE Rev. Biomed. Eng. **2**, 147–171 (2009)
4. He, K., Zhang, X., Ren, S., Sun, J.: Delving deep into rectifiers: surpassing human-level performance on imagenet classification. In: Proceedings of the IEEE International Conference on Computer Vision, pp. 1026–1034 (2015)
5. Kingma, D.P., Ba, J.: Adam: A method for stochastic optimization. arXiv preprint arXiv:1412.6980 (2014)
6. Litjens, G., et al.: A survey on deep learning in medical image analysis. arXiv preprint arXiv:1702.05747 (2017)
7. Long, J., Shelhamer, E., Darrell, T.: Fully convolutional networks for semantic segmentation. In: Proceedings of the IEEE Conference on Computer Vision and Pattern Recognition, pp. 3431–3440 (2015)
8. Loupy, A., et al.: The banff 2015 kidney meeting report: current challenges in rejection classification and prospects for adopting molecular pathology. Am. J. Transplant. **17**(1), 28–41 (2017)
9. Pedraza, A., Gallego, J., Lopez, S., Gonzalez, L., Laurinavicius, A., Bueno, G.: Glomerulus classification with convolutional neural networks. In: Valdés Hernández, M., González-Castro, V. (eds.) MIUA 2017. CCIS, vol. 723, pp. 839–849. Springer, Cham (2017). https://doi.org/10.1007/978-3-319-60964-5_73
10. Racusen, L.: Improvement of lesion quantitation for the banff schema for renal allograft rejection. Transplant. Proc. **28**, 489–490 (1996)
11. Ronneberger, O., Fischer, P., Brox, T.: U-Net: convolutional networks for biomedical image segmentation. In: Navab, N., Hornegger, J., Wells, W.M., Frangi, A.F. (eds.) MICCAI 2015. LNCS, vol. 9351, pp. 234–241. Springer, Cham (2015). https://doi.org/10.1007/978-3-319-24574-4_28
12. Temerinac-Ott, M., et al.: Detection of glomeruli in renal pathology by mutual comparison of multiple staining modalities. In: 2017 10th International Symposium on Image and Signal Processing and Analysis (ISPA), pp. 19–24. IEEE (2017)
13. van den Hoogen, M.W., et al.: Rituximab as induction therapy after renal transplantation: a randomized, double-blind, placebo-controlled study of efficacy and safety. Am. J. Transplant. **15**(2), 407–416 (2015)

Cellular Community Detection for Tissue Phenotyping in Histology Images

Sajid Javed[1(✉)], Muhammad Moazam Fraz[1,3,5], David Epstein[2],
David Snead[4], and Nasir M. Rajpoot[1,4,5]

[1] Department of Computer Science, University of Warwick, Coventry, UK
{s.javed.1,n.m.rajpoot}@warwick.ac.uk
[2] Department of Mathematics, University of Warwick, Coventry, UK
[3] SEECS, NUST, Islamabad, Pakistan
[4] UHCW NHS Trust, Coventry, UK
[5] The Alan Turing Institute, London, UK

Abstract. A primary aim of detailed analysis of multi-gigapixel histology images is assisting pathologists for better cancer grading and prognostication. Several methods have been proposed for the analysis of histology images in the literature. However, these methods are often limited to the classification of two classes i.e., tumor and stroma. Also, most existing methods are based on fully supervised learning and require a large amount of annotations, which are very difficult to obtain. To alleviate these challenges, we propose a novel community detection algorithm for the classification of tissue in *Whole-slide Images* (WSIs). The proposed algorithm uses a novel graph-based approach to the problem of detecting prevalent communities in a collection of histology images in an semi-supervised manner resulting the identification of six distinct tissue phenotypes in the multi-gigapixel image data. We formulate the problem of identifying distinct tissue phenotypes as the problem of finding network communities using the geodesic density gradient in the space of potential interaction between different cellular components. We show that prevalent communities found in this way represent distinct and biologically meaningful tissue phenotypes. Experiments on two independent *Colorectal Cancer* (CRC) datasets demonstrate that the proposed algorithm outperforms current state-of-the-art methods.

Keywords: Community detection · Tissue phenotyping

1 Introduction

The automatic recognition of phenotypes in a WSI is an important step in computational pathology [5]. It assists pathologists in better cancer grading and prognostication [8]. One of the basic approaches which is commonly used to address this problem is texture analysis [1,7,12]. Texture analysis methods compute the internal texture features of different histology images for training a classifier. The classifier then predicts distinct tissue types based on the trained

© Springer Nature Switzerland AG 2018
D. Stoyanov et al. (Eds.): COMPAY 2018/OMIA 2018, LNCS 11039, pp. 120–129, 2018.
https://doi.org/10.1007/978-3-030-00949-6_15

features. For instance, Bianconi *et al.* proposed five different kinds of perception-based texture features [1]. Linder *et al.* proposed a simple SVM classifier trained on a set of local binary patterns and contrast measure features [7]. These methods are often limited to the two class discrimination problem and they do not fully capture the biological contents of tissue types resulting in a performance degradation.

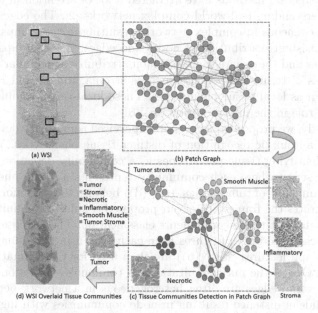

(a) WSI   (b) Patch Graph

(d) WSI Overlaid Tissue Communities   (c) Tissue Communities Detection in Patch Graph

Fig. 1. The framework of our proposed community detection algorithm for multi-component tissue classification. (a) WSI of CRC. (b) Patch-based graph construction, where each node in a graph represents a feature vector of a patch. (c) Proposed community detection algorithm where different colors represent local communities. (d) Local tissue communites overlaid on WSI.

A growing number of *Deep Learning* (DL) methods have also been used to classify WSIs into distinct tissue types [4]. Xu *et al.* proposed a fully supervised deep CNN model for the classification of tissue images [13]. Hang *et al.* proposed an unsupervised domain adaptation deep network for segmenting tissue images into meaningful regions [3]. Encouraging results were reported in these studies but they are limited to the discrimination of tumor epithelium and stroma components [1,3,7,13]. Moreover, DL methods share a common denominator which is their need for large amount of annotated histology data on which to train [4,5]. Unfortunately, the annotation of histology images is a very tedious task, especially when it comes to the annotation of distinct tissue phenotypes. CRC WSIs do not consist of only tumor and stroma components. They also contain a complex rich structure of several other tissue components including smooth muscle, inflammatory, necrotic, and tumor-associated stroma types [6].

Kather *et al.* recently proposed texture analysis methods for the identification of multiple tissue components [5]. This method trained several texture features compromising texture, local binary pattern, histogram, and gabor features etc., on several tissue images. The SVM classifier was then trained on these set of features for multi-tissue classification. However, texture features do not exploit the biological structure and hence may degrade the classification performance. Community detection methods have attracted a lot of attention in the past few years for understanding real-world complex networks [2]. The edges and nodes are often homogeneous in complex networks resulting in the groups of a higher number of consistent distribution of nodes and edges. These groups are known as communities and they share some common attributes and similar behaviours. Different types of tissue components such as stroma, tumor, and necrotic etc. are also known as local tissue phenotypes. These local tissue communities play an important role in the interpretation of WSIs.

In this study, we propose a novel semi-supervised community detection algorithm for the automatic recognition of distinct tissue phenotypes. We pose the problem of identifying tissue phenotypes as a community detection problem in histology landscape where each community represents a tissue phenotype, for instance a community of tumor (Figs. 1(c)–(d)) belongs to the tumor phenotype. Figure 1 illustrates the framework of our proposed cellular community detection algorithm for distinct tissue components classification. The proposed algorithm computes patch-based graph features from each WSI using Delaunay triangulation. These graph features (Fig. 1(b)) are then used for the estimation of proximity matrix. One of the major limitations in community detection methods is that the nodes and edges are often heterogeneous in a network because of the pairwise Euclidean distance resulting in node communities with higher number of intra-community and lower number of inter-community edges. To address this problem, we propose to map the node features into the geometric space and represent each node as a geodesic distance vector. The geodesic density gradient [9] is then computed in the space of graph features for clustering similar tissue regions into the same tissue communities (Fig. 1(c)). The nodes in the clusters of tissue communities represent distinct biologically meaningful tissue components such as tumor, stroma, necrotic, inflammatory, tumor-associated stroma, and smooth muscle (Fig. 1(d)). Nodes in a community are then used for the selection of the best exemplars. We use a deep CNN model [10] for the estimation of representative exemplars from each tissue community. Results on CRC WSIs taken from two independent datasets show that the proposed algorithm improves the visual results by segmenting tissues into six distinct classes (Fig. 1(d)) and obtained an average F-score of 92% across all tissue types. To the best of our knowledge, this is the first paper which employs community detection for the identification of distinct tissue phenotypes.

The rest of this paper is organized as follows. Section 2 describes the proposed approach in detail. Experimental results are discussed in Sect. 3, and finally conclusions and future directions are given in Sect. 4.

2   The Proposed Approach

The proposed approach consists of four main steps: (1) pre-processing, (2) color graph construction and computation of potential cell features, (3) geodesic space density gradient method for community detection, and (4) the selection of best exemplars using a CNN approach. First, we extract patches of size $200 \times 200\,\mu m^2$ (400×400 pixels at 20×) from each WSI (Fig. 2(a)) of CRC and then we use spatially constraint DL model for cell detection and classification [11]. The output of the network is a set of four different types of cells including *Epithelial* (Ep), *Spindle-shaped* (S), *Necrotic* (N), and *Inflammatory* (I) from each patch (Fig. 2(b)). Second, we employ Delaunay triangulation and estimate a feature vector of potential interaction between cells from each patch. Third, we compute a proximity matrix using a graph features and then we use a community detection algorithm known as the geodesic density gradient [9] in the space of potential interaction between cellular components for detecting network communities representing tissue classes. The nodes in each community represent patches of the same tissue type. These nodes are then used for the selection of best exemplars using [10]. In the following sections, we explain each step of the proposed algorithm in more detail.

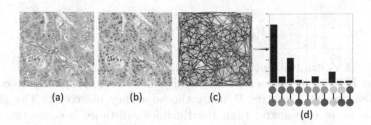

(a)   (b)   (c)

(d)

Fig. 2. Computation of potential interaction between cellular components: (a) A patch of size 400×400 extracted at 20×; (b) Results of cell detection and classification, where red, green, yellow, and blue dots show Ep, I, S, and N cells; (c) Patch-based graph construction via delaunay triangulation where each node represents cell in a graph. Color edges show similar cells, are connected together while black edges show the connectivity of cross cellular components. (d) A feature vector of potential interaction between cellular components. (Color figure online)

2.1   Graph Construction

For each patch $\mathbf{X}_i \in \mathbb{R}^{p \times p}$, we construct a color cell graph using a Delaunay triangulation such that the vertices correspond to the spatial location of cells and assign color edges between cellular components (Fig. 2(c)). For each patch, we compute a feature vector by exploiting the following 10 potential interactions between cellular components compromising Ep to Ep, Ep to I, Ep to S, Ep to N, I to I, I to S, I to N, S to S, S to N, and N to N cells (Fig. 2(c)). We compute 10 dimensional histogram feature vector $\mathbf{h}_i \in \mathbb{R}^m$ (Fig. 2(d)) for each patch $\mathbf{X}_i$ and create an input data matrix for each WSI as $\mathbf{H} = [\mathbf{h}_1, \mathbf{h}_2, ..., \mathbf{h}_n] \in \mathbb{R}^{n \times m}$, where n denotes the number of patches and $m = 10$ represents potential features.

2.2 Geodesic Space Density Gradient Community Detection

Geodesic Density Gradient (GDG) is a potential method for finding network communities [9]. The method first maps nodes in the geodesic space by defining new distances in that space to resolve the problem of compact community boundaries. It then computes density gradient in the geodesic space and finds the maximum density nodes region in that space. Finally, it employs K-means clustering algorithm to assign similar nodes to the k clusters. In this study, we formulate the problem of identifying network communities for the detection of distinct tissue classes from matrix $\mathbf{H}$.

2.2.1 Geodesic Distance Proximity Matrix

To map histology images represented by a feature vector in the geodesic space, we first construct an undirected graph $\mathbf{G} = (\mathbf{V}, \mathbf{A})$, where the vertices $\mathbf{V}$ correspond to each column of feature matrix $\mathbf{H}$ and $\mathbf{A}$ is the edge weight matrix. The adjacency matrix $\mathbf{A}$ is computed by employing chi-squared distance measure as:

$$A(i,j) = \sum_{k=1}^{m} \frac{(\mathbf{v}_{i,k} - \mathbf{v}_{j,k})^2}{\mathbf{v}_{i,k} + \mathbf{v}_{j,k}}, \tag{1}$$

where $\mathbf{v}_i$ and $\mathbf{v}_j$ are two feature vectors in matrix $\mathbf{H}$ of length m. If $A(i,j) > 0$, then there is an edge between nodes i and j in $\mathbf{G}$. From graph $\mathbf{G}$, we compute the geodesic distance matrix $\mathbf{P}$ using the adjacency matrix $\mathbf{A}$. The geodesic distance is more meaningful than the Euclidean distance because two feature vectors may have relatively smaller euclidean distance, but larger distance in the geodesic space. Let $\mathbf{p}_i(j)$ be the shortest path or geodesic distance between nodes $\mathbf{v}_i$ and $\mathbf{v}_j$. In order to assign nodes $\mathbf{v}_i$ and $\mathbf{v}_j$ to the same community and avoid the community boundary problem, GDG employs direct distance $S_{i,j}$ and indirect distance $H_{i,j}^2$ as follows:

$$S_{i,j} = \frac{\mathbf{p}_i(j) + \mathbf{p}_j(i)}{2} \text{ and } H_{i,j}^2 = \sum_{k=1}^{n-2} (\mathbf{p}_i(k) - \mathbf{p}_j(k))^2, \ s.t. \ k \neq \{i,j\}, \tag{2}$$

where $S_{i,j}$ is basically a geodesic distance between nodes $\mathbf{v}_i$ and $\mathbf{v}_j$ while $H_{i,j}$ is a distance which is induced by all other nodes in $\mathbf{G}$. By combining distances, we get

$$d_{i,j} = \sqrt{(\mathbf{y}(\mathbf{p}_i - \mathbf{p}_j))^\top (\mathbf{y}(\mathbf{p}_i - \mathbf{p}_j))}, \tag{3}$$

where $\mathbf{y}_{ii} = \mathbf{y}_{jj} = \sqrt{\alpha}$ and $\mathbf{y}_{kk} = \sqrt{\beta}$ is an $n \times n$ matrix for scaling distance using constants α and β. Equation (3) defines the distance between two nodes in the geodesic space.

2.2.2   Gradient Density Computation in Geodesic Space

The main motivation to compute the gradient density field in the geodesic space is to bring together those nodes which belong to the same community, and move far those nodes which correspond to the different communities. The purpose is the compactness of community and constraining larger inter-community gaps which improve the performance of a proposed algorithm. Let say that a node $\mathbf{v}_j$ is mapped to a point $\mathbf{p}_j$ in the geodesic space. A point $\mathbf{p}_j$ induces the Gaussian probability density function with mean $\mathbf{p}_j$ and variance b_w^2, then the density of point $\mathbf{p}$ because of a node $\mathbf{p}_j$ can be computed as:

$$K(d_{p,j}, b_w) = \frac{1}{\eta} \exp \frac{-(\mathbf{y}(\mathbf{p} - \mathbf{p}_j))^\top (\mathbf{y}(\mathbf{p} - \mathbf{p}_j))}{2b_w^2}, \tag{4}$$

where η is a normalizing factor. When a point $\mathbf{p}$ moves far from $\mathbf{p}_j$, density decreases with the increasing distance (Eq. (3)). The Gaussian density enforced by all other nodes at any point $\mathbf{p}$ can be estimated as:

$$\hat{f}(\mathbf{p}) = \frac{1}{nb_w^n} \sum_{j=1}^{n} K\left(\frac{\mathbf{y}(\mathbf{p} - \mathbf{p}_j)}{b_w^2}\right). \tag{5}$$

which is the summation of densities induced by individual nodes. To get the gradient of a density field, we take the gradient of (5), substitute the values of K from (4), and differentiate it w.r.t. $\mathbf{p}$ as

$$\nabla f(\mathbf{p}) = \frac{\mathbf{y}}{nb_w^{n+1}} \sum_{j=1}^{n} \frac{\mathbf{y}(\mathbf{p} - \mathbf{p}_j)}{b_w^2} K(d_{p,j}, b_w). \tag{6}$$

where ∇ is the gradient operator with respect to each of the n spatial dimensions. The above formulation is the density gradient known as weighted average shift and points in the maximum increase in density. By setting $\nabla f(p) = 0$, we get the approximate gradient density as:

$$\nabla p = \frac{\sum_{j=1}^{n} \mathbf{p}_j K(d_{p,j}, b_w)}{\sum_{j=1}^{n} K(d_{p,j}, b_w)}. \tag{7}$$

Eq. (7) is applied on each node in an iterative manner. This results in a drift of each node in each iteration. The iterations will stop when the l_1 norm between the new and old estimate is less than a certain threshold. By computing the density gradient of a node, it is expected that a node drifts in the positive density gradient. We consider the path followed by a node from its starting position to the stopping position as the node trajectory and the node stopping position as the trajectory end point. The GDG method computes the maximum density region by considering two trajectories of a node. Finally, the k-means clustering algorithm is performed as a post-processing step to assign closest points to the k clusters resulting in six distinct tissue components.

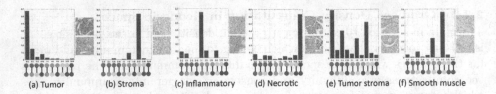

(a) Tumor   (b) Stroma   (c) Inflammatory   (d) Necrotic   (e) Tumor stroma   (f) Smooth muscle

Fig. 3. Results of the proposed algorithm for multi-component tissue classification. Representative features with corresponding images are shown for each tissue community. (a) Tumor, (b) Stroma, (c) Inflammatory, (d) Necrotic, (e) Tumor-associated stroma, and (f) Smooth muscle.

2.3   Exemplar Selection

The nodes in each community represent image patches of similar histology types. To represent each community from representative node features, we use a CNN model [10] for the selection of best five exemplars from each histology community. The model [10] is trained using CRC tissues for tumor segmentation. In case of selecting the best tumor exemplars from the tumor community, we compute the activation maps for each image from the final layer of trained CNN model during the testing phase. We deduce a single scalar value from each activation map using the last convolutional layers which shows significant importance for each patch. We draw a 2D plane of these scalar values and select first those scalars which are close to the mean values in the 2D plane (Fig. 3(a)–(f)).

3   Experimental Evaluation

We validated our proposed algorithm both quantitatively and qualitatively on two independent CRC datasets. The results are compared with seven state-of-the-art methods including B5F-SVM [5], B6F-SVM [5], *K-Medoids clustering with Chi-squared Distance* (KM-CD) [6], KM with *Bhattacharya Distance* (KM-BD) [6], KM-KLD [6], and two variants of the proposed approach by using BD and KLD measures. The original authors implementation was used for the first two methods [5], while we implemented K-Medoids clustering methods reported in [6]. The quantitative results are compared in terms of TP, FP, TN, FN, and F_1 score as a performance measures. The aim is to maximize TP, TN and F_1 score for more accurate classification of tissue components while minimizing TN and FN. We qualitatively evaluated proposed algorithm on first dataset, which consists of 10 H&E stained WSIs. Visual results of six distinct tissue components of only one WSI of CRC is shown in Figs. 4(a)–(b). The second dataset consists of total 3,750 tissue images divided into 6 categories namely, Tumor, Stroma, *Complex Stroma* (CS), *Lymphocyte* (Lympho), Debris, and *Mucosal Glands* (MG) [5]. Figures 4(c)–(h) present the visual results of the proposed approach only while Figs. 5(a)–(b) and Table 1 show the classification performance in terms of confusion matrix and average F_1 score as compared to other existing methods.

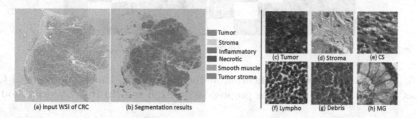

Fig. 4. Visual results of the proposed algorithm on two independent CRC datasets. (a)-(b) show the results using local university hospital CRC dataset while (c)-(h) show the results using CRC dataset reported in [5]. Different colors in (b) show distinct tissue components while (c)-(h) represent the visual results of Tumor,Stroma, CS, Lympho, Debris, and MG tissue components.

For the Tumor component, only proposed algorithms, B5F-SVM, and B6F-SVM methods produced best results (Figs. 5(a)–(b) and Table 1), while the remaining methods exhibit some discrepancies in terms of F_1 score. In stroma class, only KM-CD and proposed algorithms attained an average F_1 score of above 0.90 which is significantly larger than other compared methods. For more complicated CS component, only B5F-SVM and BF6-SVM methods performed good. In this case, F_1 score of the proposed algorithm is degraded 2% in comparison with best performing methods because the proposed algorithm is sensitive against spatial arrangements of cellular components therefore several CS tissues were incorporated into either tumor or MG types. For category Lympho, all of the compared methods (excluding KM-BD and KM-KLD) produced an average F_1 score of more than 90% (Table 1) therefore Lympho class does not pose a challenge for most of the compared methods. For Debris class, only two compared methods such as KM-BD and KM-KLD generated a large number of FP and FN resulting in low F_1 score while other methods attained an average F_1 score of more than 0.90. The MG component is one of the most difficult class among all categories because it contains heterogeneous cells. The existing methods generate a large amount of FN resulting in a low performance. Only the proposed algorithm and B6F-SVM method produced good results (Table 1). Overall, the proposed algorithm obtained an average F_1 score of 92% which is 3% greater than the second best performing method while all other compared methods showed degraded performance.

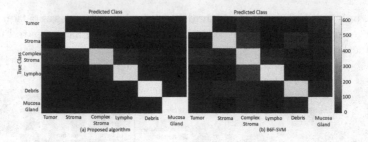

Fig. 5. Performance of the classification methods on six distinct tissue types of CRC dataset [5]. (a) Performance of the proposed algorithm and (b) B6F-SVM method [5].

Table 1. Comparison of the quantitative results with state-of-the-art classification methods in terms of average F_1 score. Bold numbers show best performing method.

Methods	Tumor	Stroma	Complex Stroma	Lympho	Debris	Mucosa Gland	Average
KM-CD [6]	0.85	0.92	0.71	0.95	0.91	0.81	0.85
KM-BD [6]	0.51	0.70	0.46	0.73	0.65	0.56	0.60
KM-KLD [6]	0.76	0.79	0.59	0.84	0.78	0.62	0.73
B5F-SVM [5]	0.87	0.86	**0.85**	0.93	0.89	0.89	0.88
B6F-SVM [5]	0.87	0.88	**0.85**	0.94	0.90	**0.90**	0.89
Proposed-CD	**0.92**	**0.94**	0.83	**0.97**	**0.96**	**0.90**	**0.92**
Proposed-KLD	0.88	0.79	0.77	0.90	0.93	0.86	0.88
Proposed-BD	0.76	0.80	0.67	0.91	0.91	0.71	0.79

4   Conclusions

In this paper, a novel tissue community detection algorithm is presented for semi-supervised classification of distinct tissue components. To classify each tissue, a potential interaction between cellular components is mapped into the geometric space and gradient density is estimated to assign two similar tissue image patches to the same community. The representative exemplars are selected to represent each community of tissue using the CNN. Experiments on two datasets demonstrated that the proposed algorithm achieved the best performance compared to existing methods. In future, we plan to further investigate this study for the classification of more than six types of CRC and other cancer types.

Acknowledgement. This work was supported by the Medical Research Council [MR/P015476/1].

References

1. Bianconi, F., Álvarez-Larrán, A., Fernández, A.: Discrimination between tumour epithelium and stroma via perception-based features. Neurocomputing **154**, 119–126 (2015)
2. Harenberg, S., et al.: Community detection in large-scale networks: a survey and empirical evaluation. Wiley Interdisc. Rev.: Comput. Stat. **6**(6), 426–439 (2014)
3. Huang, Y., Zheng, H., Liu, C., Ding, X., Rohde, G.K.: Epithelium-stroma classification via convolutional neural networks and unsupervised domain adaptation in histopathological images. IEEE J-BHI **21**(6), 1625–1632 (2017). https://doi.org/10.1109/JBHI.2017.2691738
4. Janowczyk, A., Madabhushi, A.: Deep learning for digital pathology image analysis: a comprehensive tutorial with selected use cases. JPI **7** (2016)
5. Kather, J.N., et al.: Multi-class texture analysis in colorectal cancer histology. Sci. Rep. **6**, 27988 (2016)
6. Korsuk, S., et al.: Novel digital tissue phenotypic signatures of distant metastasis in colorectal cancer. Arch. Pathol. Lab. Med. **140**(1), 41–50 (2015)
7. Linder, N., et al.: Identification of tumor epithelium and stroma in tissue microarrays using texture analysis. Diagn. Pathol. **7**(1), 22 (2012)
8. Louis, D.N., et al.: Computational pathology: a path ahead. Arch. Pathol. Lab. Med. **140**(1), 41–50 (2015)
9. Mahmood, A., Small, M., Al-Maadeed, S.A., Rajpoot, N.: Using geodesic space density gradients for network community detection. IEEE T-KDE **29**(4), 921–935 (2017). https://doi.org/10.1109/TKDE.2016.2632716
10. Qaiser, T., et al.: Fast and accurate tumor segmentation of histology images using persistent homology and deep convolutional features. arXiv preprint arXiv:1805.03699 (2018)
11. Sirinukunwattana, K., Raza, S.E.A., Tsang, Y.W., Snead, D.R., Cree, I.A., Rajpoot, N.M.: Locality sensitive deep learning for detection and classification of nuclei in routine colon cancer histology images. IEEE T-MI **35**(5), 1196–1206 (2016)
12. Tamura, H., Mori, S., Yamawaki, T.: Textural features corresponding to visual perception. IEEE T-SMC **8**(6), 460–473 (1978)
13. Xu, J., Luo, X., Wang, G., Gilmore, H., Madabhushi, A.: A deep convolutional neural network for segmenting and classifying epithelial and stromal regions in histopathological images. Neurocomputing **191**, 214–223 (2016)

Automatic Detection of Tumor Budding in Colorectal Carcinoma with Deep Learning

John-Melle Bokhorst[1,2](✉), Lucia Rijstenberg[2], Danny Goudkade[3],
Iris Nagtegaal[2], Jeroen van der Laak[1,2], and Francesco Ciompi[1,2]

[1] Diagnostic Image Analysis Group, Radboud University Medical Center,
Nijmegen, Netherlands
john-.melle.bokhorst@radboudumc.nl
[2] Department of Pathology, Radboud University Medical Center,
Nijmegen, Netherlands
[3] Department of Pathology, Maastricht University Medical Center,
Maastricht, Netherlands

Abstract. Colorectal cancer patients would benefit from a valid, reliable and efficient detection of Tumor Budding (TB), as this is a proven prognostic biomarker. We explored the application of deep learning techniques to detect TB in Hematoxylin and Eosin (H&E) stained slides, and used convolutional neural networks to classify image patches as containing tumor buds, tumor glands and background. As a reference standard for training we stained slides both with H&E and immunohistochemistry (IHC), where one pathologist first annotated buds in IHC and then transferred the obtained annotations to the corresponding H&E image. We show the effectiveness of the proposed three-class approach, which allows to substantially reduce the amount of false positives, especially when combined with a hard-negative mining technique. Finally we report the results of an observer study aimed at investigating the correlation between pathologists at detecting TB in IHC and H&E.

Keywords: Deep learning · Computational pathology
Colorectal carcinoma · Tumor budding

1 Introduction

Tumor budding is defined as the presence of detached single epithelial cells or small clusters of up to 5 cells at the invasive front of colorectal cancer. It can also be found within the tumor mass, which is typically organized in irregular clusters of long stretched tumor glands. Tumor budding (TB) has received increasing attention by gastrointestinal pathologists as a promising adverse prognostic factor of lymph node and distant metastasis for colorectal carcinoma (CRC) patients. Incorporation of the phenomenon into the currently used staging system would contribute to more effective risk stratification [5]. Unfortunately, there

© Springer Nature Switzerland AG 2018
D. Stoyanov et al. (Eds.): COMPAY 2018/OMIA 2018, LNCS 11039, pp. 130–138, 2018.
https://doi.org/10.1007/978-3-030-00949-6_16

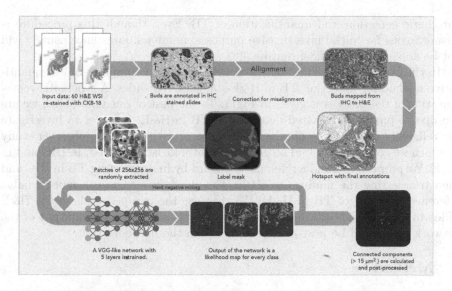

Fig. 1. Schematic overview of the proposed approach.

is no established procedure for the detection of TB so far, mainly due to the fact that there has been no reproducible method of assessment.

One of the main obstacles to the reproducibility of TB quantification has been the process of choosing the fields in which tumor budding is most intensive. In 2016, it was decided as a standard to assess the extent of TB on Hematoxylin and Eosin (H&E) in a $0.785\,\text{mm}^2$ hot-spot with highest TB density [5].

Microscopic identification of tumor buds in H&E by a pathologist can be difficult because of the resemblance of buds to surrounding stromal cells, fragmented glands and the concealment of buds in the setting of a peritumoral inflammatory reaction. Most studies of tumor budding have provided little detail regarding the morphologic criteria, used to include or exclude a potential bud [6]. Although tumor budding can be assessed in standard Hematoxylin and Eosin (H&E) in unproblematic cases, immunohistochemistry (IHC) with cytokeratin antibodies facilitates the detection. IHC staining highlights all epithelium and can be used for the identification of most adenocarcinomas. As IHC does not stain stromal components or tumor infiltrating lymphocytes in colon cancers, this staining can be helpful.

While manual evaluation of histological slides is still essential in clinical routine, automated image processing can provide high-throughput analysis of tumor tissue and assist the pathologist, by performing tasks such as the segmentation, classification and detection of phenomena. In recent years, Deep Learning has been leveraged to successfully address this kind of tasks. Recent developments in the field of computational pathology with Convolutional Neural Networks (CNN) are demonstrated by [1,8]. In the area of tumor budding, seminal work was done in Caie et al. [2], where immunofluorescence was employed for the automatic detection and quantification of TB. Even though this procedure is advantageous for initial investigative purposes, immunofluorescence usually will not be applicable to clinical routine [2].

In this study we will focus on the development of a computer aided, quantitative method for detecting TB in H&E-stained CRC slides. A schematic representation of the process is given in Fig. 1. To the best of our knowledge, we are the first to pursue automated detection of TB in H&E. In order to investigate the reliability of manually obtained annotations, we conduct an observer study in which we compare the bud scores of the pathologists involved, in IHC and in H&E. We propose to build a reference standard by first detecting TB in IHC and then transferring the findings to H&E. This procedure ensures a more reliable reference standard of TB in H&E. We then use the mapped buds in the H&E slides to train a CNN for multiple class patch classification. The output of the network finally will be post-processed to obtain the bud detection.

2   Method

Materials - Data from 60 CRC patients with presence of tumor budding reported during the initial sign out were included in this study. Tissue slides were prepared from tissue blocks on which the invasive front was clearly visible, which were stained with H&E, digitized, and re-stained with CK8-18. This procedure ensured having two digitized slides of exactly the same tissue section with two different stains. Glass slides were digitized using the Pannoramic P250 Flash II scanner from 3D-Histech, at 20X magnifiction (spatial resolution of 0.24 μm/px). Following the aforementioned hot-spot driven procedure, the invasive fronts of tissue sections were visually established from low-resolution images and the one hot-spot per image was chosen. After the hot-spot was selected, buds were manually annotated by one pathologist by clicking a point in the centre of the bud and drawing a circle around it automatically, based on a pre-calculated, average bud area of 600 μm². In this way, an average of 5 TB per hot-spot was obtained. To obtain a reference standard, we transferred the buds annotated in the IHC slides to the H&E slides. Due to re-staining, deformations can occur, which in some cases results in misalignment in the tissue images. For this reason, we performed a semi-automatic image alignment process. This process was done by software from 3D-Histech after selection of corresponding points in the H&E and IHC slides. The transferred annotations in H&E were corrected for false positives after visual assessment by the pathologist.

Tumor buds are not only located at the invasive tumor front (peritumoral budding), but are also found within the tumor mass, namely in the stroma between the tumor glands (intratumoral budding). For this reason, having a differentiation between buds and tumor glands is beneficial, because it allows the discrimination of small tumor areas from large tumor areas, and also potentially identifies small groups of tumor cells that are part of the tumor mass, and therefore no TB. This motivated us to make annotations of tumor glands (TG) as well by delineating the tumor glands in the H&E hotspot images, which we used in the development of our method.

Table 1. Architecture of the CNN used in this project. MP = max-pool layer, D = dropout layer with 0.5 drop-probability, convA-B is a convolutional layer with B filters of size A $\times$ As. The last convolutional layer has C filters, where C indicates the number of classes (C = 2 or C = 3 in this paper).

conv5-32	conv5-32	MP	conv3-64	conv3-64	MP	D	conv3-128	conv3-128	MP	conv3-256	conv3-256	MP	D	conv3-512	conv3-512	avg-pool	conv4-1024	conv1-512	conv1-C	soft-max

The set of input images and corresponding annotations was randomly divided into a training (36 images, 194 buds), a validation (14 images, 73 buds in total) and a test set (10 images, 38 buds in total).

Convolutional Networks for Tumor Bud Detection - Inspired by the VGG16-net architecture [7], which was ranked at the top of ILSVRC challenge 2014, a VGG-like network was developed with two configurations: one with 2 output classes (TB versus Background) and one with 3 output classes (TB, TG, Background), as shown in Table 1. The input of both network configurations is a RGB patch of 256×256 px. This size was chosen in order to contain the surface of the area equivalent to the largest TB in our dataset, i.e., $\approx 2500 \, \mu m^2$. For training purposes, class balanced patches were randomly sampled within the hot-spot. In order to sample TB patches, all pixels within the circle around the manually annotated TB location were considered. Training data were augmented by random flipping, rotating, elastic deformation, blurring, brightness (random gamma) and contrast changes. This artificially increased the number of samples and is known to optimize the network's robustness to variations in real samples of the images. Because of the relatively small amount of data and related risk of overfitting, we applied L2 regularization ($\lambda = 0.00009$), and dropout layers were added after the 2nd and 4th max-pool layer, with a drop-probability of 0.5. The densely connected layers were replaced by convolutional layers with 1024, 512 filters respectively to enable classification of arbitrary size inputs during inference. The final convolutional layer has C 1×1 filters with C representing the number of output classes. The training procedure involved optimizing the multinomial logistic regression objective (softmax), using stochastic gradient descent with Nesterov momentum. The batch size was set to 15, momentum to 0.9. We used an adaptive learning rate scheme, where the learning rate was initially set to 0.00001 and then multiplied by a factor of 0.7 after every 5 epoch if no increase in performance was observed on the validation set. The weights of the network were initialized as proposed in He et al. [3]. The networks were trained for 100 epochs.

We investigated the effect of doing hard-negative mining (HNM), applied exclusively to the output of the network with the three output classes. After classifying the training set with the trained network, we trained this network with the same settings again from scratch. In contrast to the procedure followed earlier during training, in which balanced mini-batches were used, we reduced the

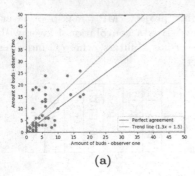 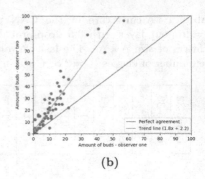

(a)   (b)

Fig. 2. Scatterplots of the amount of scored buds per image in (a) H&E, (b) IHC. Perfect agreement line and trend-line have been plotted.

Table 2. Corr. matrix of budding score in (left) H&E and (right) CK8-18. Classes are in line with the ITBCC 2016; Bd1 (0–4 buds), Bd2 (5–10 buds) and Bd3 (10+ buds)

	Obs. 1	Bd1	Bd2	Bd3		Obs. 1	Bd1	Bd2	Bd3
Obs. 2					Obs. 2				
Bd1		31	4	0	Bd1		18	0	0
Bd2		3	4	0	Bd2		3	2	0
Bd3		4	6	8	Bd3		4	8	25

number of patches of the third class (Background) by 50% and replaced them by hard-negatives: false positives with a probability of 0.7 (empirically determined) or higher, obtained from the likelihood maps. The number of patches of the TB and TG class were kept equal. The used 50-50 ratio was empirically observed to produce better results compared to other settings.

The output of all networks is in the form of C likelihood maps. In order to obtain the final detection of TB, we computed the center of mass of all connected components obtained after thresholding the output map of the TB class. The threshold was determined based on the validation-set and set to a probability of 0.7. Automatic detections at an euclidean distance of hand-identified buds, smaller or equal to 26 μm (which corresponds to the equavialent diameter, i.e., a single CRC tumor cell) were labeled as bud. In order to reduce TB false positives in the Tumor Glands, we applied a post-processing step to the results of the 3-class network configuration with HNM. For this purpose, we used the likelihood maps of the TG class to extract the contour of each classified glands. Bud candidates with a center or mass within the TG border and half the estimated equivalent diameter of a TB (i.e., 13 μm) were removed, as it was assumed that this implied an incomplete detachment from the tumor gland and therefore indicating a group of tumor cells that still belonged to the tumor gland itself.

In order to compute quantitative detection performance, the final detections were compared to the hand-annotated TB, in terms of F1-score and free receiver-operation characteristics (FROC) curve.

3   Observer Study

An observer study was conducted to assess inter-observer variability at TB detection on the 60 H&E and on the corresponding 60 IHC hot-spot images. In addition to the pathologist who, as described, annotated the buds in an earlier phase for the reference standard, a pathologist from another hospital was now involved. The pathologists annotated 747 buds in total in H&E. Among these detections only 143 (20%) were detected by both observers. Points closer than 26 μm from each other were considered as belonging to the same bud. On IHC a total amount of 2092 buds was identified, 570 (27%) by both pathologists. The amount of TB counted in each image by the two observers is depicted as a scatter plot in Fig. 2, where also the correlations are shown between the observers. As can be seen, the second pathologist significantly annotated more buds (x1.3 in H&E, x1.8 in IHC).

In order to get further insights on the interobserver agreement, Intraclass- and Spearman Correlation coefficients as well as Kappa values were calculated. Intraclass correlations (Two-way mixed single measures with Absolute agreement) of r = 0.664 (95% CI 0.442–0.798) in H&E and r = 0.679 (95% CI 0.233–0.847) in IHC were found. Note the relatively large 95% confidence intervals (CK range even greater than H&E range). Spearman correlation coefficients were found for H&E r = 0.706 and CK r = 0.907.

For calculation of the Kappa scores raw bud counts were classified according to ITBCC 2016 classes, see Table 2. Kappa values of 0.46 (H&E) and 0.55 (IHC) were determined.

4   Experimental Results

We evaluated the performance of the different network configurations as described in Sect. 2.2, via FROC analysis, as shown in Fig. 3. For this purpose, we first compared the performance of the networks with 2 classes and with 3 classes without HNM. As can be seen in Fig. 3 the FROC performance of the network with 3 classes is substantially better than the one with 2 classes, achieving higher sensitivity with less false positives per hot-spot, even when HNM is not used. Secondly we assessed the perfor-

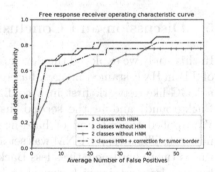

Fig. 3. FROC curves presenting the performance for all CNN's.

mance of the CNN trained with three classes when training examples were uniformly randomly sampled, and when hard-negative mining was used. It can be observed that the network with hard-negative mining gives consistently better FROC performance. Finally we compared the results of the CNN with HNM, with and without false positive reduction based on the distance from TG. As shown, the proposed post-processing technique improves the results slightly. The

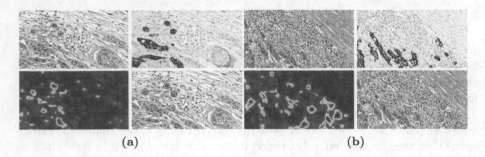

(a)   (b)

Fig. 4. Example of details of two hot-spots (a) and (b). The original H&E stained image (top left), the CK stained image (top right), the likelihood map (bottom left) and the final output (yellow) with the manual annotations (blue) are depicted. (Color figure online)

network with 2 classes and 3 classes, reached a F1-score of 0.16 and 0.20 respectively, with recalls of 0.68 and 0.72. With HNM without post-processing a F1-score of 0.31 was reached and with HNM plus post-processing a F1-score was reached of 0.36, with a recall of 0.72 for both networks.

The likelihood maps of predicted hot-spots of the CNN with HNM and post-processing are shown in Fig. 4. A closer inspection of the accompanying maps seems to indicate that this CNN is better able to distinguish TB and stroma components from each other compared to the former CNN's, although it is also visible that it still contains false positives in the stroma area.

5   Discussion and Conclusions

In this study we explored the development of a computer aided tool for detection of TB in H&E stained images, based on convolutional neural networks. We used a VGG-like network first in a configuration with two output classes (TB and Background) and in the second instance with TG as an extra output class. We applied the method of hard-negative mining to the results of the 3-class network. The ratio for HNM was set at 50-50. We also tested with different higher ratios (more hard-negatives less Background) however, we saw that the network became less certain with regard to labeling of the class TB (lower sensitivity), a phenomenon possibly due to TB incompleteness in the reference standard. In connection with the persistent problem of the false positives in the tumor glands, we eventually applied a post-processing step to the last results. With this step, buds detected by the network in the immediate vicinity of the outline of the tumor glands were removed. However, it is clear that this procedure carries the risk that buds in the immediate vicinity of the gland contour are missed. We have analyzed the results after the post-processing step. We mainly investigated residual false positives. The analysis indicates that also the presence of larger buds (potentially poorly differentiated clusters; small clusters of >5 tumor cells) is problematic in the task to be performed, which is discriminating between TB and TG. We therefore propose to use not only TB and TG but also PDCs as separate output class in future work.

We calculated the degree of agreement between the scores of the two pathologists using both the Spearman Correlation Coefficients and ICCs. The ICC takes into account how many buds in an image have been annotated by both pathologists, whereas the Spearman Correlation Coefficient only reflects the relationship between the number of annotated buds, and thus also gives a high correlation when the same number is scored, but not –more specifically– the same buds. Several TB investigators have included an assessment of interobserver variation in prognostic studies [4]. Based on scores from 2 or more observers, reported Kappa values for tumor budding scores range from 0.41 (moderate) to 0.938 (very good), depending on methodological factors, but also for example on the experiance of the participants. The level of agreement found in our study is moderate, although fairly in line with numbers reported by others. In view of this a better reference standard may be obtained by considering majority voting of buds detected by several pathologists in a pool of experienced observers. The reference standard can also be improved by a more precise TB annotation. In this seminal work we marked buds in the dataset by clicking and then creating an artificial outline (circle; surface equal to calculated average bud-surface). As a result, the smaller, usually single buds have been presented for analysis in conjunction with much surrounding stroma. In the future, this step could be replaced with delineating the real outlines (basement membranes) of the TB.

As our results confirm, generally a plurality of TB are found in CK, so apparently many buds in H&E are withdrawn from human perception. For this reason in our study previously marked equivalent IHC stained images were used for annotating the H&E training data set. Although this procedure will have contributed to the reliability of the reference standard in H&E, this may not have been sufficient. This conclusion is supported by our observations on the testing of several ratios for the HNM process, as we have noticed that increasing the ratio (more hard-negatives, less Background) led to a lower sensitivity. In connection with these findings, future work could focus on detection of TB in IHC staining first, whereby a procedure for better consensus on TB status between pathologists will be sought.

Acknowledgement. This project was funded by a research grant from the Dutch Cancer Society, project number 10602/2016-2. The authors would like to thank Irene Otte-Holler and Rob van de Loo for staining and scanning the WSI's.

References

1. Bejnordi, B.E., et al.: Deep learning-based assessment of tumor-associated Stroma for diagnosing breast cancer in histopathology images. In: Biomedical Imaging (ISBI 2017), pp. 929–932. IEEE (2017)
2. Caie, P.D., et al.: Quantification of tumour budding, lymphatic vessel density and invasion through image analysis in colorectal cancer. J. Transl. Med. **12**(1), 156 (2014)

3. He, K., Zhang, X., Ren, S., Sun, J.: Delving deep into rectifiers: surpassing human-level performance on imagenet classification. In: Proceedings of the IEEE International Conference on Computer Vision, pp. 1026–1034 (2015)
4. Koelzer, V.H., Zlobec, I., Lugli, A.: Tumor budding in colorectal cancerready for diagnostic practice? Hum. Pathol. **47**(1), 4–19 (2016)
5. Lugli, A., et al.: Recommendations for reporting tumor budding in colorectal cancer based on the international tumor budding consensus conference (ITBCC) 2016. Mod. Pathol. **30**, 1299–1311 (2017)
6. Mitrovic, B., Schaeffer, D.F., Riddell, R.H., Kirsch, R.: Tumor budding in colorectal carcinoma: time to take notice. Mod. Pathol. **25**(10), 1315 (2012)
7. Simonyan, K., Zisserman, A.: Very deep convolutional networks for large-scale image recognition (2014)
8. Wang, D., Khosla, A., Gargeya, R., Irshad, H., Beck, A.H.: Deep learning for identifying metastatic breast cancer. arXiv preprint arXiv:1606.05718 (2016)

Significance of Hyperparameter Optimization for Metastasis Detection in Breast Histology Images

Navid Alemi Koohbanani[1,2]([⊠]), Talha Qaisar[1], Muhammad Shaban[1], Jevgenij Gamper[1], and Nasir Rajpoot[1,2]

[1] Department of Computer Science, University of Warwick, Coventry, UK
n.alemi-koohbanani@warwick.ac.uk
[2] The Alan Turing Institute, London, UK

Abstract. Breast cancer (BC) is the second most leading cause of cancer deaths in women and BC metastasis accounts for the majority of deaths. Early detection of breast cancer metastasis in sentinel lymph nodes is of high importance for prediction and management of breast cancer progression. In this paper, we propose a novel deep learning framework for automatic detection of micro- and macro- metastasis in multi-gigapixel whole-slide images (WSIs) of sentinel lymph nodes. One of our main contributions is to incorporate a Bayesian solution for the optimization of network's hyperparameters on one of the largest histology dataset, which leads to 5% gain in overall patch-based accuracy. Furthermore, we present an ensemble of two multi-resolution deep learning networks, one captures the cell level information and the other incorporates the contextual information to make the final prediction. Finally, we propose a two-step thresholding method to post-process the output of ensemble network. We evaluate our proposed method on the CAMELYON16 dataset, where we outperformed "human experts" and achieved the second best performance compared to 32 other competing methods.

1 Introduction

Breast cancer (BC) is the second most common type of cancers and the primary cause of cancer mortality in women. Majority of deaths from BC are due to its metastasis to other organs in the body [1]. Therefore, early stage detection is important for the diagnosis and prognosis of BC. The sentinel lymph node (SLN) biopsy is the most pragmatic way of attaining BC metastasis. One of the challenging aspects of this problem is that a lymph node tissue contains some other cells (histiocytes) and regions (follicles and medullary sinus) having morpholgical resemblance to tumor cells and tumor regions as shown in Fig. 1.

Feature engineering to capture the discriminative attributes of each region is a non trivial task. Therefore, learning robust and discriminative features from the data in an automated manner using convolutional neural networks (CNNs) is a captivating choice for the problem at hand. Here, we proposed a framework

© Springer Nature Switzerland AG 2018
D. Stoyanov et al. (Eds.): COMPAY 2018/OMIA 2018, LNCS 11039, pp. 139–147, 2018.
https://doi.org/10.1007/978-3-030-00949-6_17

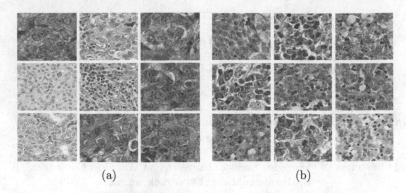

(a)   (b)

Fig. 1. (a) Tumor regions, exhibiting non-uniformity in chromatin texture (b) histio-cyte regions resembling tumor regions in terms of their morphological appearance.

for the detection of metastasis in whole-slide images (WSIs). Firstly, extracted patches from WSIs are categorized based on the presence of metastasis using CNNs. In this paper, modified CNNs are utilized and we show the high impact of these modification for boosting CNNs performance. To this end, we used Gaussian process for optimization of hyperparameters. Secondly, to enhance the final prediction on WSI level, cellular and contextual information are combined to get tumour likelihood map. This is achieved by fusing the outputs of two modified CNNs trained on patches of different resolutions and sizes. Finally, we propose a simple yet effective algorithm to convert the tumor likelihood map into a binary map. The proposed two-step thresholding method removes the highly uncertain regions from the likelihood map. Confidence score of each WSI for containing a metastases region is calculated using area of highly certain metastasis regions.

We have made three major contributions in this paper. First, we show that following a principled Bayesian approach to hyperparameter optimization can significantly improve performance for histology images with standard state-of-the-art CNNs instead of using new complex network architectures. Second, we propose an ensemble strategy to mimic the routine clinical practice where pathologist examines a WSI at different magnification levels ($40\times$, $20\times$, etc.) under the microscope. Our proposed method combines the predictions of two different resolutions ($40\times$ and $20\times$) to make final prediction which integrates cellular and contextual information. Finally, with the above principled approach, we achieve competitive results; 2nd only to the current winner in the leaderboard and better than the pathologists' performance on one of the largest publicly available histology image datasets (CAMELYON16[1]), according to the criterion used by the challenge organizers.

---

[1] https://camelyon16.grand-challenge.org.

2   Related Work

Czerniecki *et al.* [2] proposed IHC biomarkers to assist the pathologist in breast metastasis detection but it requires more time, cost and increases the number of slides required for analysis. An automated computer-based system was also developed for detecting micro-metastasis in lymph node biopsies [3] based on cell detection. Instead of cell detection, our method detects metastasis using robust multi-resolution features automatically learned from the very large dataset.

Recently, deep learning techniques have been used for a variety of histology image analysis problems. Some early work using CNNs was done for mitosis counting for primary breast cancer [4]. Recently, Wang *et al.* [5] assign a prediction value to each patch using CNNs and then make decision based on probability map of WSIs. Overall, the CAMELYON16 challenge [6] has shown the utility of deep learning algorithms for automatic tissue analysis, outperforming the pathologists in terms of detecting tumors within the WSIs. Since feeding large high resolution patches into deep learning model is computationally infeasible; one should consider a trade off between patch size and the amount of information lying within that patch. Most existing methods, consider using patches at one resolution. Here we train two separate networks with high and low resolution patch size, then merge probability maps generated by these networks.

The rest of this paper is organized as follows. Section 3 introduces the methodology details, Sect. 4 demonstrate the experimental results and comparison. Finally, Sect. 5 draws conclusion about the paper.

3   Methodology

Our framework (Fig. 2) is based on ensemble of two networks where one network encodes on cellular structures and the other captures the context. We find optimal architecture of classification networks and training hyperparameters through Bayesian optimization method. WSI probability map is created by merging all probabilities of small patches.

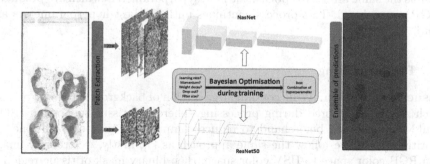

Fig. 2. An overview of the proposed approach.

3.1   Bayesian Optimization for Boosting Network Performance

Selection of the best network for a specific task is not trivial due to the stochastic nature of deep learning networks and their dependencies on various hyperparameters. These hyperparameters range from learning rate, momentum to more complicated variables like selection of number of layers, filter size, etc. Optimal selection of these hyperparameters is a known challenge in computer vision community. Non-optimal values of hyperparameters may lead to poor overall performance of a network. Search over all possible combination of hyperparameter values (grid search or random search) is computationally infeasible and extremely time consuming.

One way to overcome this problem is using Bayesian Optimization (BO) methods [7]. Gaussian process (GP) is one of BO approaches that can be used to predict optimal hyperparameter. A GP is a supervised learning method, which addresses the problem of leaning input-output mappings from training data. It utilizes kernel functions to learn these relations. In GP having observed N input vectors (hyperparameters) and their corresponding output variables (accuracies), we wish to make assumption about unobserved parameters. Acquisition function using this set of information suggest the next set of parameters. Here we used Expected Improvement [8] as an acquisition function to find the optimal settings for network and optimization level hyperparameters.

Convolutional filter size controls the receptive field for subsequent layers. Larger filter size for initial layers works well for natural images as classification of these images does not require high resolution information. However, tumor classification highly depends on high resolution images of tumor cells. Therefore, we consider filter size of first convolutional layer along with l2 weight regularization as hyperparameters to give flexibility to the network to choose the best receptive field. The standard architectures of ResNet and NASNet does not contains any dropout layer. However, dropout layer is useful to reduce over-fitting. Therefore, it is considered as another hyperparameters along with learning rate and momentum to be chosen by Bayesian optimization. The best hyperparameters are selected during training with early stopping. It means where the validation remains the same after two epochs, the new hyperparameters which are predicted by GP are replaced. This process continues until the best hyperparameters are found.

3.2   Pre-processing

Tissue Localization. WSIs contain large section of background (white) regions which should be ignored during processing. Therefore, tissue separation (ROI) should be performed beforehand to reduce computation time and efficiency of algorithm. Here, we follow the same approach as [5], firstly, we transfer image from RGB color space to HSV color space, then binary mask of tissue region is obtained by applying Otsu adaptive thresholding algorithm on S channel.

3.3   Network Selection and Ensembling

The proposed framework is an ensemble of two different CNNs. One network learns the cell level representation of tumour and normal patches at 40× resolution. However, since context information is also important along with high resolution cell appearance, cell level information is not enough to predict the label of a patch. We used second network to capture the context information from a larger patch at 20× resolution.

Different networks have been proposed in the literature for classification of natural images (ImageNet) including Inception-V3 [9], ResNet [10], DenseNet [11], and NASNet [12]. We choose ResNet for cellular feature based classification of patches due to its better performance. Moreover, use of smaller filter size in the first convolution layer and dropout before last fully connected layer results in improved performance. We explored these modification along with weight optimization level hyperparameters through BO.

We feed large patches(448 × 448) at lower resolution to the context based prediction models. Having the context information as input, NasNet has the best performance. It is based on seperable convolution with different filter sizes that helps to learn the representation at different resolution. Additionally, inclusion of dropout with hyperparameter tuning results in 5% increase in patch level performance as shown in Table 1.

Finally, we construct the probability map from each network (ResNet and NasNet) for each WSI. Then we fuse them together to build the final probability map that reflect both high and low resolution information. Afterwards, the probability maps are post-processed by morphological operations to achieve tumor localization and WSI classification. The post-processing procedure is described in Sect. 3.4.

Table 1. Validation accuracy of different networks trained on the CAMELYON16 dataset. Networks trained on both 448 × 448 at 20× resolution and 224 × 224 at 40× resolution with default hyperparameter settings and after applying Bayesian Optimization (BO)

Networks	40×	40× (BO)	20×	20× (BO)
ResNet [10] (%)	97.32	99.12	86.15	90.56
InceptionV3 [9] (%)	95.10	95.10	85.67	86.57
DenseNet [11] (%)	95.35	96.67	87.68	89.15
NasNet [12] (%)	96.28	97.01	86.75	91.91

3.4   Post-processing

Our careful inspection of the tumor probability shows that regions with high probability values but characterized by abrupt changes in the values (i.e. high uncertainty) generally correspond to false positive decisions. Therefore, to eliminate these high uncertainty regions, we use two different threshold values t_{low}

Table 2. Dropout, learning rate, momentum, Weight decay for 1st layer and filter size for 1st layer

Networks	Dropout	learning rate	momentum	weight decay	filter size
ResNet (40×)	✓	0.001	0.95	12	3×3
ResNet (20×)	✓	0.001	0.97	12	3×3
InceptionV3 (40×)	✓	0.010	0.80	-	3×3
InceptionV3 (20×)	✓	0.001	0.85	12	3×3
DenseNet (40×)	-	0.001	0.90	12	3×3
DenseNet (20×)	-	0.010	0.87	12	3×3
NasNet (40×)	✓	0.001	0.95	12	3×3
NasNet (20×)	-	0.010	0.99	12	3×3

and t_{high}. First, we obtain sets of regions $\mathcal{B}(t_{\text{low}})$ and $\mathcal{B}(t_{\text{high}})$ by thresholding the tumor probability image at t_{low} and t_{high}, respectively. From the construction, each region in $\mathcal{B}(t_{\text{high}})$ is a shrink version of some region in $\mathcal{B}(t_{\text{low}})$, and there can be multiple regions in $\mathcal{B}(t_{\text{high}})$ that correspond to the same region in $\mathcal{B}(t_{\text{low}})$. For each region $C^{(i)} \in \mathcal{B}(t_{\text{low}})$, let $\left\{ \tilde{C}_1^{(i)}, ..., \tilde{C}_{N_i}^{(i)} \right\} \in \mathcal{B}_{\text{high}}(t)$ be a set of N_i regions corresponding to $C^{(i)}$. We eliminate each region $C^{(i)} \in \mathcal{B}(t_{\text{low}})$ such that

$$\frac{\left| \cup_{j=1}^{N_i} \tilde{C}_{N_i}^{(i)} \right|}{|C^{(i)}|} < \alpha, \tag{1}$$

where α is the area threshold ratio, and $|\cdot|$ denotes the cardinality of a set. We set $t_{\text{low}} = 0.3$, $t_{\text{high}} = 0.9$, and $\alpha = 0.5$. For each remaining candidate region, we calculate the confidence of being tumor using the minimum probability found in that region weighted by its area. The candidate regions are further removed based on this weighted confidence.

After clearing probability map, we binarize all probability maps by applying a threshold (threshold = 0.6). We dilate the tumor regions as much as 75 um, since locations within the 75 um of tumor areas are also considered as true positive for FROC validation criteria that is used in CAMELYON 16 challenge. Finally, the coordinate of those regions along with the maximum value on probability map are recorded for FROC. For plotting ROC curve, the maximum probability of largest tumor area is reported as probability of WSI being tumor.

4 Experimental Results

Our experiments were conducted on the CAMELYON16 dataset to evaluate the proposed framework for cancer metastasis detection in WSIs. This dataset contains 400 WSIs: 270 WSIs for training and remaining 130 WSIs for evaluation purpose. The cancer metastasis regions were exhaustively annotated under the supervision of expert pathologists. The WSIs were stored at different magnification levels with the highest magnification level of 40× and 0.243 um/pixel

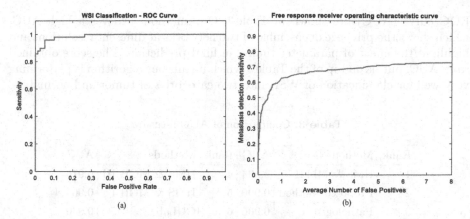

Fig. 3. (a) ROC curve and (b) FROC curve which shows sensitivity against average number of false positives

resolution. Our experiments were performed at both 40× and 20× magnifications to have both cellular and contextual information. Table 2 shows the values that have been chosen for hyper parameters after applying BO. Filter size should be 3 × 3 for most of networks as the objects in images are small and l2 regularization in first layer can lead to better performance. Moreover we use drop out in our settings according to BO estimation.

We experimented with different networks on our data in order to choose optimal achitecture. To improve the performance of the networks we used hyper-parameter optimization method (Sect. 3.1) for increasing the accuracy. Table 1 shows the performance of different networks on our dataset. We achieve considerable increase in accuracy after hyperparameter optimization. To this end, we chose Resnet50 for patches of 40× and NasNet for 20× patches as they have higher accuracy on their corresponding dataset after BO optimization. The accuracy increased 2% and 5% for ResNet (with 40× patches) and NasNet (with 20× patches) respectively which is a high improvement with regarding high variability in dataset and low inter class variability.

After first iteration of training, we process the WSIs in patch based manner to generate the probability maps. We observed that the probability maps of WSIs have many false positive regions that affect the final result. To this end, by recognizing the false positive regions, corresponding patches from WSIs are extracted and fed back into the networks for fine-tuning. Therefore, we come up with cleaner probability map with very few false positive regions.

We followed the same evaluation criteria (ROC and FROC) as explained in [6]. As shown in Fig. 3, ROC curve clearly depicts that our method predicts large number of tumor slides with very few false positives. In fact, it classifies 83% of tumor slides without throwing any false positive. The FROC curve also shows that the algorithm is capable of localizing tumor regions with various mean number of false positives in per whole slide image. The values for area under

ROC curve (AUC) are shown in Table 3. Our method achieves a high AUC, which shows the privilege of ensemble of two networks at different resolutions and highlight the effect of parameter tuning in final prediction. The score obtained from AUC, put us on top of the Table 3 which means our algorithm is performing very well for classification of WSIs into two categories of tumor and normal.

Table 3. Comparison of AUC measure

Rank	Methods	AUC	Rank	Methods	AUC
1	HMS & MITII [5]	0.994	4	HMS & MGH III	0.976
2	Proposed Method	0.990	5	HMS & MGH I	0.965
3	Pathologist	0.966	6	CULab	0.940

5   Conclusions

In this paper, we investigated the impact of hyperparameter optimization on network performance. Tuning hyperparameters with the Gaussian process could increase the validation accuracy on average by 5%. Furthermore, a multi-resolution network for detecting breast cancer metastasis from sentinel lymph node WSIs was proposed. Therefore, with combined contextual and cell level information and also optimizing hyperparameter, we achieve AUC and average sensitivity of 0.990 and 0.6583 respectively. This results in competitive performance of our framework applied on the CAMELYON16 dataset.

References

1. American Cancer Society: Cancer Facts & Figures (2015)
2. Czerniecki, B.J., et al.: Immunohistochemistry with pancytokeratins improves the sensitivity of sentinel lymph node biopsy in patients with breast carcinoma. Cancer **85**(5), 1098–1103 (1999)
3. Weaver, D.L., et al.: Comparison of pathologist-detected and automated computer-assisted image analysis detected sentinel lymph node micrometastases in breast cancer. Mod. Pathol. **16**(11), 1159 (2003)
4. Cireşan, D.C., Giusti, A., Gambardella, L.M., Schmidhuber, J.: Mitosis detection in breast cancer histology images with deep neural networks. In: Mori, K., Sakuma, I., Sato, Y., Barillot, C., Navab, N. (eds.) MICCAI 2013 Part II. LNCS, vol. 8150, pp. 411–418. Springer, Heidelberg (2013). https://doi.org/10.1007/978-3-642-40763-5_51
5. Wang, D., et al.: Deep learning for identifying metastatic breast cancer. arXiv preprint arXiv:1606.05718 (2016)
6. Bejnordi, B.E., et al.: Diagnostic assessment of deep learning algorithms for detection of lymph node metastases in women with breast cancer. JAMA **318**(22), 2199–2210 (2017)
7. Snoek, J., et al.: Practical Bayesian optimization of machine learning algorithms. In: Advances in Neural Information Processing Systems, pp. 2951–2959 (2012)

8. Bergstra, J.S., et al.: Algorithms for hyper-parameter optimization. In: Advances in Neural Information Processing Systems, pp. 2546–2554 (2011)
9. Szegedy, C., et al.: Rethinking the inception architecture for computer vision. In: Proceedings of the IEEE Conference on Computer Vision and Pattern Recognition, pp. 2818–2826 (2016)
10. He, K., et al.: Deep residual learning for image recognition. In: Proceedings of the IEEE Conference on Computer Vision and Pattern Recognition, pp. 770–778 (2016)
11. Huang, G., et al.: Densely connected convolutional networks. In: Proceedings of the IEEE Conference on Computer Vision and Pattern Recognition, vol. 1 (2017)
12. Zoph, B., et al.: Learning transferable architectures for scalable image recognition. arXiv preprint arXiv:1707.07012 (2017)

Image Magnification Regression Using DenseNet for Exploiting Histopathology Open Access Content

Sebastian Otálora[1,2]([✉]), Manfredo Atzori[2], Vincent Andrearczyk[2], and Henning Müller[1,2]

[1] University of Geneva (UNIGE), Geneva, Switzerland
[2] University of Applied Sciences Western Switzerland (HES-SO), Sierre, Switzerland
juan.otaloramontenegro@hevs.ch

Abstract. Open access medical content databases such as PubMed Central and TCGA offer possibilities to obtain large amounts of images for training deep learning models. Nevertheless, accurate labeling of large-scale medical datasets is not available and poses challenging tasks for using such datasets. Predicting unknown magnification levels and standardize staining procedures is a necessary preprocessing step for using this data in retrieval and classification tasks. In this paper, a CNN-based regression approach to learn the magnification of histopathology images is presented, comparing two deep learning architectures tailored to regress the magnification. A comparison of the performance of the models is done in a dataset of 34,441 breast cancer patches with several magnifications. The best model, a fusion of DenseNet-based CNNs, obtained a kappa score of 0.888. The methods are also evaluated qualitatively on a set of images from biomedical journals and TCGA prostate patches.

1 Introduction

Pathologists analyze biopsies looking for structural patterns such as nuclei and gland deformations to grade various types of cancer and to describe the structures in the images for later writing the pathology report. These visual patterns are traditionally inspected using a light microscope at a certain magnification level but also increasingly through digital biopsy slides, namely Whole Slide Images (WSIs).

Deep Learning (DL) models and, in particular, Convolutional Neural Networks (CNNs) learn high–level discriminative features for digital pathology tasks [2,4,7] such as classification and content–based image retrieval [5]. Most supervised DL models require thousands or even hundreds of thousands of manually annotated patches when building a model from scratch, which is extremely difficult to obtain for medical data. Given the availability of open access data repositories such as the cancer genome atlas (TCGA[1]), digital teaching files and

---

[1] http://cancergenome.nih.gov/.

D. Stoyanov et al. (Eds.): COMPAY 2018/OMIA 2018, LNCS 11039, pp. 148–155, 2018.
https://doi.org/10.1007/978-3-030-00949-6_18

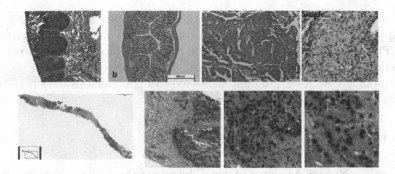

Fig. 1. Top row: PubMed Central images from the category *Light Microscope*. In their captions, the magnifications were identified as 5×, 5×, 10× and 20× respectively. Bottom row: Three Prostate biopsy patches extracted at a 5×, 10× and 20×, respectively.

PubMed Central (PMC[2]), an open question is how to use these datasets for leveraging useful knowledge from them effectively, since they offer an attractive possibility to obtain large amounts of relevant medical images for training models, and ultimately solving concrete medical inquiries. In TCGA, the available WSIs lack of local annotations, but the magnification information is provided in the WSI file. In PMC, the challenge is bigger since there is a wide variety of organs and species (humans, macaques and mice), staining procedures and slide preparation methods and also unknown magnification levels of the images. Example images are shown in Fig. 1. All these factors vary strongly among digital pathology images and even more after figure editing, for example when writing a scientific publication or after publishing an article. Raw data of the WSIs from where the images come are never available.

Several authors have studied the influence of the magnification level for WSI classification, nuclei detection and segmentation with interesting findings. Bayramoglu et al. [1] trained a multitask CNN to predict both malignancy and image magnification level simultaneously, showing that the network trained with multiple magnifications outperforms the single magnification one, they also encourage to regress the magnification level instead of limiting a classifier to a discrete set of levels. Janowczyk et al. [4], trained a standard CNN for nuclei segmentation based on the AlexNet architecture, forcing the network to learn better boundaries, they discuss the need for re-training of models for each magnification level. Kumar et al. [6] designed a CNN that outputs a 3-class probability for each pixel (background, boundary and inside nuclei) and evaluate their method on several tissue types outperforming CellProfiler and Fiji in a fixed magnification level. Otálora et al. [8] trained a deep CNN to predict three fixed levels of magnification and evaluated in a single type of tissue, their results show that a pretrained network has better overall performance; however, in content–based retrieval tasks, where the query pattern could be in any type of tissue at any magnification, this classifier is of limited usability.

---

[2] https://www.ncbi.nlm.nih.gov/pmc/, URLs as of 3 January 2018.

The objective of this paper is to tackle the variability in scale using a regression approach based on deep CNNs tailored to regress directly the magnification level. The proposed approach is tested on different type of tissues in open access datasets showing the generalization of the method, an exploration of the combination of different regression approaches led to a good quantitative performance of magnification prediction.

2   Methods

Regressing Nuclei Average Area: The average nuclei area in terms of pixels can provide an estimate of the magnification of an image, this regressor can be used for computing differences between nuclei areas of different kind of tissues as shown in the results section, nevertheless this depends on the cell type and disease. This regression has the advantage that bypasses the problem of nuclei segmentation at test time, even though the annotated masks are still needed for computing the average area ground truth. In both architectures, the last layer is designed to output only a real number, i.e., the nuclei average area, in order to minimize the mean squared error between the ground truth and predicted average areas. Predicting the magnification with an average area is done by computing the closest magnification mean area using the mean of the nucleus areas in the training set patches and then assigning its correspondent magnification. i.e. if the predicted area is 650 pixels, the magnification assigned will be 30×.

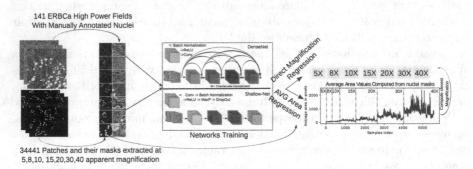

Fig. 2. Overall schema of our magnification regression approach. For the regression of the average area, the segmentation masks are used to compute the average area of nuclei. Instead of having a unit layer at the end of each architecture that regres to the magnification (top branch), a unit outputs the average area (bottom branch).

A comparison of two different CNN architectures is done in the two scenarios of direct magnification and average area nuclei regression, as shown in Fig. 2. The first architecture is the state–of–the–art DenseNet architecture [3] that features a dense connectivity pattern among its layers. DenseNet introduces direct connections between any two subsequent layers with the same feature map size. The main advantage of DenseNet over other very deep architectures is that it reuses information at multiple levels without drastically increasing the number of parameters in the model, particularly with the inclusion of bottleneck and compression layers. For the second architecture, a relatively shallow network, named ShallowNet, is designed. It consists of 4 consecutive blocks of convolution, batch normalization, rectified linear units and dropout with a probability of 0.25. The comparison of the two architectures assesses the performance gain in deeper and more complex architectures versus a more parameter–efficient one. In the case of direct regression of the magnification, the last output unit of the two networks is set to predict the magnification value of the patch directly, without computing the area of the nuclei in the segmentation mask. The regressed magnification is mapped to the closest magnification by calculating the minimum absolute value between the prediction and the magnification classes. The details of the two DL architectures are:

DenseNet-BC 121: The chosen architecture is the 121-layer variation of DenseNet with 7 million parameters and perform experiments fine–tuning all the layers from pre–trained ImageNet weights and training the weights from scratch.

ShallowNet: A 4–layer CNN consisting of 3×3 convolutional kernels, followed by batch normalization, ReLU activation, dropout of 0.25 and max–pooling of a 2×2 neighborhood, ending in a dense layer with a linear activation that is expected to output the average area of the nuclei in the patch. This designed network has 2.7 million parameters.

As baseline for the area regression the DL nuclei segmentation method of Kumar et al. [6] is choosen. Since the calculated average nuclei area is needed for comparing it with the regression approach, we added the first and second output probability maps of their network that corresponds to the probabilities of pixels belonging to the inner nuclei and their boundary. An Otsu threshold is computed from this output to obtain a binary mask from which the average area of the nuclei is calculated using the resulting blobs. All the nuclei that were on the edge of the patches where removed to have a more robust prediction. Also, detected areas of less than 20 pixels are not taken into account since in the ground truth the minimum nuclei area at $5\times$ was 24 pixels. Even though this was not a fair comparison, since the model of Kumar was trained for a single magnification, this highlights the advantage of having a flexible area regressor.

2.1   Datasets

The data used for training in our approach is the publicly available dataset used for nuclei segmentation in [4], that allows to confidently estimate via manual annotations the ground–truth nuclei average area, and also downsampling the original image and masks to obtain the different magnification levels. This dataset consists of 141 images and masks of 2000×2000 pixels @$40\times$ ROIs of

Table 1. Number of ERBCa patches extracted per magnification and partition

Partition/#patches	5X	8X	10X	15X	20X	30X	40X	*Total*
Train	47	1087	2026	3646	4501	5368	5819	22494
Validation	4	294	585	1066	1302	1527	1575	6353
Test	18	214	440	864	1099	1401	1558	5594
Total	69	1595	3051	5576	6902	8296	8952	34441

estrogen receptor-positive breast cancer (ERBCa). The images contain a subset (not all nuclei in the images were annotated) of 12,000 manually annotated nuclei. We extracted 34441 patches for 5, 8, 10, 15, 20, 30, and 40× magnifications. The number of patches per magnification was kept within the same ranges when possible, i.e., for 5× and 8× is not possible to extract as many patches as in 30 or 40× due to the large area covered by the lower magnifications. The patches are separated into training (94), validation (27) and test (20) partitions checking that all the patches from a single image were in the same partition. In each patch, the condition that at least 3 complete nuclei were present is ensured. The complete distribution of patches is shown in Table 1 and example patches with masks are displayed in the input of the networks in Fig. 2.

For assessing the generalization of the approach, we tested it on two external open access databases: TCGA patches and PubMed Central histopathology images. The best trained model was tested on the test partition of 99125 patches used in the evaluation of the method reported in [5]. The patches corresponds to areas of low (45081) and high (54044) grade prostate cancer, with reported Gleason scores 6-7 and 8-9-10 respectively, at 20× magnification. For the PMC set, a total of 5,764,238 images with captions were crawled. A standard multi-modal CNN architecture was used for the captions and images to identify the image modality, e.g. light microscopy, x-ray, MRI, etc. The classification process led to a total of 291 prostate histopathology images.

3   Results

In Table 2 the magnification prediction results in the ERBCa test patches set are summarized. The DenseNet architecture trained from scratch to regress magnification led to a better classification performance than any other method separately, this is likely due to two factors: First, since is not computing an intermediate area, the network is less prone to introduce noise of overlapping classes measured by the area, and secondly, since it doesn't start with pre-trained weights from Imagenet, it has more flexibility to learn appropriate filters for histopathology images. Three combinations of both approaches were explored: Concatenation of the feature vectors, using the average-area learned weights to then fine-tune to regress magnification, and linearly combining the magnifications predicted by the area and the direct approaches, i.e.: $\alpha \times \text{Densenet}_{area} + 1 - \alpha \times \text{Densenet}_{magnif}$. From this experiments, the first two

Table 2. Left: classification results in the test set for all the compared methods. Right: confusion matrix for the best model: linear mixture of DenseNet models ($\alpha = 0.2$).

Method	Kappa	F1-Score		5×	8×	10×	15×	20×	30×	40×
ShallowNet	0.720	0.779	5×	13	1	4	0	0	0	0
DenseNet	0.888	0.911	8×	0	169	45	0	0	0	0
Pre-trained Densenet	0.616	0.693	10×	0	46	331	60	0	0	0
(Area) Kumar	0.079	0.037	15×	0	0	3	748	113	0	0
(Area) ShallowNet	0.506	0.619	20×	0	0	0	1	1097	1	0
(Area) DenseNet	0.713	0.772	30×	0	0	0	0	15	1364	22
(Area) Pre-trained Densenet	0.734	0.789	40×	0	0	0	0	0	177	1381
(Area) GT Area	0.615	0.692								
Fusion (α)area + (1-α)direct	**0.888**	**0.912**								
Fusion (PMC 3-Class subset)	0.16	0.392								

did not show any significant improvement in the test set over the two approaches separately, thus not reported here. The third one led to a slightly better performance than the direct approach, using an alpha value of 0.2, and is the model which is reported in the confusion matrix for Table 2. In the area regression scenario, the two DL regressors presented are consistently closer to the ground truth average area than the baseline method. The baseline works very well on the lower-medium magnifications but fails at capturing the changes in big nuclei. Examples of patches are presented in Fig. 4. The class-activation maps are computed using the Grad-CAM method in the last dense layer as implemented in Keras-vis[3]. Both networks were implemented in Keras and optimized using Adam with initial learning rates explored logarithmically between 0.01 and 10^{-9}. The best learning rates were found to be 0.01, 0.0001, and 0.0001 for the ImageNet pre–trained and trained from scratch DenseNet and ShallowNet respectively. The best performing model on the test ERBCa patches was also evaluated on TCGA-PRAD and PMC databases.

PMC Histopathology Prostate Images: Since the PMC images are directly from articles the size and resolution of the images varies widely. Only the central 224×224 pixels in RGB channels were considered as this is the input size for our network. The predictions for most image patches are accurate as shown in Fig. 4, even with unknown stainings as the first two examples show. The lower magnifications (very small nuclei) are more challenging and, as a result, some of the predictions for those images are not correct as shown in the bottom–right images. A random selection of 55 images from were the magnification are available from the captions were selected to perform a quantitative test. In Fig. 3 a t-SNE embedding shows how the images at 20× tend to cluster in a single part of the feature space, whereas the 10 and particularly 5× images are more spread across since their differences with closer magnifications are more subtle, as also seen in the quantitative results in the ERBCa patches.

---

[3] keras-vis https://github.com/raghakot/keras-vis(2017).

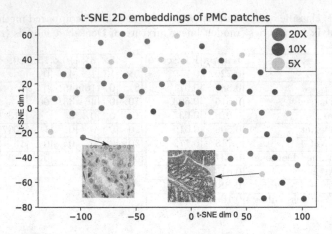

Fig. 3. 2D t-SNE embedding from the 1024-dimensional representation of 55 randomly selected PMC prostate images.

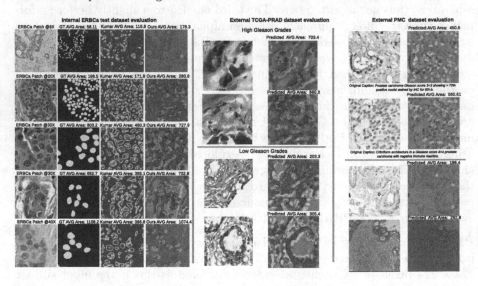

Fig. 4. Predictions of the regressor in all the datasets. Left: ERBCa test patches at several magnifications, our model is more robust in higher magnifications than the automatic segmentation. Middle: TCGA-PRAD patches from low and high cancer grades, showing dissimilar predicted average areas. Right: PMC prostate images: the first two predictions are consistent with manual annotations, the bottom ones are incorrect.

In the TCGA-PRAD dataset 92% of the patches were classified correctly at 20× using the area-magnification fusion approach.

4   Conclusion

In this paper, two CNN architectures were trained to regress the magnification level in histopathology images, using a direct regression approach or by first learning the average area of the nuclei. For internal evaluation, the best magnification regressor was a linear combination of two DenseNets: One trained to regress the area and the other to regress the magnification, this model had the best performance in terms of Kappa and F1-score, suggesting that a complementarity between the two models exists. In the case of the area regression a comparison is done with a state–of–the–art DL segmentation method, showing better overall performance as measured in MAE and F1-scores. Finally, the predictions of our model on the TCGA and PMC databases were accurate for a subset of filtered prostate images. Our model was able to generalize to several tissue types and provides useful information for exploiting the content in open access databases of histopathology images.

References

1. Bayramoglu, N., Kannala, J., Heikkilä, J.: Deep learning for magnification independent breast cancer histopathology image classification. In: 2016 23rd International Conference on Pattern Recognition (ICPR), pp. 2440–2445. IEEE (2016)
2. Cruz-Roa, A., et al.: Accurate and reproducible invasive breast cancer detection in whole-slide images: a deep learning approach for quantifying tumor extent. Sci. Rep. **7**, 46450 (2017)
3. Huang, G., Liu, Z., Weinberger, K.Q., van der Maaten, L.: Densely connected convolutional networks. In: Proceedings of the IEEE Conference on Computer Vision and Pattern Recognition (2017)
4. Janowczyk, A., Madabhushi, A.: Deep learning for digital pathology image analysis: a comprehensive tutorial with selected use cases. J. Pathol. Inf. **7**, 29 (2016)
5. Jimenez-del-Toro, O., Otálora, S., Atzori, M., Müller, H.: Deep multimodal case–based retrieval for large histopathology datasets. In: Wu, G., Munsell, B.C., Zhan, Y., Bai, W., Sanroma, G., Coupé, P. (eds.) Patch-MI 2017. LNCS, vol. 10530, pp. 149–157. Springer, Cham (2017). https://doi.org/10.1007/978-3-319-67434-6_17
6. Kumar, N., Verma, R., Sharma, S., Bhargava, S., Vahadane, A., Sethi, A.: A dataset and a technique for generalized nuclear segmentation for computational pathology. IEEE Trans. Med. Imaging **36**(7), 1550–1560 (2017)
7. Litjens, G., et al.: Deep learning as a tool for increased accuracy and efficiency of histopathological diagnosis. Sci. Rep. **6**, 26286 (2016)
8. Otálora, S., Perdomo, O., Atzori, M., Andresson, M., Hedlund, M., Müller, H.: Determining the scale of image patches using a deep learning approach. In: IEEE 15th International Symposium on Biomedical Imaging, ISBI 2018. IEEE, April 2018

Uncertainty Driven Pooling Network for Microvessel Segmentation in Routine Histology Images

M. M. Fraz[1,2,4(✉)], M. Shaban[1], S. Graham[1], S. A. Khurram[5],
and N. M. Rajpoot[1,2,3]

[1] Department of Computer Science, University of Warwick, Coventry, UK
moazam.fraz@warwick.ac.uk
[2] The Alan Turing Institute, London, UK
[3] University Hospitals Coventry and Warwickshire, NHS Trust, Coventry, UK
[4] National University of Sciences and Technology, Islamabad, Pakistan
[5] The University of Sheffield, Sheffield, UK

Abstract. Lymphovascular invasion (LVI) and tumor angiogenesis are correlated with metastasis, cancer recurrence and poor patient survival. In most of the cases, the LVI quantification and angiogenic analysis is based on microvessel segmentation and density estimation in immuno-histochemically (IHC) stained tissues. However, in routine H&E stained images, the microvessels display a high level of heterogeneity in terms of size, shape, morphology and texture which makes microvessel segmentation a non-trivial task. Manual delineation of microvessels for biomarker analysis is labor-intensive, time consuming, irreproducible and can suffer from subjectivity among pathologists. Moreover, it is often beneficial to account for the uncertainty of a prediction when making a diagnosis. To address these challenges, we proposed a framework for microvessel segmentation in H&E stained histology images. The framework extends DeepLabV3+ by using an improved dice coefficient based custom loss function and also incorporating an uncertainty prediction mechanism. The proposed method uses an aligned Xception model, followed by atrous spatial pyramid pooling for feature extraction at multiple scales. This architecture counters the challenge of segmenting blood vessels of varying morphological appearance. To incorporate uncertainty, random transformations are introduced at test time for a superior segmentation result and simultaneous uncertainty map generation, highlighting ambiguous regions. The method is evaluated using 1167 images of size 512 × 512 pixels, extracted from 13 WSIs of oral squamous cell carcinoma (OSCC) tissue at 20x magnification. The proposed net-work achieves state-of-the-art performance compared to current semantic segmentation deep neural networks (FCN-8, U-Net, SegNet and DeepLabV3+).

Keywords: Microvessel detection · Tumor angiogenesis
Lymphovascular invasion · Separable convolution
Pyramid pooling based neural network · Uncertainty quantification

© Springer Nature Switzerland AG 2018
D. Stoyanov et al. (Eds.): COMPAY 2018/OMIA 2018, LNCS 11039, pp. 156–164, 2018.
https://doi.org/10.1007/978-3-030-00949-6_19

1   Introduction

The ability for cancer to spread to distant or adjacent tissues is a key indicator of poor patient prognosis. The tumor cells can gain access to blood or lymphatic vessels by intravasation allowing them to circulate through the intravascular stream. This lamphovascular invasion (LVI) can lead to the proliferation of tumor cells at another site in the body. This phenomena is more commonly referred to as metastasis. The lymphatic or vascular invasion by the primary tumor is considered as a sign of aggressive disease and is usually accompanied by metastases to the regional lymph nodes and to distant sites. The formation of new blood vessels is important for growth, survival and metastatic spread of tumor cells [2]. Tumor neoangiogenesis leads to formation of new blood vessels in the tumor tissues, with an initial purpose to facilitate the transport of nutrients and oxygen to help the tumor cells survive [2]. In diagnostic clinical pathology, most of the currently existing tissue datasets and pathways recommend commenting on the presence or absence of LVI however the degree of angiogenesis is not routinely examined or reported. In the existing research literature, there is strong evidence that microvessel density in tumor tissue is directly correlated with an increased risk of cancer spread, an increased incidence of disease recurrence and poor patient survival [10]. Recent studies have re-ported LVI detection and quantification as an important risk factor in disease progression particularly in breast, cervical and lung cancers [10]. However, most of the results in these studies are based on manual localization of microvessels in the subjectively defined regions of tissue whole slide image (WSI). The subjective identification of LVI and angiogenic regions in the tissues is irreproducible, time consuming and often requires clinical knowledge. In routine pathological practice, accurate segmentation of microvessels can assist pathologists in identification of LVI which would otherwise be very time-consuming. Furthermore, objective quantification of LVI and tumor angiogenesis from multi giga pixel histopathology images will provide extremely valuable 'big data' aiding prediction of tumor behavior and prognosis.

The appearance of microvessels in Hematoxylin and Eosin (H&E) stained histology WSIs is characterized by the presence of endothelial cells forming a closed structure that surrounds the red blood cells, as illustrated in Fig. 1. The heterogeneity in the visual appearance of microvessels makes their segmentation a non-trivial task. Immunohistochemical (IHC) staining for markers such as CD31 and CD34 can be used for microvessel identification, which is comparatively expensive than H&E staining. Due to this cost, it is not commonly used in routine clinical practice. Likewise, the IHC-stained histology images are rarely available in public datasets, which is a major hindrance in the development of automated methods to investigate the role of microvessels in tumor prognosis and therapy response.

Deep learning has recently been successfully used for the automated analysis of histopathology images. Specifically, deep learning based architectures have been proposed for the detection, segmentation and classification of histopathological structures in WSI images including deep learning based architectures are

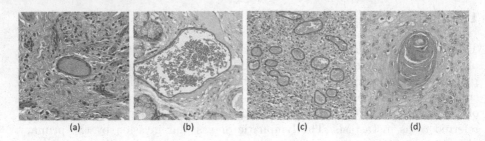

(a)   (b)   (c)   (d)

Fig. 1. Representative images of microvessels in H&E stained histology images of OSCC tissue, illustrating their shape variability. The microvessel boundary annotation is shown in red color. (a–c) Microvessel with varying red cells density, (c) Microvessels at different sizes, (d) Keratinization, which appears similar to microvessel (Color figure online)

proposed for detection and classification of nuclei [14], mitoses [12], lymphocytes [13], tumor and stromal regions [15] and glandular structure segmentation [7]. A few studies have explored automatic quantification of tumor angiogenic hotspots by the detection of microvessels in IHC stained histology images only [8]. Most recently, a fully convolutional neural network (FCN) based method for microvessel detection in H&E stained images is presented [16]. However, the methods for microvessel segmentation particularly in H&E stained images are limited.

In this paper, we present a framework for precise segmentation of microvessels in H&E stained histology images at multiple scales and resolutions by using an uncertainty aware spatial pyramid pooling deep neural network architecture. The Deeplabv3+ [4] architecture is extended by using an improved dice coefficient minimization based custom loss function and by accounting for the uncertainty of a prediction. The proposed network aims to solve the key challenges posed by automated microvessel segmentation. The method uses a modified Aligned Xception model [5] followed by an atrous spatial pyramid pooling (ASPP) unit [3] for feature extraction at multiple scales. This overcomes a major challenge of segmenting vessels of various sizes. Moreover, despite achieving state-of-the art performance in semantic segmentation, the deep networks typically do not inherently model the segmentation uncertainty. For this purpose, we apply random transformations to the images during test time, as a method to generate the approximate predictive distribution. Taking the average of these predictions of transformed images yields a superior segmentation and enables us to observe ambiguous areas, where the network is uncertain in a decision. The methodology is evaluated on 1167 images of size 512×512 pixels, extracted from 13 WSIs of oral squamous cell carcinoma (OSCC) tissue at 20x magnification, and demonstrated promising results. Moreover, the proposed network achieves state-of-the-art performance compared to current semantic segmentation deep neural networks (FCN-8 [9], U-Net [11], SegNet [1] and DeepLabv3+ [4]).

2   Proposed Method

We propose a framework for the segmentation of microvessels in H&E stained histology images. The framework extends the DeepLabV3+ network architecture by utilizing a modified Aligned Xception with ASPP and a custom loss function for precise segmentation. Moreover, the framework also incorporates random transformations at test time to account for the uncertainty of a prediction, as shown in Fig. 2. A substantially deep network is needed for meaningful feature extraction. Traditional convolutional neural network architectures (AlexNet, VGG, Google Net, ResNet e.t.c.) used for image classification has inherited limitations to model geometric transformations due to fixed geometric structures in their building modules. Moreover, these networks use a hierarchical combination of maxpooling and convolutions to increase the receptive field size. This results in loss of image information which may be very significant for precise object segmentation. In order to deal with these issues, the feature extraction process should be invariant to geometric transformations and retain the low level image information. The Xception model [5] has demonstrated promising performance in image classification task on ImageNet in terms of speed and accuracy. Xception has been modified to incorporate geometric transformations modeling capability for feature extraction. Further to this, Chen *et al.* [4] proposed the replacement of all max pooling operations with depthwise separable convolutions with striding. This allows the application of atrous separable convolutions for multiscale feature extraction at arbitrary resolution. Atrous convolution is an extension of the standard convolution operation, which provide us with the ability to explicitly control the resolution of features computed by deep convolutional neural networks and adjust filter's receptive field for capturing multiscale information. The use of separable convolutions reduces the number of convolutional parameters, hence increasing the computational efficiency. Subsequntly, this is more suitable for processing of multi giga pixel WSIs.

Inspired from [4], we have use modified Aligned Xception model for microvessel feature extraction at multiple scales and resolution with geometric deformation invariance. Incorporating multiscale and geometric transformation invariant features allow us to perform accurate microvessel segmentation. Atrous spatial pyramid pooling [3] with varying dilation rates (6, 12 and 18) is applied at the end of the encoder, to aggregate the multilevel features. This pooling module allows our proposed network to segment microvessels of varying shape and sizes. Global average pooling has been used to incorporate the global level context. Moreover, a 1×1 convolution is performed before each operation, followed by a dropout layer and another 1×1 convolution for dimensionality reduction. The features from each dilation operation is concatenated to give a powerful representation of high level image contextual information. The low level image information for precise delineation of microvessel boundaries is taken from the shallow layers of the deep network and concatenated with the feature map obtained after bilinear upsampling by a factor of 4. The feature map size is illustrated at each block level. The output is upsampled twice by a factor of 4 to obtain the final output after applying the softmax layer. We have used the loss function τ_{DSC} based

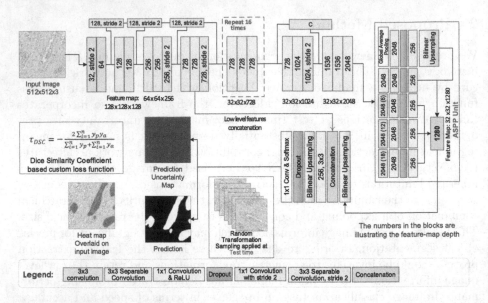

Fig. 2. The Proposed Framework: The legend represents the color coding of various types of convolutional layers. The values inside the blocks represents the depth of corresponding feature map. In ASPP Unit, the value in brackets denotes the dilation rates of 6, 12 and 18. To obtain the prediction uncertainty map, Random Transformation Sampling at test time is applied. The network uses custom loss function based on minimization of –ve of Dice Similarity Coefficient.

on minimizing the negative of dice similarity coefficient (DSC) for training the network. The custom loss function is explained in Eq. 1

$$\tau_{DSC} = -\frac{2 \sum_{i=1}^{n} y_p y_a}{\sum_{i=1}^{n} y_p + \sum_{i=1}^{n} y_a} \tag{1}$$

where, y_p represents the value of softmax predicted segmentation $map \in [0, \dots, 1]$, and y_a is the ground at each pixel i.

Traditional deep learning models are capable of learning discriminative features and have the ability to accurately map the high dimensional input data to expected output. However, the models do not quantify that how certain the model is its prediction. A Bayesian approach in machine learning can model the uncertainty, but current deep learning models do not represent the prediction uncertainty. Recently, a number of methods for uncertainty quantification by estimating the posterior distribution have been proposed [6]. We estimate the model uncertainty by applying random transformations [7] to the test input images. With this we are able to capture the noise inherent in the input data and visualize the regions where the segmentation network is uncertain in its prediction. To obtain the predictive distribution, we apply a random transformation $\delta(x)$ to a set of m images, where δ performs median blur, Gaussian blur, rotation, flipping or Gaussian noise. Taking the mean of this ample gives the refined prediction and the variance within the sample gives the uncertainty in prediction. The prediction and the uncertainty can be defined as;

$$\mu = -\frac{1}{m} \sum_i^m f(\delta_i(x); W); \quad \sigma = -\frac{1}{m} \sum_i^m f(\delta_i(x); W - \mu)^2 \qquad (2)$$

where, μ is the prediction of microvessel segmentation, σ is the prediction uncertainty and m in the number of applied transformations. δ_i denotes the random transformation applied to the input image x. Taking the average of the prediction of transformed images give better segmentation.

3   Experiments and Results

3.1   Materials

We have used a set of 1167 image tiles of size 512 × 512 pixels taken from 13 H&E stained WSIs of OSCC tissue at 20× magnification. The ground truth for microvessels is validated by two independent pathologists. The dataset is split into training, validation sets such that 686 training and 226 validation images are obtained from 10 WSIs. The test set is comprises of 255 images taken from remaining 3 WSIs. The training and validation images are augmented with random rotation, elastic distortion, random flip, median blur and Gaussian blurring.

3.2   Experimental Settings

The framework is implemented in Keras 2.2 with TensorFlow backend and trained on workstation equipped with Nvidia GeForce GTX 1080 Ti for 175 epochs (35000 iterations). We have used Adam optimizer, the learning rate was initialized at 10-4, the input image to the network is 512 × 512 × 3 and the batch size is 2. As explained in Sect. 2, we have used a custom loss function based on minimizing the dice score.

3.3   Evaluation

The model is quantitative evaluated using Jaccard index, Dice Similarity Coefficient (DSC), Accuracy, Sensitivity, Specificity, Precision and Recall. Furthermore, several state-of-the-art segmentation methods including FCN-8 [9], U-Net [11], SegNet [1] and DeepLabv3+ [4] are implemented for comparative analysis, which is presented in Table 1.

Table 1. Quantitative performance measures of miscrovessel segmentation, compared with FCN-8 [9], U-Net [11], SegNet [1] and DeepLabv3+ [4].

	Jaccard index	DSC	Accuracy	Sensitivity	Specificity	Precision	Recall
FCN-8	0.8562	0.9225	0.9612	0.9086	0.9791	0.9368	0.9086
U-Net	0.8561	0.9225	0.9616	0.9010	0.9818	0.9442	0.9017
SegNet	0.8431	0.9148	0.9569	0.9100	0.9729	0.9524	0.8854
DeepLabv3+	0.8741	0.9329	0.9667	0.9085	0.9834	0.9540	0.9089
Proposed	**0.8851**	**0.9390**	**0.9694**	**0.9261**	**0.9862**	**0.9558**	**0.9225**

4   Discussion and Conclusion

Microvessels in H&E stained histology images display a high level of heterogeneity with respect to their size, shape, texture, and luminal red cells density. Figure 3 shows the visual results of four challenging cases in microvessel segmentation. Figure 3(a, b) shows a vessel partially filled with red cells. U-Net is unable to segment the microvessel region where red cells are not present whereas SegNet, FCN-8 and DeepLabV3+ manage to segment the microvessel but with low confidence. In contrast, the proposed method segments the complete vessel with high confidence. The misdetections of variable sized microvessels and the segmentation microvessels located in close proximity are the other challenging case, illustrated in Fig. 3c and d respectively. FCN-8, U-Net and DeepLabv3+ are unable to segment small vessels Fig. 3c and the SegNet merged the two closely located by vessels into one large vessels (Fig. 3d). The proposed method successfully segments the vessels belonging to these challenging cases. The proposed method achieves a superior performance for all evaluation measures as illustrated in Table 1. Although, the quantitative performance gain may not appear notably significant, the visual results illustrated in (Fig. 3) show that the proposed framework successfully localizes and segments the microvessels of different shapes and sizes with mercurial density of red cells. The segmentation of microvessels in the histology images is the first step in automated quantification of LVI and estimation of tumor angiogenesis. In routine pathological practice, the microvessels can be identified using IHC stained histology images with associated time and cost implications. We have present a method for the precise segmentation of microvessels in H&E stained histology images. The proposed method uses a modified aligned Xception model, atrous spatial pyramid pooling and a customized dice coefficient minimization based loss function to segment microvessels of various shapes and size. Random transformations at test time are used to incorporate the predictive uncertainty. Taking the average of these predictions gives a superior segmentation. The visual results and quantitative performance measures illustrate that the proposed method is able to precisely segment the microvessels in challenging cases.

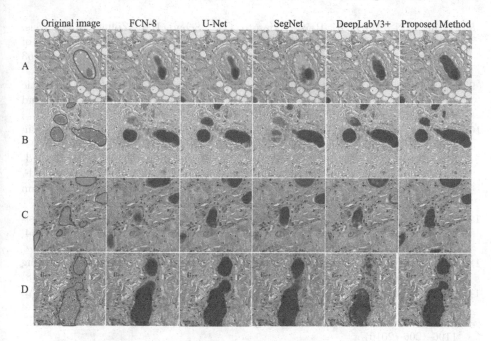

Fig. 3. Visual illustration of microvessel segmentation results shown as miscrovessel prediction heatmap obtained by FCN-8 [9], U-Net [11], SegNet [1], DeepLabv3+[4] and the proposed approach, overlaid on the original images. The miscrovessel boundary is marked on the original images in the 1st column. (Color figure online)

References

1. Badrinarayanan, V., Kendall, A., Cipolla, R.: SegNet: a deep convolutional encoder-decoder architecture for image segmentation. arXiv preprint arXiv:1511.00561 (2015)
2. Carmeliet, P.: Angiogenesis in life, disease and medicine. Nature **438**(7070), 932 (2005)
3. Chen, L.C., Papandreou, G., Kokkinos, I., Murphy, K., Yuille, A.L.: DeepLab: semantic image segmentation with deep convolutional nets, atrous convolution, and fully connected CRFs. IEEE Trans. Pattern Anal. Mach. Intell. **40**(4), 834–848 (2018)
4. Chen, L.C., Zhu, Y., Papandreou, G., Schroff, F., Adam, H.: Encoder-decoder with atrous separable convolution for semantic image segmentation. arXiv preprint arXiv:1802.02611 (2018)
5. Chollet, F.: Xception: deep learning with depthwise separable convolutions. arXiv preprint arXiv:1610–02357 (2017)
6. Gal, Y., Ghahramani, Z.: Dropout as a Bayesian approximation: representing model uncertainty in deep learning. In: International Conference on Machine Learning, pp. 1050–1059 (2016)
7. Graham, S., Chen, H., Dou, Q., Heng, P.A., Rajpoot, N.: MILD-Net: minimal information loss dilated network for gland instance segmentation in colon histology images. arXiv preprint arXiv:1806.01963 (2018)

8. Kather, J.N., Marx, A., Reyes-Aldasoro, C.C., Schad, L.R., Zöllner, F.G., Weis, C.A.: Continuous representation of tumor microvessel density and detection of angiogenic hotspots in histological whole-slide images. Oncotarget **6**(22), 19163 (2015)

9. Long, J., Shelhamer, E., Darrell, T.: Fully convolutional networks for semantic segmentation. In: Proceedings of the IEEE Conference on Computer Vision and Pattern Recognition, pp. 3431–3440 (2015)

10. Noma, D., et al.: Prognostic effect of Lymphovascular Invasion on TNM staging in stage i non-small-cell lung cancer. Clin. Lung Cancer **19**(1), e109–e122 (2018)

11. Ronneberger, O., Fischer, P., Brox, T.: U-Net: convolutional networks for biomedical image segmentation. In: Navab, N., Hornegger, J., Wells, W.M., Frangi, A.F. (eds.) MICCAI 2015 Part III. LNCS, vol. 9351, pp. 234–241. Springer, Cham (2015). https://doi.org/10.1007/978-3-319-24574-4_28

12. Saha, M., Chakraborty, C., Racoceanu, D.: Efficient deep learning model for mitosis detection using breast histopathology images. Comput. Med. Imaging Graph. **64**, 29–40 (2018)

13. Saltz, J., et al.: Spatial organization and molecular correlation of tumor-infiltrating lymphocytes using deep learning on pathology images. Cell Rep. **23**(1), 181 (2018)

14. Sirinukunwattana, K., Raza, S.E.A., Tsang, Y.W., Snead, D.R., Cree, I.A., Rajpoot, N.M.: Locality sensitive deep learning for detection and classification of nuclei in routine colon cancer histology images. IEEE Trans. Med. Imaging **35**(5), 1196–1206 (2016)

15. Xu, J., Luo, X., Wang, G., Gilmore, H., Madabhushi, A.: A deep convolutional neural network for segmenting and classifying epithelial and stromal regions in histopathological images. Neurocomputing **191**, 214–223 (2016)

16. Yi, F., et al.: Microvessel prediction in H&E stained pathology images using fully convolutional neural networks. BMC Bioinf. **19**(1), 64 (2018)

5th International Workshop on Ophthalmic Medical Image Analysis, OMIA 2018

Ocular Structures Segmentation from Multi-sequences MRI Using 3D Unet with Fully Connected CRFs

Huu-Giao Nguyen[1,2,3(✉)], Alessia Pica[1], Philippe Maeder[4],
Ann Schalenbourg[5], Marta Peroni[1], Jan Hrbacek[1], Damien C. Weber[1],
Meritxell Bach Cuadra[3,4,6], and Raphael Sznitman[2]

[1] Proton Therapy Center, Paul Scherrer Institut, ETH Domain, Villigen, Switzerland
[2] Ophthalmic Technology Laboratory, ARTORG Center,
University of Bern, Bern, Switzerland
huu.nguyen@artorg.unibe.ch
[3] Medical Image Analysis Laboratory, CIBM, University of Lausanne,
Lausanne, Switzerland
[4] Radiology Department, Lausanne University Hospital (CHUV),
Lausanne, Switzerland
[5] Adult Ocular Oncology Unit, Jules-Gonin Eye hospital, Lausanne, Switzerland
[6] Signal Processing Laboratory, Ecole Polytechnique Fédérale de Lausanne,
Lausanne, Switzerland

Abstract. The use of 3D Magnetic Resonance Imaging (MRI) has attracted growing attention for the purpose of diagnosis and treatment planning of intraocular ocular cancers. Precise segmentation of such tumors are highly important to characterize tumors, their progression and to define a treatment plan. Along this line, automatic and effective segmentation of tumors and healthy eye anatomy would be of great value. The major challenge to this end however lies in the disease variability encountered over different populations, often imaged under different acquisition conditions and high heterogeneity of tumor characterization in location, size and appearance. In this work, we consider the Retinoblastoma disease, the most common eye cancer in children. To provide automated segmentations of relevant structures, a multi-sequences MRI dataset of 72 subjects is introduced, collected across different clinical sites with different magnetic fields (3T and 1.5T), with healthy and pathological subjects (children and adults). Using this data, we present a framework to segment both healthy and pathological eye structures. In particular, we make use of a 3D U-net CNN whereby using four encoder and decoder layers to produce conditional probabilities of different eye structures. These are further refined using a Conditional Random Field with Gaussian kernels to maximize label agreement between similar voxels in multi-sequence MRIs. We show experimentally that our approach brings state-of-the-art performances for several relevant eye structures and that these results are promising for use in clinical practice.

© Springer Nature Switzerland AG 2018
D. Stoyanov et al. (Eds.): COMPAY 2018/OMIA 2018, LNCS 11039, pp. 167–175, 2018.
https://doi.org/10.1007/978-3-030-00949-6_20

1   Introduction

Retinoblastoma (RB) is the most common form of intraocular cancer with high morbidity and mortality rates in children. To diagnose and treat this cancer, several imaging modalities are typically necessary to properly characterize the tumor, its growth and for any follow-up care. While traditionally 2D Fundus imaging, 2D ultrasound or 3D Computed Tomography (CT) [1,2] were the modalities of choice, 3D Magnetic Resonance Imaging (MRI) has gained increased interest within the ophthalmic community thanks to the high spatial resolutions, multiplanar capabilities and high intrinsic contrast [4,5]. In effect, 3D MRI thus allows for clear overall improved discrimination between anatomical structures and different pathological regions such as the gross tumor volume, retinal detachment, and intraocular bleeding as illustrated in Fig. 1. As such, automatic and effective segmentation of tumors and healthy eye anatomy would be of great value for both disease diagnosis and treatment planning. For example, having reliable MR imaging biomarkers would open the door to eye cancer radiogenomics [3] (the association of radiological image features with gene expression profile) which can support in prognosis and patient selection for targeted treatment, thereby contributing to precision medicine.

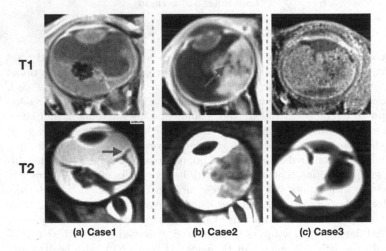

Fig. 1. Illustration of the major challenge of our RB dataset with different sizes, textures, locations and irregular/ill-defined shapes of tumors with the presence of retinal detachment (blue arrow), subretinal hemorrhage (yellow arrow) and tumor necrosis (red arrow) in MR images. (Color figure online)

Towards this goal, previous methods based on geometrical and statistical models had addressed ocular segmentation in 3D medical imaging (e. g., MRI or CT). For instance, a parametric model allowed for coarse eye structure segmentations [6], while [7] introduced a 3D shape model of the retina to study abnormal shape changes and peripheral vision. Similarly, 3D mesh construction with morphologic parameters such as distance from the posterior corneal pole and deviation from sphericity have also been proposed [8], as well as using Active

Shape Models (ASM) to analyze eye shape information [9,10]. Unfortunately, a major limitation of the aforementioned methods is that they focus solely on healthy eyes, while the characterization of tumors it self has by and large not been addressed. Actually, ocular tumor segmentation is challenging because of small amount of data but images acquired under different conditions and huge variability of tumors in location, size and appearance.

To this end, we propose a fully automated framework capable of delineating both healthy structures and RB tumors from multi-contrast 3D MRI. It includes data pre-processing for normalization, multi-sequence MRI registration, an effective coarse segmentation and output post-processing to improve localization accuracy. Our approach is based on the popular UNet Convolutional Neural Network (CNN) architecture [11], whereby we segment different healthy eye structures and the tumor in a single step. From this multi-class 3D segmentation, we further refine our estimate by using a Gaussian edge potential Conditional Random Field (CRF) to maximize label agreement between similar voxels in the multi-sequence MRIs. Although we applied here the original implementations of above methods, their combination and the application context is novel with the possibility to easily be extended to other types of tumors. We compare our proposed framework with state-of-the-art techniques on a large *mixed* dataset, including both healthy and pathological eyes as well as children and adult data from different magnetic fields and MR sequences. Our method allows simultaneous segmentation of both healthy and tumor regions to be identified and outperforms existing approaches used a mixture of ASM [9,10] and deep learning [12].

2 Methodology

2.1 Dataset

Originating from two clinical centers, our study contains 72 eyes consisting of 32 RB, 16 healthy children (HC) eyes, and 24 healthy adult (HA) eyes. All MRI examinations were performed with a Siemens scanner (SIEMENS Magnetom Aera, Erlangen, Germany) with both T1-weighted (T1w) and T2-weighted (T2w) contrasts. A 3T MR with a head coil was used to image asleep children aged 4 months to 8 years old (mean age of 3.29 ± 2.15 y.o.), with a cohort eye diameter size mean of 12.9 ± 1.3 mm (range [10.4–15.9] mm). A 1.5T MR with

Table 1. MR imaging acquisition parameters: children imaging was done asleep at 3T head coil while adult were awake and imaged at 1.5T with a surface coil.

	Children		Adults	
	T1-VIBE	T2-Spin Echo	T1-VIBE	T2-SPACE
Repetition time (ms)	19	1000	6.55	1400
Echo time (ms)	3.91	131	2.39	185
Flip angle	12°	120°	12°	150°
Voxel size (mm³)	$0.4 \times 0.4 \times 0.4$ and $0.48 \times 0.48 \times 0.5$	$0.45 \times 0.45 \times 0.45$	$0.5 \times 0.5 \times 0.5$	$0.5 \times 0.5 \times 0.5$ and $0.82 \times 0.82 \times 0.8$
FOV (Voxels)	$256 \times 256 \times 120$	$256 \times 256 \times 160$	$256 \times 256 \times 80$	$256 \times 256 \times 80$

surface coil was used for awake adults aged 28.4 ± 5.2 years old (range [23–46]y.o.), with a cohort eye size mean of 24.7 ± 0.6 mm (range [23.3–26] mm). The study was approved by the Ethics Committee of the involved institutions and all subjects provided written informed consent prior to participation. All subject information in our study was anonymized and de-identified. Table 1 shows the different parameters used for the two MRI acquisition protocols. Whenever children and adult images are used together we denote *mixed cohort* (MC).

MRI Data Normalization: Clinical image quality is affected by many factors such as noise, low varying intensity, variations due to non-uniform magnetic field, imperfections of coils, magnetic susceptibility at interfaces, all of which can be influenced by different imaging parameters such as signal-to-noise ratio or acquisition time. In order to compensate for such effects, all MRI volumes were pre-processed with an anisotropic diffusion filtering [14], to reduce noise without removing significant image content. We applied the N4 algorithm [15] to correct for bias field variations and performed histogram-based intensity normalization [16] to build an intensity profile of the dataset. In order to improve the performance in segmentation and computation time, we defined a volume of interest (VOI) of the eye by retaining a $72 \times 72 \times 64$ volume centered on the eye such that the optic nerve was always included. Rigid registration was applied to move T2w images into T1w image space.

Manual Segmentation: For training and validation purposes, manual delineations of the eye lens, sclera, tumors and optic nerve were done by radiation oncologist expert. First, segmentation for sclera and tumor was individually by intensity thresholding. Then, manual editing was done to refine borders and remove outlier regions. For small structures such as the lens and the optic nerve, manual segmentations were performed directly using a stylus.

2.2 Automated Anatomical Structure Segmentation

Coarse segmentation: Similar to the original UNet method presented in [11], we consider an encoder and decoder network that takes as input multiple image channels for each of the imaging sequence types (see Fig. 2). Each encoding and decoding pathway contains 4 layers that effactually changes the feature dimension (i.e., 32, 64, 128, 256, 512). The same architecture accounts for the decoding pathway. In each case, $3 \times 3 \times 3$ Convolutions are used with a Batch normalization and parametric rectified linear unit (PRelu) operations.

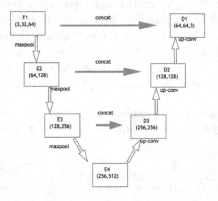

Fig. 2. 3D Unet architecture used.

Between two layers in the encoder pathway, $2 \times 2 \times 2$ max pooling with strides of two in each dimension are used. In the decoder pathway, the connection of two subsequent layers is performed with an up-convolution of $2 \times 2 \times 2$ with strides of two in each dimension. Concatenation is performed to connect the output tensors of two layers of the encoder and decoder pathways at same level. To train our network, we used the Adam optimizer and the Dice loss function. At inference time, softmax is used to extract probability maps for each class.

3D Fully Connected CRF with Gaussian Kernels: To provide a more refined segmentation, we use a 3D CRF [17] to maximize label agreement between similar voxels (or patches) in the multi-sequences MRI. The 3D CRF incorporates unary potentials of individual voxels and pair-wise potentials (in terms of appearance and smoothness) on neighboring voxels to provide more accurate eye structure segmentation.

Considering an input image I and a probability map P (i. e., provided by the above network), the unary potential is defined to be the negative log-likelihood $\psi_u(z_i) = -logP(z_i|I)$, where z_i the predicted label of voxel i. The pair-wise potential has the form $\psi_p(z_i, z_j) = \mu(z_i, z_j)k(f_i, f_j)$, where μ is a label compatibility function, $k(f_i, f_j)$ and is characterized by integrating two Gaussian kernels of appearance (first term) and smoothness (second term), as follows

$$k(f_i, f_j) = w_1 \exp\left(-\frac{|p_i - p_j|^2}{2\theta_1^2} - \frac{|I_i - I_j|^2}{2\theta_2^2}\right) + w_2 \exp\left(-\frac{|p_i - p_j|^2}{2\theta_3^2}\right),$$

where p_i are voxel locations, I_i are voxel intensities, f_i are voxel feature vectors as described in [13], w_j are weight factor between the two terms, and the θ's are tunable parameters of the Gaussian kernels. The Gibbs energy of CRF model is then given by $\sum (\psi_u(z_i), \psi_p(z_i, z_j))$ [17].

3 Experiments

We performed leave-one-out cross-validations to quantitatively compare the results of the proposed segmentation scheme (i. e., iteratively chose one eye as a validation case, while the remaining subjects are used as the training set). The quality of the segmentations were evaluated by computing the predicted and true volume overlap using the Dice similarity coefficient (DSC) and the Hausdorff distance (HD). For the training step, we crop volumes to size $72 \times 72 \times 64$. We report the best performance result obtained from the different parameter settings detailed as follows. For 3D Unet: regularisation type {L1,L2}; number of samples per volume {8, 16, 32}; volume padding size {8, 16, 32}; learning rate for the optimiser {0.001, 0.005}; maximum iterations {0.001, 0.005}. For CRF: neighborhood size {[3,3,3], [5,5,5]}; intensity-homogeneous distance {[5,5,5], [10,10,10]}; kernel weights of appearance and smoothness terms {[1,1], [3,1], [1,3]}. The performance of the proposed method compared to two baselines algorithms found in the literature: an ASM method [9,10] and a 3D CNN [12].

Table 2. Comparison of eye structures segmentation performances. Results are shown in terms of average Dice (DSC%) and Hausdorff distance (HD mm) scores. ¶Average results of tumor is computed with 32RB only. ‡The $p - value < 0.005$ of Wilcoxon test was obtained between these DCS values.

Dataset	Method			Sclera	Lens	Optic-nerve	RB tumor¶
16HC + 24HA	ASM	DSC		93.7 ± 3.8	**87.1 ± 4.9**	77.5 ± 5.7	N/A
		HD		1.87 ± 0.58	**0.69 ± 0.23**	1.36 ± 0.61	N/A
	Unet + CRF	DSC		**94.8 ± 3.8**	86.9 ± 2.5	**78.9 ± 4.8**	N/A
		HD		**1.37 ± 0.37**	0.82 ± 0.35	**1.21 ± 0.45**	N/A
32RB + 16HC	Unet + CRF	DSC		94.71 ± 2.2	85.9 ± 4.61	77.2 ± 6.1	‡57.9 ± 13.2
		HD		1.25 ± 0.32	0.71 ± 0.27	1.7 ± 0.39	5.54 ± 2.65
32RB + 16HC + 24HA	Unet + CRF	DSC		**94.75 ± 2.1**	**86.5 ± 4.52**	**77.8 ± 5.2**	‡**59.1 ± 12.4**
		HD		**1.22 ± 0.31**	**0.72 ± 0.25**	**1.63 ± 0.3**	**5.33 ± 2.54**

First, we compare the proposed method with that of an ASM [10] on 40 healthy eyes (24 HA subjects with a 1.5T MRI system and 16HC subjects in a 3T MRI system). Table 2 (top half) reports quality measures and indicates that our approach performs slightly better on the sclera and optic nerve but not on the lens. Indeed, the sclera and the optic nerve have large anatomical variability, due in part to large differences in eye size. The ASM is limited in its ability to take these large variations into account (see Fig. 3 first column, where the smallest healthy eye was used as testing input).

Second, we evaluate the segmentation accuracy of healthy structures in presence of RB tumors (see Table 2 (botom half)). Two training scenarios are considered: (1) 48 children eyes from 3T MR images (i. e., 32RB + 16HC) and, (2) the mixed cohort, MC (i. e., 32RB + 16HC + 24HA) described in Sect. 2.1. Segmentation results on the sclera, lens and optic nerve are superior when using the MC. However, no statistical differences (Wilcoxon signed rank test) were found.

RB segmentation results are presented as function of its size in Fig. 4. The mean DSC and HD using a MC training is of 59.1 ± 12.4% and 5.33±2.54 mm, respectively. When compared with a training set using children eyes only, our approach yields gains of 1.2% DSC and 0.21 mm HD (mean DSC of 57.9±13.2% and mean HD of 5.54 ± 2.65 mm). Similarly to healthy structures, these results indicate that our approach benefits from using the MC dataset and that healthy and children eyes, regardless of what scanner used to image them, can be used jointly to improve segmentation performances. Let us note that differences in DSC were statistically significant ($p < .005$). Qualitative results are shown in second and third column of Fig. 3. Finally, as regards similar approaches in the literature [12], our method provide slight improvements (reported results in [12] on 16 RB were of the sclera (94.62 ± 1.9%), lens (85.67 ± 4.68%), optic nerve (absent) and RB tumor (62.25 ± 26.27%) for DICE overlap). However, given that their results are achieved using a different and smaller dataset, a direct comparison would not be fair.

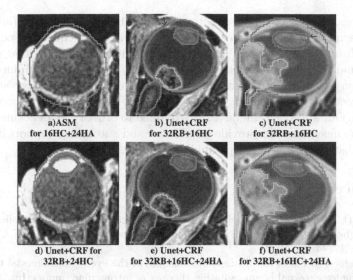

Fig. 3. First column shows the smallest healthy eye in training set. Second and third column are RB segmentation results with different training scenarios. The red narrows point improvements of our mixed dataset.

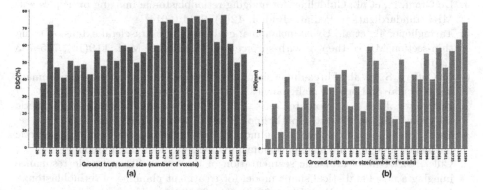

Fig. 4. Tumors segmentation performance as a function of tumor size (voxels). (a) DCS results (mean: $59.15 \pm 12.43\%$); (b) HD results (mean: $5.33 \pm 2.54\,\text{mm}$).

4 Conclusion

In this paper, we have explored the problem of simultaneous segmentation of eye structures from multi-sequences MR images to support clinicians in their need of precise tumor characterization and their progression. We proposed a thorough segmentation pipeline consisting of a combination of data quality normalization and a 3D Unet CNN segmentation model with a Gaussian kernel CRF framework. Effectively, the proposed method embeds the probability maps, the output of 3D Unet architecture, with respect to the analysis of pair-wise appearance and smoothness on neighborhood voxels using CRF model. We validated our method with a heterogeneous eye dataset consisting of a diverse population (adults and children) acquired over multiple sites with different MRI acquisition conditions. Differing from state-of-the-art ASM methods, the proposed method offers an accurate and fully automatic segmentation without any prior computations of statistics on the shape of the eye and its structures. Surprisingly, we show here that our method is also largely robust to the eye size and imaging acquisition conditions. Our approach can be easily extended to other types of occular tumors (e. g., Uveal melanoma) to provide an effective and automated support in clinical practice (diagnosis, treatment planning and follow-up).

References

1. Kook, D., et al.: Variability of standardized echographic ultrasound using 10 mHz and high-resolution 20 mHz B scan in measuring intraocular melanoma. Clin. Ophthal. **5**, 477–482 (2011)
2. Ruegsegger, M.D., et al.: Statistical modeling of the eye for multimodal treatment planning for external beam radiation therapy of intraocular tumors. Int. J. Radiat. Oncol. Biol. Phys. **84**(4), 541–547 (2012)
3. Jansen, R. et al.: MR imaging features of retinoblastoma: association with gene expression profiles. Radiology (2018)
4. De Graaf, P., et al.: Guidelines for imaging retinoblastoma: imaging principles and MRI standardization. Pediatr. Radiol. **42**(1), 2–14 (2014)
5. Tartaglione, T., et al.: Uveal melanoma: evaluation of extrascleral extension using thin-section MR of the eye with surface coils. La Radio. Med. **119**(10), 775–783 (2014)
6. McCaffery, S., et al.: Three-dimensional high-resolution magnetic resonance imaging of ocular and orbital malignancies. Archiv. Ophthal. **120**, 747–754 (2002)
7. Beenakker, J., et al.: Automated retinal topographic maps measured with magnetic resonance imaging. Invest. Ophthalmol. Vis. Sci. **56**, 1033–1039 (2015)
8. Singh, K., et al.: Three-dimensional modeling of the human eye based on magnetic resonance imaging. Invest. Ophthalmol. Vis. Sci. **47**, 2272–2279 (2006)
9. Ciller, C., et al.: Automatic segmentation of the eye in 3D magnetic resonance imaging a novel statistical shape model for treatment planning of retinoblastoma. Int. J. Radiat. Oncol. Biol. Phys. **92**(4), 94–802 (2015)
10. Nguyen H.-G., et al.: Personalized anatomic eye model from T1-weighted VIBE MR imaging of patients with Uveal melanoma. J. Radiat. Oncol. Biol. Phys. (2018)
11. Çiçek, Ö., Abdulkadir, A., Lienkamp, S.S., Brox, T., Ronneberger, O.: 3D U-net: learning dense volumetric segmentation from sparse annotation. In: Ourselin, S., Joskowicz, L., Sabuncu, M.R., Unal, G., Wells, W. (eds.) MICCAI 2016 Part II. LNCS, vol. 9901, pp. 424–432. Springer, Cham (2016). https://doi.org/10.1007/978-3-319-46723-8_49
12. Ciller, C., et al.: Multi-channel MRI segmentation of eye structures and tumors using patient-specific features. PLoS ONE **12**(3), e173900 (2017)
13. Kamnitsas, K., et al.: Efficient multi-scale 3D CNN with fully connected CRF for accurate brain lesion segmentation. Med. Image Anal. **36**, 61–78 (2017)
14. Perona, P., et al.: Scale-space and edge detection using anisotropic diffusion. IEEE Trans. Pattern Anal. Mach. Intell. **12**(7), 629–639 (1990)
15. Tustison, N., et al.: N4ITK: improved N3 bias correction. IEEE Trans. Med. Imaging **29**(6), 1310–1320 (2010)

16. Nyul, L., et al.: New variants of a method of MRI scale standardization. IEEE Trans. Med. Imaging **19**(2), 143–50 (2000)
17. Krähenbühl, P.: Efficient inference in fully connected CRFs with Gaussian edge potentials. Adv. Neural Inf. Process. Syst. **24**, 109–117 (2011)

Classification of Findings with Localized Lesions in Fundoscopic Images Using a Regionally Guided CNN

Jaemin Son[1], Woong Bae[1], Sangkeun Kim[1], Sang Jun Park[2], and Kyu-Hwan Jung[1(✉)]

[1] VUNO Inc., Seoul, Korea
{woalsdnd,iorism,sisobus,khwan.jung}@vuno.co
[2] Department of Ophthalmology, Seoul National University Bundang Hospital and Seoul National University College of Medicine, Seongnam, Korea
sangjunpark@snu.ac.kr

Abstract. Fundoscopic images are often investigated by ophthalmologists to spot abnormal lesions to make diagnoses. Recent successes of convolutional neural networks are confined to diagnoses of few diseases without proper localization of lesion. In this paper, we propose an efficient annotation method for localizing lesions and a CNN architecture that can classify an individual finding and localize the lesions at the same time. Also, we introduce a new loss function to guide the network to learn meaningful patterns with the guidance of the regional annotations. In experiments, we demonstrate that our network performed better than the widely used network and the guidance loss helps achieve higher AUROC up to 4.1% and superior localization capability.

1 Introduction

Fundoscopic images provide comprehensive visual clues about the condition of the eyes. For the analysis, ophthalmologists search for abnormal visual features called *findings* from the images and make decisions on *diagnoses* based on the findings discovered. For instance, the severity of diabetic retinopathy (DR) is clinically judged by the existence and the extent of relevant findings (microaneurysm, hemorrhage, hard exudate and cotton wool patch, etc) [14].

In recent years, Convolutional Neural Networks (CNN) have achieved the level of professional ophthalmologists in diagnosing DR and diabetic macular edema (DME) [5,13]. However, CNNs in the literature are trained to make decisions on the diagnoses directly without localizing lesions. There exist several studies that visualize the lesions that contribute to the decision on diagnoses [3,12], though, types of findings were not identified for the lesions.

© Springer Nature Switzerland AG 2018
D. Stoyanov et al. (Eds.): COMPAY 2018/OMIA 2018, LNCS 11039, pp. 176–184, 2018.
https://doi.org/10.1007/978-3-030-00949-6_21

In the past, segmentation methods with hand-crafted feature-extractors had been proposed for the detection of hemorrhage [1], hard Exudate [11], drusen deposits [9] and cotton wool patch [8]. However, since the heuristic feature-extractors embed biases of the human designer regarding visual properties of the target findings, unexpected patterns are not well detected severely constraining the performance in real world applications. CNN for segmentation [2] or detection [10] would improve the performance, however, manual annotation of lesions is labor-intensive especially when they are spread in the images, thus, renders the process of data collection highly expensive.

In this paper, we demonstrate an inexpensive and efficient approach to collecting regional annotation of findings and propose a CNN architecture that classifies the existence of a target finding and localizes the lesions. We show that training with the guidance of regional cues not only helps localize the lesions of findings more precisely but also improves classification performance in some cases. This is possible because the regional guidance encourages the network to learn right patterns of findings instead of biases in the images.

2   Proposed Methods

2.1   Data Collection

We collected regional annotations of findings for macular-centered images obtained at health screening centers and outpatient clinics in Seoul National University Bundang Hospital using a data collection system (Fig. 1). Annotators chose a type of finding on the right panel and selected the corresponding regions on the left panel. When the eye was normal, no finding was annotated to the image.

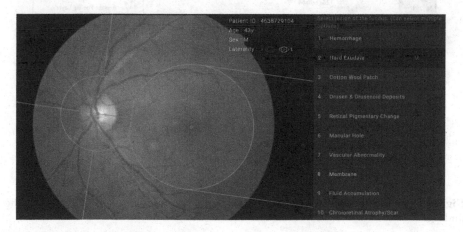

Fig. 1. Data collection system developed to retrieve regional annotations of findings in fundoscopic images.

We divided an image into 8 regions in a way that each region reflects the anatomical structure of the eyes and the regional characteristics of findings. When the distance between the optic disc and fovea is D, circles are drawn at the centers of the optic disc and fovea with the radius of $\frac{2}{5}D$ and $\frac{2}{3}D$ and the intersections of the two circles are connected with a line segment. Then, a half-line passing through the optic disc and fovea (L) cuts the circle of the optic disc in half and two half-lines parallel to L and tangent to the circle of fovea are drawn in a direction away from the optic disc. Finally, a line perpendicular to L is drawn to pass through the center of the optic disc.

We also separated annotations into training and test sets based on the expertise of the annotators. Training set was annotated by 27 board-certified ophthalmologists and the test set was annotated by 16 certified retina specialists and 9 certified glaucoma specialist. Each fundoscopic image was annotated by 3 ophthalmologists in total. Training and test set amount to 66,473 and 15,451 images respectively. This study was approved by the institutional review board at Seoul National Bundang Hospital (IRB No. B-1508-312-107) and conducted in accordance with the tenets of the Declaration of Helsinki.

2.2 Network Architecture

As shown in Fig. 2, our network architecture consists of residual layer (feature maps after residual unit [6]), reduction layer (feature maps after 3×3 conv with stride 2, batch-norm, ReLU), average pooling layer, atrous pyramid pooling layer [2] and 1×1 conv (depth $= 1$) layer.

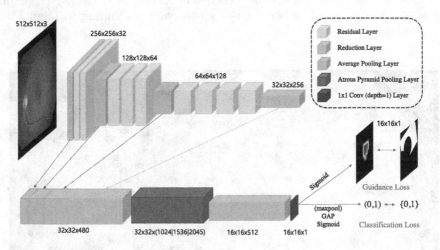

Fig. 2. Proposed network architecture for localization and classification of a specific finding.

The depth of layers doubles when the height and width halve. First four reduction layers with different sizes are concatenated after average pooling to exploit both low level and high level features. The concatenated feature maps are atrous-pyramid-pooled with the dilation rates of $1, 2, 4, 8$ (findings with large size), or the rates of $1, 2, 4$ (findings with medium size) or the rates of $1, 2$ (findings with small size). We employed atrous pyramid pooling to aggregate features with various size of receptive fields [2].

Note that the 1×1 conv layer is a linear combination of the previous feature maps as in class activation map(CAM) [15]. With a sigmoid function, values in the layer are normalized to $(0, 1)$, thus, the consequent layer can be considered as normalized activation map. Our activation map differs from CAM in that additional loss function guides the activation to appear only in desirable areas. Also, the 1×1 conv layer is globally average-pooled and normalized with sigmoid function to yield the prediction regarding the existence. Therefore, the activation map is directly related to the prediction in our network architecture and does not require external operations for visualization. In case of findings with small lesions (hemorrhage, hard exudate, cotton wool patch, drusen and retinal pigmentary change), max-pooling layer is inserted before the GAP layer to compensate for the low GAP values to compensate for low GAP values.

2.3 Objective Function

For a fundoscopic image ($I \in R^{W_I \times H_I}$), the existence of a target finding in the image I is encoded to $y_{true} \in \{0, 1\}$ and the probability of the existence $y_{pred} \in (0, 1)$ is the output from the network. When k images are given as a mini-batch, binary cross entropy for classification loss in Fig. 2 is given by

$$L_{class}(\mathbf{y}_{true}, \mathbf{y}_{pred}) = \frac{1}{k} \sum_{i=1}^{k} \left[-y_{true}^i \log y_{pred}^i - (1 - y_{true}^i) \log(1 - y_{pred}^i) \right] \quad (1)$$

where $\mathbf{y}_{true} = \{y_{true}^1, \ldots, y_{true}^k\}$ and $\mathbf{y}_{pred} = \{y_{pred}^1, \ldots, y_{pred}^k\}$.

When the last feature maps are size of $W_F \times H_F$, a region mask for a target finding ($M \in \{0, 1\}^{W_F \times H_F}$) is given as label and the activation map ($A \in (0, 1)^{W_F \times H_F}$) is generated from the network. With a mini-batch of size k, guidance loss in Fig. 2 is given by

$$L_{guide}(\mathbf{A}, \mathbf{M}) = \frac{1}{kW_F H_F} \sum_{i=1}^{k} \sum_{l=1}^{W_F H_F} (1 - m_l^i) \log(\max(a_l^i, \epsilon)) \quad (2)$$

where $\mathbf{A} = \{A^1, \ldots, A^k\}$ and $\mathbf{M} = \{M^1, \ldots, M^k\}$ and m_l^i and a_l^i are values at lth pixel in M^i and A^i for $l = 1, \ldots, W_F H_F$. Note that $\epsilon > 0$ is added inside the logarithm for numerical stability when $a_l^i \approx 0$. In a nutshell, the guidance loss suppresses any activation (a_l^i) in regions where the value of the mask is 0 ($m_l^i = 0$) and has no effect for activation inside the mask ($m_l^i = 1$).

Then, total loss is given by combining the classification loss and the guidance loss,

$$L_{total} = L_{class}(\mathbf{y}_{true}, \mathbf{y}_{pred}) + \lambda L_{guide}(\mathbf{A}, \mathbf{M}) \tag{3}$$

where λ balances two objective functions.

3   Experiments

3.1   Experimental Setup

We selectively show results of clinically important findings - findings associated with DR and DME [14] (hemorrhage, hard exudate, drusen, cotton wool patch (CWP)), macular hole, membrane and retinal nerve fiber layer defect (RNFL defect). The training set is split into derivation set (90%) and validation set (10%). Model was optimized with the derivation set until the validation loss stagnates and exacerbates. The model with the lowest validation loss is tested on the test set which we regard as gold standards. We defined that a target finding is absent when no ophthalmologists annotated and present when more than 2 out of 3 ophthalmologists annotated. The union of annotated regions is provided as regional cues during training.

We aim to measure the effectiveness of the guidance loss by experimenting with our CNN architecture (Fig. 2) with/without the regional guidance and comparing the results in terms of Area Under Receiver Operating Characteristic curve (AUROC), specificity, sensitivity and activations in the regional cues (AIR). AIR is defined as the summation of activations inside the regional cues divided by the summation of all activations. AIR is measured for both true positive and false negative in classification where the regional cues are available. We used the same network architecture (Fig. 2) to implement the networks with or without the regional guidance by changing λ in Eq. 3 ($\lambda = 0$ without the guidance).

Original color fundoscopic images are cropped to remove black background and resized to 512×512 for the network input. The resized images are randomly augmented by affine transformation (flip, scaling, rotation, translation, shear) and random re-scaling of the intensity. An image is normalized to $[0, 1]$ by dividing by 255. Weights and biases are initialized with Xavier initialization [4]. As an optimizer, we used SGD with *Nesterov* momentum 0.9 and decaying learning rate. Batch size is set to 32 following the recommendation that small batch size leads to better generalization [7]. We set $\epsilon = 10^{-3}$ in Eq. 2 to obtain numerical stability and $\lambda = 1$ in Eq. 3 to treat classification loss and guidance loss equally.

3.2   Experimental Results

Comparison of the performance between inception-v3 [5] and our two models (with/without guidance loss) is summarized in Table 1. We can observe positive effects of the guidance loss to AIR for TP and FN throughout all findings. This is desirable, since it means that the network attends inside the regional cues for

Table 1. Comparison of inception-v3 and our two models (with/without the guidance) on the test set with respect to AUROC. Activations in the regional cues (AIR) on true positive (TP) and false negative (FN) in classification. Among multiple operating points, specificity and sensitivity that yield the best harmonic mean are chosen.

Findings	AUROC			AIR (TP)		AIR (FN)	
	Inception-v3	With	Without	With	Without	With	Without
Macular hole	0.9592	**0.9870**	0.9676	**0.9999**	0.3156	**0.9999**	0.2611
Hard exudate	0.9889	**0.9938**	0.9910	**0.8089**	0.5999	**0.5750**	0.4614
Hemorrhage	0.9760	0.9862	**0.9895**	**0.8890**	0.6388	**0.4899**	0.3857
Membrane	0.9654	**0.9831**	0.9795	**0.9699**	0.3696	**0.8446**	0.2499
Drusen	0.9746	**0.9811**	0.9786	**0.8292**	0.5611	**0.5931**	0.4012
Cotton wool patch	0.9633	**0.9792**	0.9741	**0.8058**	0.5450	**0.4672**	0.4538
RNFL defect	0.9037	**0.9263**	0.8870	**0.7233**	0.4024	**0.4801**	0.2838

classification, thus the network is less likely to learn biases of datasets. Also, difference in AIR is larger in the cases of TP between the two models than those of FN. This is reasonable since FN consists of hard cases for the networks to classify, while TP is relatively easy to be classified with high confidence.

When it comes to AUROC, only macular hole and RNFL defect showed significant improvements. It is interesting to notice that these findings are observed in specific regions. This can be explained by the fact that learning becomes easier as the network is guided to attend to important regions for classification. On the other hand, findings that spread over the extensive areas such as hemorrhage, hard exudate and drusen took less or no advantage of regional cues for classification. We suspect that this happens because these findings would have wide regional cues that the guidance is marginal and the lesions are small that guidance would be more difficult. It is observed that when AUROC is higher, sensitivity is also higher and specificity is lower. However, significant difference is seen only for macular hole and RNFL defect.

In Fig. 3, we qualitatively compare activation maps of networks with/without the guidance loss. Before superimposed onto the original image, activation maps are upscaled through bilinear interpolation and blurred with 32×32 Gaussian filter for natural visualization and normalized to $[0, 1]$. As obvious in the figure, the network generates much more precise activation maps when trained with the regional cues. Though it is not as salient as is segmented pixelwise and includes few false positives in some cases, our activation maps provide meaningful information about the location of the findings which would be beneficial to clinicians. Without the guidance loss, activation maps span far more than the surroundings of lesions and sometimes highlight irrelevant areas.

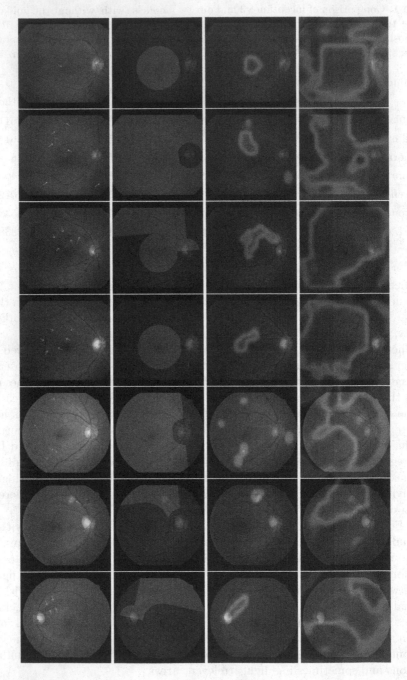

Fig. 3. (From left to right) original fundus image, mask, activation map with and without guidance loss. (From top to bottom) Macular Hole, Hard Exudate, Hemorrhage, Membrane, Drusen, Cotton Wool Patch, RNFL Defect.

4   Conclusion and Discussion

In this paper, we introduced an approach to exploiting regional information of findings in fundoscopic images for localization and classification. We developed an efficient labeling tool to collect regional annotations of findings and proposed a network architecture that classifies findings with localization of the lesions. When trained with the guidance loss that makes use of the regional cues, our network generates more precise activation maps with better attention to the relevant areas for classification. Also, the proposed regional guide also improves the classification performance of findings that occur only at specific regions.

References

1. Bae, J.P., Kim, K.G., Kang, H.C., Jeong, C.B., Park, K.H., Hwang, J.M.: A study on hemorrhage detection using hybrid method in fundus images. J. Dig. Imaging **24**(3), 394–404 (2011)
2. Chen, L.C., Papandreou, G., Kokkinos, I., Murphy, K., Yuille, A.L.: DeepLab: semantic image segmentation with deep convolutional nets, atrous convolution, and fully connected CRFs. arXiv preprint arXiv:1606.00915 (2016)
3. Gargeya, R., Leng, T.: Automated identification of diabetic retinopathy using deep learning. Ophthalmology **124**(7), 962–969 (2017)
4. Glorot, X., Bengio, Y.: Understanding the difficulty of training deep feedforward neural networks. In: Proceedings of the Thirteenth International Conference on Artificial Intelligence and Statistics, pp. 249–256 (2010)
5. Gulshan, V., et al.: Development and validation of a deep learning algorithm for detection of diabetic retinopathy in retinal fundus photographs. JAMA **316**(22), 2402–2410 (2016)
6. He, K., Zhang, X., Ren, S., Sun, J.: Deep residual learning for image recognition. In: Proceedings of the IEEE Conference on Computer Vision and Pattern Recognition, pp. 770–778 (2016)
7. Keskar, N.S., Mudigere, D., Nocedal, J., Smelyanskiy, M., Tang, P.T.P.: On large-batch training for deep learning: generalization gap and sharp minima. arXiv preprint arXiv:1609.04836 (2016)
8. Köse, C., ŞEvik, U., İKibaş, C., Erdöl, H.: Simple methods for segmentation and measurement of diabetic retinopathy lesions in retinal fundus images. Comput. Methods Programs Biomed. **107**(2), 274–293 (2012)
9. Rapantzikos, K., Zervakis, M., Balas, K.: Detection and segmentation of drusen deposits on human retina: potential in the diagnosis of age-related macular degeneration. Med. Image Anal. **7**(1), 95–108 (2003)
10. Ren, S., He, K., Girshick, R., Sun, J.: Faster R-CNN: towards real-time object detection with region proposal networks. In: Advances in Neural Information Processing Systems, pp. 91–99 (2015)
11. Sasaki, M., Kawasaki, R., Noonan, J.E., Wong, T.Y., Lamoureux, E., Wang, J.J.: Quantitative measurement of hard exudates in patients with diabetes and their associations with serum lipid levels. Invest. Ophthalmol. Vis. Sci. **54**(8), 5544–5550 (2013)
12. Takahashi, H., Tampo, H., Arai, Y., Inoue, Y., Kawashima, H.: Applying artificial intelligence to disease staging: deep learning for improved staging of diabetic retinopathy. PLoS One **12**(6), e0179790 (2017)

13. Ting, D.S.W., et al.: Development and validation of a deep learning system for diabetic retinopathy and related eye diseases using retinal images from multiethnic populations with diabetes. JAMA **318**(22), 2211–2223 (2017)
14. Wilkinson, C., et al.: Proposed international clinical diabetic retinopathy and diabetic macular edema disease severity scales. Ophthalmology **110**(9), 1677–1682 (2003)
15. Zhou, B., Khosla, A., Lapedriza, A., Oliva, A., Torralba, A.: Learning deep features for discriminative localization. In: Proceedings of the IEEE Conference on Computer Vision and Pattern Recognition, pp. 2921–2929 (2016)

Segmentation of Corneal Nerves Using a U-Net-Based Convolutional Neural Network

Alessia Colonna, Fabio Scarpa$^{(\boxtimes)}$, and Alfredo Ruggeri

Department of Information Engineering, University of Padova, Padua, Italy
fabio.scarpa@unipd.it

Abstract. In-vivo confocal microscopy provides information on the corneal health state. In particular, images taken at a specific depth allow the visualization of the nerves fibers. The correlation between corneal nerves morphology and pathology has been shown several times. However, the difficulty in obtain an accurate tracing of the nerves (manually or automatically) and the execution times limit the widespread use of this technique in clinical practice. In this work, we propose a U-Net-based Convolutional Neural Network (CNN) for the fully automatic tracing of corneal nerves. The proposed CNN's architecture consists of a contracting path, which captures nerve descriptors, and a symmetric expanding path, which enables precise nerve localization. The proposed algorithm provides nerve segmentation with sensitivity higher than 95% with respect to manual tracing, and improves the results obtained by a previous fully automatic technique. Furthermore, corneal nerve representation obtained in the proposed CNN provides an improvement in the image automatic classification between healthy subjects and subjects with diabetic neuropathy, demonstrating the potential of CNN in identifying clinically useful features.

Keywords: Corneal nerves · Convolutional neural network · Deep learning

1 Introduction

The cornea is a multi-stratified avascular membrane with peculiar optic-physical transparency and light refraction. It is the most important component of the ocular diopter, and it is one of the most densely innervate tissues in the human body [1, 2].

Thanks to in-vivo confocal and specular microscopy, images of various layers of the cornea can be acquired rapidly and in a non-invasive way. Analyzing these images is very important to obtain clinical information about the health of the cornea and/or systemic diseases (e.g. diabetes). In particular, images taken at a specific depth (sub-basal nerve plexus) allow the visualization of the nerves fibers. Over the last decades, many studies demonstrated a correlation between morphometric parameters of the corneal nerves (density, branching, tortuosity, etc.) and a wide group of ocular and systemic diseases (dry eye syndrome, keratoconus, diabetic neuropathy, etc.) [3–8].

D. Stoyanov et al. (Eds.): COMPAY 2018/OMIA 2018, LNCS 11039, pp. 185–192, 2018.
https://doi.org/10.1007/978-3-030-00949-6_22

In order to provide robust corneal nerve descriptors, it is essential to trace/segment the corneal nerves accurately. Image analysis software for nerve tracing have been proposed over the last years, but they still require a manual refinement, and/or a long execution time [9–12]. Thus, to the best of our knowledge, none of them is currently used in clinical practice.

In this work, we propose a U-Net-based Convolutional Neural Network (CNN) for the fully automatic tracing of corneal nerves. We also investigated if the corneal nerves representation obtained in the proposed CNN provides an improvement in the automatic classification between healthy subjects and subjects with diabetic neuropathy. CNNs, which are arguably the most popular deep learning architectures, are playing an increasing role in the field of computer vision and image processing, with promising results especially in medical image analysis [13–17]. This study investigates if CNNs can also successfully be used for corneal nerves image analysis.

2   Material

In this study, we used a dataset of 9039 confocal images of sub-basal corneal nerve plexus from healthy and diabetic (type 1 or 2) subjects. Images were acquired using the Heidelberg Retina Tomograph (HRT-II) with the Rostock Cornea Module (Heidelberg Engineering GmbH, Heidelberg, Germany), covering a field of $400 \times 400 \ \mu m^2$ (384 × 384 pixels). Image acquisition was performed in different clinical centers and each image was anonymized eliminating any patient information. We divided these images into three subsets.

Subset 1 is composed by 8909 images from healthy (6151 images, 85 subjects) and pathological (2758 images, 424 subjects) subjects. We used this subset to develop the proposed algorithm.

Subset 2 is composed by 30 images, acquired from 30 different subjects (10 healthy and 20 pathological). A clinic expert manually segmented all images from this subset tracing the centerline of each nerves. We used this subset to evaluate the automatic tracing of the nerves.

Subset 3 is also used in [7] and is composed of 100 images: five non-overlapped images, acquired from the central part of cornea, from each of 10 healthy and 10 pathological subjects. We used this subset to investigate the capability to classify the correct health state of each individual.

3   Method

3.1   Preprocessing

Due to the curvature of cornea layers and the possible inaccurate alignment of the instrument on the corneal apex during image acquisition, the most external area of the images appears less illuminated and often blurred, sometimes showing corneal structure belonging to layers adjacent to the sub-basal nerves plexus. Thus, the first step of the proposed automatic procedure consists in a crop of the most external area (10 pixels) of the analyzed image. Then, image is resized by a factor of 0.7 to achieve a size of 256×256 pixels, useful for the next step of the proposed method. The bi-cubic function used to resize the image also allowed a partial noise reduction.

3.2 Convolutional Neural Network

CNN can be used to obtain a label classification on every single pixel of an image [14, 15], or to obtain a single label classification of the whole image [16, 17]. We used both these approaches: the first for the segmentation of the corneal nerves, the second for the classification of healthy and pathological subject. The second task is achieved using the corneal nerves representation obtained in the CNN developed in the first task.

3.2.1 Corneal Nerves Segmentation

As shown in Fig. 1, the convolutional network used to segment corneal nerves is based on the U-net [13] and is made of a contracting-encoder part and an expanding-decoder part, which allows obtaining a label classification on every single pixel.

The contracting part, as shown, starts with a 256×256 image (i.e. the prepro-cessed image), applies a first convolution 3×3 with stride of two (which down-samples the image) and, at the end, presents four blocks. Each block consists of two convolutional units (convolution 3×3, Batch Normalization, Rectified Linear Unit - ReLu) followed by 2×2 max pooling layer (to down-sample the image).

The expanding part of the proposed net also consists in four blocks, each of them concatenating the up-sampled image with the correspondent one in the encoder part and then applying two convolutional unit (convolution 3×3, Batch Normalization, Rectified Linear Unit - ReLu). After the four blocks, we implement one more trans-posed convolution, in order to get back to the initial dimensions, and a 1×1 convolution.

We trained the U-net on randomly sampled images from **subset 1**, leaving out 30% of this set as a validation set. We trained the network with mini batches of 128 images for 6 epochs and shuffled the training data before each training epoch.

Since classes are unbalanced (pixels of nerves are under-represented with respect to background pixels), we balanced them using inverse class frequency as weight.

Due to the fact that very few public datasets are available with the manual tracking of the corneal nerves, we used as label set the tracing of the nerves obtained with the algorithm proposed in [10]. Each pixel of the image was labeled as 0 (background) or 1 (nerve fiber). To avoid overfitting our label set and in order to obtain a better nerve segmentation, we trained our network for 6 epochs only.

The network was trained in less than 5 h and it was able to generate a full nerve tracing in less than 1 s on a PC equipped with Nvidia GeForce GTX GPU, without any code optimization.

To set the best hyper-parameters, we used the leave-out set as validation set: we tried different values of hyper-parameters (e.g. mini batches, max number of epochs, momentum, learning rate) based on the visual inspection on the results performed on that set.

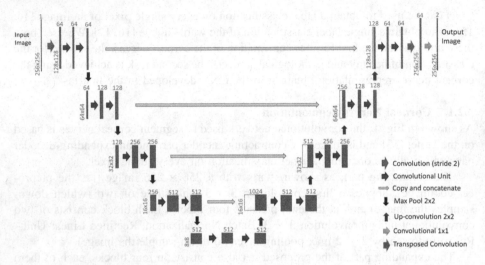

Fig. 1. Neural Network architecture: U-net. Each blue box corresponds to a 3D tensor. The number on top of each box denotes the number of channels (number of filters). The number at the lower left edge of the box provide the x-y-size. White boxes in the decoder part represent concatenated tensor correspondent in the encoder part. The arrows denote the different operations. (Color figure online)

3.2.2 Neuropathy Classification

In order to classify subjects as healthy or pathological, we made a single label classification for each image. As shown in Fig. 2, we used the features extracted in the previous net lower layer as inputs, applied another convolution (1 × 1), down-sampled with a max-pool and applied fully connect layers.

During the tracing of the corneal nerves, the U-net down-samples the image and to extracts some descriptors of the nerves. We used the descriptors, extracted in the U-net lower layer as features for the classification (healthy versus pathological subject).

From **subset 1**, we randomly sample 2500 pathological subject's images and 2500 healthy subject's images. We used 70% of those as training set and the remaining 30% as validation set. Each image was labeled as healthy or pathological.

We trained the network with mini batches of 256 images for 15 epochs and shuffled the training data before each training epoch.

We trained both networks end-to-end using stochastic gradient descent (SGD). We used a 0.01 learning rate and L2 regularization.

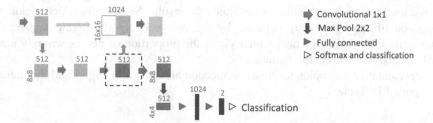

Fig. 2. Neural Network architecture for healthy/pathologic classification. The dashed box represents lower layer in previous net, which would be used as input in new task. Each blue box corresponds to a 3D tensor, while green ones represent the fully connected layers. The number of channels (number of filters/connected node) is denoted on top of the box. The x-y-size is provided at the lower left edge of the box. The arrows denote the different operations. (Color figure online)

4 Results

4.1 Corneal Nerve Segmentation

We evaluated the CNN's performance on **subset 2**, using the manual tracing as reference ground-truth. As the clinical expert was asked to trace the centerline of the identified nerve, a tolerance must be set during the comparison. A pixel recognized to be a nerve (Positive) was considered True if it was within three pixels from the reference nerve in the ground-truth. In addition, to allow a comparison with one of the literature technique, we also traced corneal nerves in images of **subset 2** with the fully automated algorithm proposed in [10].

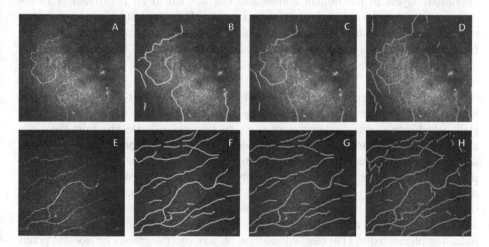

Fig. 3. From left to right: original corneal confocal images (A, E), manual tracing (B, F, yellow), automatic tracing with [10] (C, G, light blue), and automatic tracing with proposed method (D, H, green), respectively, for two representative images. (Color figure online)

We used two different indices to evaluate the results. Sensitivity, which yields the proportion of nerve correctly detected by the proposed algorithm (the higher, the better), and false discovery rate, which yields the proportion of nerves wrongly identified as such (the lower, the better).

Representative examples for visual inspection are shown in Fig. 3 and the indices are reported in Table 1.

Table 1. Sensitivity and FDR results (average ± standard deviation of 30 images)

	Sensitivity	FDR
Proposed	0.9720 ± 0.225	0.1840 ± 0.0722
Guimarães *et al.* [10]	0.9051 ± 0.818	0.0913 ± 0.0501

4.2 Neuropathy Classification

In order to discriminate between healthy subjects and individuals with diabetic neuropathy, we used **subset 3** as test set. The proposed technique provided a very good capability to distinguish between healthy and pathologic subjects (accuracy 83%), when using one image only in each subject. If all 5 images per subject are considered, we reached an accuracy of 100% (i.e., 3 or more images of the subject are labelled correctly in every subject).

5 Discussion and Conclusion

In-vivo confocal microscopy allows acquiring, in a fast and non-invasive way, images of various layers in the human cornea and this has led to an improvement in the diagnosis and monitoring of this important structure. Over the last years, many studies revealed correlations between nerve parameters and ocular and/or systemic diseases. However, the difficulty in obtaining an accurate tracing of the nerves (manually or automatically) and the execution times, limit the widespread use of this technique in clinical practice.

We propose the segmentation of corneal nerves using a U-Net-based convolutional neural network. We trained the CNN using the automatic nerve tracings obtained by [10], which sometimes contain errors (e.g., missing nerves). Despite this, the CNN appears capable of extracting the features that correctly describe the nerves, since it was able to provide a better nerve tracing (see example in Fig. 3). The proposed algorithm provided nerve segmentation with sensitivity higher than 95%. When compared with the results provided by [10], the proposed approach achieved better results in sensitivity, while in false discovery rate produced worse results. This is mainly due to the fact that the proposed approach do not involve any post processing procedure that was indeed implemented in [10] (e.g. elimination of very short segments or isolated pixels classified as nerves).

Furthermore, the proposed U-net seems to be capable of tracing even nerves not traced manually, as shown in Fig. 4. Indeed, manual tracing of corneal nerves is difficult because nerves, or part of them, are often barely visible. The manual tracing requires a long time and is not fully reproducible. Indeed, as shown in [10], the difference in sensitivity between manual tracings provided by two experts is about 8%.

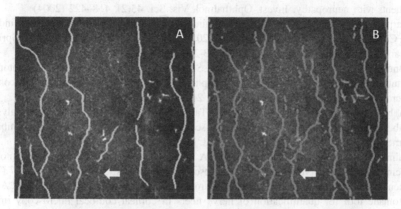

Fig. 4. Example of nerve tracing a noisy image provided by the manual procedure (A) and by the proposed algorithm (B). The arrow points to a nerve segment that was missing in the manual tracing but is detected by the proposed algorithm.

The sensitivity and the processing time obtained with the proposed method are very good and encouraging, but further work (e.g. development of a post-processing step to eliminate the shorter part labelled as nerve) is needed to provide a better nerve segmentation.

In this work, we also investigated the capability of the proposed CNN to identify diabetic neuropathy. We made a single label classification for each image, using the features extracted in the previous net lower layer as inputs. Compared to [7], we achieved better results considering for a subject both a single image and the entire set (average of 5 images), demonstrating the potential of CNN in identifying clinically useful features.

References

1. Krachmer, J.H., Mannis, M.J., Holland, E.J.: Cornea, Fundamentals, Diagnosis and Management, vol. 1. Elsevier Mosby, Philadelphia (2005)
2. Marfurt, C.F., Cox, J., Deek, S., Dvorscak, L.: Anatomy of the human corneal innervation. Exp. Eye Res. **90**(4), 478–492 (2010)
3. Benítez del Castillo, J.M., Wasfy, M.A., Fernandez, C., Garcia-Sanchez, J.: An in vivo confocal masked study on corneal epithelium and subbasal nerves in patients with dry eye. Invest. Ophthalmol. Vis. Sci. **45**(9), 3030–3035 (2004)

4. Parissi, M., Randjelovic, S., Poletti, E., et al.: Corneal nerve regeneration after collagen cross-linking treatment of keratoconus: a 5-year longitudinal study. JAMA Ophthalmol. **134** (1), 70–78 (2016)
5. De Cillà, S., Ranno, S., Carini, E., et al.: Corneal subbasal nerves changes in patients with diabetic retinopathy: an in vivo confocal study. Invest. Ophthalmol. Vis. Sci. **50**(11), 5155–5158 (2009)
6. Kallinikos, P., Berhanu, M., O'Donnell, C., et al.: Corneal nerve tortuosity in diabetic patients with neuropathy. Invest. Ophthalmol. Vis. Sci. **45**(2), 418–422 (2004)
7. Scarpa, F., Ruggeri, A.: Development of clinically based corneal nerves tortuosity indexes. In: Cardoso, M.J., et al. (eds.) FIFI/OMIA-2017. LNCS, vol. 10554, pp. 219–226. Springer, Cham (2017). https://doi.org/10.1007/978-3-319-67561-9_25
8. Annunziata, R., Kheirkhah, A., Aggarwal, S., Hamrah, P., Trucco, E.: A fully automated tortuosity quantification system with application to corneal nerve fibres in confocal microscopy images. Med. Image Anal. **32**, 216–232 (2016)
9. Dabbah, M., Graham, J., Petropoulos, I., Tavakoli, M., Malik, R.: Automatic analysis of diabetic peripheral neuropathy using multi-scale quantitative morphology of nerve fibres in corneal confocal microscopy imaging. Med. Image Anal. **15**, 738–747 (2011)
10. Guimarães, P., Wigdahl, J., Ruggeri, A.: A fast and efficient technique for the automatic tracing of corneal nerves in confocal microscopy. Transl. Vis. Sci. Technol. **5**(5), 7 (2016)
11. Chen, X., Graham, J., Dabbah, M.A., Petropoulos, I.N., Tavakoli, M., Malik, R.A.: An automatic tool for quantification of nerve fibers in corneal confocal microscopy images. IEEE Trans. Biomed. Eng. **64**(4), 786–794 (2017)
12. Al-Fahdawi, S., Qahwaji, R., Al-Waisy, A.S., et al.: A fully automatic nerve segmentation and morphometric parameter quantification system for early diagnosis of diabetic neuropathy in corneal images. Comput. Methods Programs Biomed. **135**, 151–166 (2018)
13. Ronneberger, O., Fischer, P., Brox, T.: U-Net: convolutional networks for biomedical image segmentation. In: Navab, N., Hornegger, J., Wells, W.M., Frangi, A.F. (eds.) MICCAI 2015. LNCS, vol. 9351, pp. 234–241. Springer, Cham (2015). https://doi.org/10.1007/978-3-319-24574-4_28
14. Shankaranarayana, S.M., Ram, K., Mitra, K., Sivaprakasam, M.: Joint optic disc and cup segmentation using fully convolutional and adversarial networks. In: Cardoso, M.J., et al. (eds.) FIFI/OMIA-2017. LNCS, vol. 10554, pp. 168–176. Springer, Cham (2017). https://doi.org/10.1007/978-3-319-67561-9_19
15. Giancardo, L., Roberts, K., Zhao, Z.: Representation learning for retinal vasculature embeddings. In: Cardoso, M.J., et al. (eds.) FIFI/OMIA-2017. LNCS, vol. 10554, pp. 243–250. Springer, Cham (2017). https://doi.org/10.1007/978-3-319-67561-9_28
16. Sun, J., Wan, C., Cheng, J., Yu, F., Liu, J.: Retinal image quality classification using fine-tuned CNN. In: Cardoso, M.J., et al. (eds.) FIFI/OMIA-2017. LNCS, vol. 10554, pp. 126–133. Springer, Cham (2017). https://doi.org/10.1007/978-3-319-67561-9_14
17. Ohsugi, H., Tabuchi, H., Enno, H., Ishitobi, N.: Accuracy of deep learning, a machine-learning technology, using ultra–wide-field fundus ophthalmoscopy for detecting rhegmatogenous retinal detachment. Nat. Sci. Rep. **7**, 9425 (2017)

Automatic Pigmentation Grading of the Trabecular Meshwork in Gonioscopic Images

Andrea De Giusti[1], Simone Pajaro[1(✉)], and Masaki Tanito[2,3]

[1] Nidek Technologies Srl, Albignasego, PD, Italy
{andreadegiusti,simonepajaro}@nidektechnologies.it
[2] Division of Ophthalmology, Matsue Red Cross Hospital, Matsue, Japan
tanito-oph@umin.ac.jp
[3] Department of Ophthalmology, Shimane University, Faculty of Medicine, Izumo, Japan

Abstract. Gonioscopy is essential to make a correct diagnosis of glaucoma. However, it requires a skilled examiner for being performed, and it may provide subjective information. The assessment of the iridocorneal angle by means of currently available modalities (such as ultrasound biomicroscopy or anterior segment OCT) gives anatomical quantification of angle structures without providing any chromatic information. In this study, image analysis was carried out in the pictures acquired by the prototype of a recently developed gonioscopy device, capable to automatically acquire 360° images of the iridocorneal angle, to verify the feasibility of performing an automatic Scheie's pigmentation grading of the trabecular meshwork.

Keywords: Glaucoma · Gonioscopy · Iridocorneal angle Machine learning · Scheie's pigmentation grading system

1 Introduction

Glaucoma refers to a group of eye diseases associated to optic nerve damage (i.e. glaucomatous optic neuropathy), a common pathology which possibly leads to blindness.

The guidelines of the International Council of Ophthalmology (ICO) recommend that all patients should be checked for glaucoma risk factors during ophthalmic examinations [1]. Among them, a high intra-ocular pressure (IOP) is considered one of the most indicative parameters to be monitored [2], and the structure mainly responsible for its regulation, the trabecular meshwork (TM), one of the major targets of clinical assessment.

The iridocorneal angle (or simply angle) is the region between the periphery of the iris and the sclera-cornea junction. The TM is a layered structure located on the side wall of the angle between the scleral spur (a collagen ridge of scleral

© Springer Nature Switzerland AG 2018
D. Stoyanov et al. (Eds.): COMPAY 2018/OMIA 2018, LNCS 11039, pp. 193–200, 2018.
https://doi.org/10.1007/978-3-030-00949-6_23

tissue), and the Schwalbe's line (a ring demarcating the border between the trabecula and the cornea) as illustrated in left of Fig. 1.

IOP is mainly regulated by the outflow of aqueous humor from the anterior chamber, through the TM into the Schlemm's canal, and finally into the episcleral veins [3].

Depending on the patient's specific conformation of the angle, the iris can be separated from the trabecula (open angle) or folded onto the trabecula (closed angle), respectively, permitting or limiting the normal outflow of aqueous humor and determining the overall capacity of controlling the IOP. According to this anatomical categorization, glaucomatous eyes are mainly classified as open-angle or closed-angle [2].

Secondary Open-Angle Glaucoma (SOAG) is related to the dispersion of pigment particles originated by the rubbing of the iris epithelium against the lens zonules [4,5]. These particles follow the aqueous humor flow, and accumulate on the TM, yielding the pigment dispersion syndrome (PDS) [6] which is typical of pathologies like the pigmentary glaucoma. Pigmentation may also result from the deposition of membrane-like material which is distinctive of the pseudoexfoliation syndrome.

Gonioscopy permits the inspection of the iridocorneal angle, and is indicated as a mandatory procedure by ICO for glaucoma assessment. Other techniques for the evaluation of the anterior chamber angle, like ultrasound biomicroscopy or anterior segment OCT, give anatomical quantification of the angle width. However, they do not provide any chromatic information which is also extremely important for a comprehensive and correct diagnosis.

The pigmentation of the TM was firstly described in 1957 by Scheie [7] who also proposed a pigmentation grading system consisting of five categories numbered from grade 0 (or None) when no TM pigmentation is perceivable, up to a grade IV when the TM is severely pigmented. A graph of the Scheie's pigmentation grading system is in the right of Fig. 1. Usually, pigmentation is not constant, and tends to be higher in the inferior part of the eye.

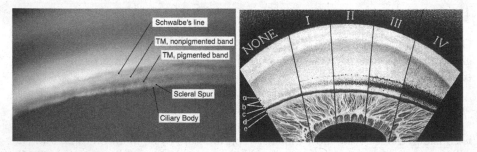

Fig. 1. Left: Iridocorneal angle in a gonioscopic image. The TM is located between the Schwalbe's line and the scleral spur and is divided into pigmented (or functional) and nonpigmented (or nonfunctional) band. Right: Scheie's pigmentation grading system of the TM. Reprinted with permission. Arch Ophthalmol. 1957;58:510–512. Copyright © 1957, American Medical Association. All Rights Reserved.

In this paper, a study on an automatic pigmentation grading system of the trabecular meshwork is presented. The proposed algorithm makes use of machine learning concepts in order to provide a classification comparable with the Scheie's pigmentation grading system.

The paper is organized as follows. Section 2 presents some insights on the data preparation. Section 3 introduces the processing algorithm. Results are presented in Sect. 4. Finally, conclusions are given in Sect. 5 together with some ideas for the future development of this work.

2   Data Preparation

This study was conducted on a dataset of 317 images of Japanese eyes already diagnosed with glaucoma. All images were collected at the Matsue Red Cross Hospital (Matsue, Japan), in 2016, during one of the clinical trials of the prototype of a newly introduced gonioscopy device (GS-1, Nidek Co., Ltd.). The study adhered to the tenets of the Declaration of Helsinki. The institutional review boards of Matsue Red Cross Hospital, Matsue (Japan), reviewed and approved the research. All subjects provided written informed consent.

Acquired images present a high variability of illumination conditions. This depends on several factors such as the illumination LED power used during the data acquisition and the actual reflectivity of the patient's iridocorneal angle. In addition, the pigmentation pattern of the TM may change significantly between different sectors of the same eye. Hence, an initial preprocessing is mandatory for reducing the input data variability.

In each image, a region of interest (ROI) was manually defined by a skilled operator in order to include the scleral spur and the TM up to the Schwalbe's line. This ROI (marked in red in Fig. 2(a)) contains the information considered during the manual assessment of the Scheie's pigmentation grade which was also given by the clinician and assumed as the ground truth.

Given the relatively small size of the available dataset, it was decided to reduce the original Scheie's five-class system to a simplified three-class system by grouping together grades 0 and I, and grades III and IV, respectively. On the other hand, grade II was kept unchanged and common to both grading systems. According to this grouping, the reduced system has grades labeled 0/I, II, and III/IV.

The rationale behind this decision was that clinical differences between a grade 0 or I, or between a grade III or IV given by an expert are often very subjective because they strongly depend on the illumination conditions during the examination. On the contrary, grade II, having a characteristic pattern with a central heavy-pigmented band, is easily identifiable and thus more reliable.

During the selection of the dataset, images were excluded either when too many iris processes or extended synechiae were covering the TM (thus preventing the assessment of the pigmentation grade), or when the image quality was considered inadequate because of insufficient brightness or for the presence of saturated regions (thus preventing any reliable color analysis).

3   Algorithm Description

The proposed algorithm relies on a machine learning (ML) classifier for attributing the pigmentation grade to the TM portion inside the selected ROI. As usual in ML, a first image preprocessing Matlab script is applied for extracting the set of relevant features which are then fed into the classifier.

3.1   Image Preprocessing

Acquired images present a high variability in brightness and pigmentation patterns. Thus, an image preprocessing step is required for reaching an illumination-independent representation and for simplifying the set of features that will be processed by the ML algorithm.

As a preliminary step, RGB colors are converted to the Lab color space. This operation separates the lightness component L from the two chromatic opponents a and b.

For addressing the dependency on the illumination variability, an auxiliary ROI (aux ROI) is considered over the cornea just anterior to the Schwalbe's line. Consequently, the auxiliary ROI can be automatically selected given the position of the operator-selected ROI (main ROI). Figure 2(a) shows some examples of the displacement of the main ROI (in red) and the aux ROI (in blue).

More importantly, the cornea is not pigmented, and has a rather homogeneous color (white or slightly yellowish) suitable for defining a reference point in the Lab color space. It was decided to consider the aux ROI median value, $(L_{aux}, a_{aux}, b_{aux})$ as this reference point. As it will be seen later on, this value is used for gaining an illumination-independent representation of colors in the main ROI.

In order to cope with the pattern variability, a fundamental observation from right of Fig. 1 is that, as a first approximation for assessing the pigmentation grade, it may suffice to characterize the color gradient along the radial direction with respect to the pupil's center instead of characterizing any specific pigmentation patterns of the TM.

For taking into consideration this geometrical behavior, pixels in the main ROI are grouped in a discrete number of arched stripes according to their distance from the estimated pupil's center. Then, each pixel within a cluster is substituted by the median Lab value of the cluster it belongs to. The effect of this operation is represented in Fig. 2(b).

The radially-dependent median operator is able to simplify the pattern still maintaining the color gradient along the radial direction. In addition, the median operator may also filter out some sporadic pigmentation spots as in the top of Fig. 2(b), or even isolated iris processes possibly entering the main ROI.

Once that the color description of the ROIs has been reduced, it is possible to proceed with the extraction of the radial color gradient (or "profile") from the main ROI, and with the creation of an illumination-independent representation exploiting the reference point given by the median value of the aux ROI previously defined.

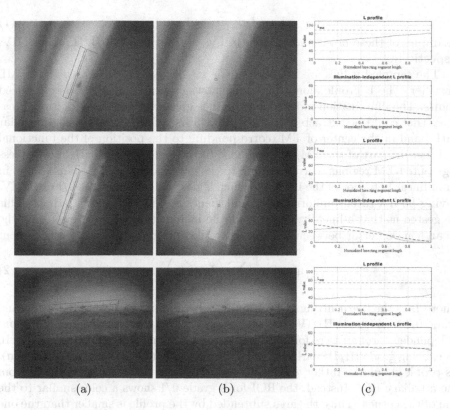

Fig. 2. Preprocessing results for a grade 0/I TM (top), a grade II TM (center), a grade III-IV TM (bottom). (a) Main ROI (in red) and aux ROI (in blue) before preprocessing. (b) Resulting main ROI gradient after preprocessing. Beginning and ending of the bisecting segment are represented by the red and blue squares, respectively. (c) (Above) illumination profile in the original image, with the L_{aux} marked as a dashed line. (Below) the corresponding illumination-independent profile and its linear interpolation (dashed line). The X axis is normalized so that "0" represents the beginning of the bisecting segment close to the iris, whereas "1" represents the end of the bisecting segment toward the cornea. Observe that illumination-independent profiles of grade 0/I (top) are characterized by a linear behavior and a small subtended area, while profiles of grade II (center) are characterized by a highly nonlinear behavior, and those of grade III/IV (bottom) are characterized by a linear behavior and a large subtended area. (Color figure online)

3.2 Illumination-Independent Profiles

Consider a segment bisecting the processed main ROI (i.e. the one connecting the red square to the blue square in Fig. 2(b)).

Let $\{L_i\}$, $i = 1, \ldots, N$ be the sampled values of L along the segment in N points. Also let L_{aux} be the median L value in the auxiliary ROI. Then, an illumination-independent description of the information carried by L is given by:

$$\hat{L}_i = L_i - L_{aux} \quad i = 1, \ldots, N. \tag{1}$$

Examples of these profiles are shown in Fig. 2(c) for grades 0/I, II, and III/IV, respectively.

The sampled values are included in the set of features forwarded to the ML classifier. The L profile along this segment was sampled in $N = 16$ equispaced points. The illumination-independent profiles give some hints on additional features that could be extracted from the data. First of all, grade II has a peculiar dark band in the center of TM (corresponding to the position of the functional TM, i.e. the part of the TM located ahead of the Schlemm's canal), and two less pigmented TM regions (corresponding to the nonfunctional TM). This results in a nonlinear behavior of the L profile which is not typical for grades 0/I or III/IV, having both a rather uniform shade and, hence, an almost linear gradient. This suggested us that a linearity index could be included in the feature set, and the mean squared error between the actual profile and its linear fitting was chosen:

$$\text{mse} = \frac{1}{N} \sum_{i=1}^{N} \left(\hat{L}_i - \bar{L}_i \right)^2, \tag{2}$$

where $\bar{L}_i$ is the linear fitting of $\hat{L}_i$. Finally, other indices for better discriminating grade 0/I from grade III/IV (both having a linear behavior of the illumination-independent profile) were found to be the area subtended by the profile itself. In fact, being grade III/IV darker than the cornea (due to the heavy pigmentation), its profile should be rather distant from the reference value L_{aux} extracted from the auxiliary ROI. Instead, the ROI for a grade 0/I shows a color similar to the one of the cornea. Thus, the area subtended by the profile is smaller than the one for a grade III/IV. Hence, the following integral indices were also introduced:

$$\text{Area}(\hat{L}) = \sum_{i=1}^{N} \hat{L}_i \ , \ \text{Area}(\hat{a}) = \sum_{i=1}^{N} \hat{a}_i \ , \ \text{Area}(\hat{b}) = \sum_{i=1}^{N} \hat{b}_i. \tag{3}$$

3.3 Classification Using Random Forests

The correspondence between the reduced pigmentation grading system and the extracted image indices is obtained by means of the random forest classifier [8] implemented in the R statistical framework [9].

The dataset containing 317 manually-graded images was divided into a training set with 80% of the available images, and a validation set with 20% of images. The three classes were reasonably distributed, accounting 25.2% for grade 0/I, 36.9% for grade II, and 37.9% for grade III/IV, respectively. A total of 2000 trees were grown in the forest during the training phase obtaining the error rates reported in Fig. 3(a).

4 Results

Performance of the trained random forest was tested on a validation set having 64 images never used during the training. The confusion matrices for the training and validation are reported in Table 1.

Table 1. Confusion matrix during (a) ML training (b) ML validation

Actual Grades	Predicted Grades 0-I	II	III-IV	Errors
0-I (27.7%)	63	5	2	10.00%
II (36.0%)	5	84	2	7.69%
III-IV (36.3%)	5	5	82	10.87%

(a)

Actual Grades	Predicted Grades 0-I	II	III-IV	Errors
0-I (15.6%)	10	0	0	0.00%
II (40.6%)	0	23	3	11.54%
III-IV (43.8%)	0	2	26	7.14%

(b)

During validation, the maximum class error was 11.54% for grade II, whereas the average grade error was 7.81%, corresponding to a success rate of 92.19%. Interestingly, grade 0/I was always correctly predicted, and wrong predictions were always between the grade II and III/IV.

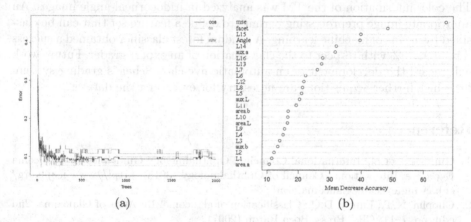

(a)   (b)

Fig. 3. (a) Error rates of the random forest computed during the training phase. (b) Variable importance for the trained random forest: prism facet, angle of the bisecting segment with respect to the horizontal axis of the eye, L, a, and b median values of the auxiliary ROI, sampled illumination-independent L profile along the ROI bisecting segment as illustrated in Sect. 3.2, the integral indices given in (3), mean squared error between the illumination-independent L profile and its linear fitting as given by (2). The most significant variable for discriminating among the grades was the linearity index (mse).

A relevant outcome of the random forest training is the characterization of the importance of input features, intended as their discriminant capability in producing the correct classification.

Results obtained in this study are reported in Fig. 3(b), where variables are sorted in descending order according to their estimated importance.

This figure suggests that the best discriminant feature is related to the linear fitting index (mse). This is not completely unexpected since there is an evident change of behavior for the illumination-independent L profile when passing from grade 0/I to grade II, and from grade II to grade III/IV.

Other relevant features are the facet and angle indices. These two variables identify the eye sector from which the image of the iridocorneal angle was acquired. Again, this does not surprise much since pigmentation tends to be more evident in the inferior part of the eye due to the gravity force that conveys more aqueous humor, and hence pigment, toward the bottom part of the anterior chamber. Interestingly, this fact statistically arises during the training of the random forest. Finally, also the terminal section of the processed L profiles (namely, L15, L14, L16 and L13) were important. These indices characterize the nonfunctional TM near the sclera and this is the TM region that changes more significantly as a function of the pigmentation grade.

5   Conclusions

The color information of the TM was analyzed in iridocorneal-angle images. An appropriate image preprocessing permits to define a feature set that can be classified by means of machine learning. A random forest classifier obtained a success rate of 92.19% with respect to the classification of an expert grader. Future work will regard the development of an automatic five-class Scheie's grading system, for which further acquisitions are necessary for extending the dataset.

References

1. Gupta, N., et al.: International Council of Ophthalmology Guidelines for Glaucoma Eye Care. International Council of Ophthalmology (2015). http://www.icoph.org/enhancing_eyecare/glaucoma.html
2. Choplin, N.T., Lundy, D.C.: Classification of glaucoma. In: Atlas of Glaucoma, 2nd edn, pp. 7–11. CRC Press, Boca Raton (2007)
3. Faschinger, C., Hommer, A.: Anatomical structures of the chamber angle. In: Faschinger, C., Hommer, A. (eds.) Gonioscopy, 2nd edn, pp. 11–16. Springer, Berlin (2012). https://doi.org/10.1007/978-3-642-28610-0_3
4. Choplin, N.T., Lundy, D.C.: Secondary open-angle glaucomas. In: Atlas of Glaucoma, 2nd edn, pp. 133–136. CRC Press, Boca Raton (2007)
5. Alward, W.L.M., Longmuir, R.A.: Abnormalities associated with an open angle. In: Color Atlas of Gonioscopy, 2nd edn, pp. 87–93. Wolfe Publishing, Prescott (1994)
6. Scheie, H.G., Cameron, J.D.: Pigment dispersion syndrome: a clinical study. Br. J. Ophthalmol. **65**, 264–269 (1981)
7. Scheie, H.G.: Width and pigmentation of the angle of the anterior chamber: a system of grading by gonioscopy. Am. Med. Assoc. Arch. Ophthalmol. **58**(4), 510–512 (1957)
8. Breiman, L.: Random forests. Mach. Learn. **45**(1), 5–32 (2001)
9. R Core Team: R: A Language and Environment for Statistical Computing. R Foundation for Statistical Computing (2017). https://www.R-project.org/

Large Receptive Field Fully Convolutional Network for Semantic Segmentation of Retinal Vasculature in Fundus Images

Gabriel Lepetit-Aimon$^{(\boxtimes)}$, Renaud Duval, and Farida Cheriet

Ecole Polytechnique de Montréal, Montreal, QC H3T 1J4, Canada
gabriel.lepetit-aimon@polymtl.ca

Abstract. Analysis of the retinal vasculature morphology from fundus images, using measures such as arterio-venous ratio, is a promising lead for the early diagnosis of cardiovascular risks. The accuracy of these measures relies on the robustness of the vessels segmentation and classification. However, algorithms based on prior topological knowledge have difficulty modelling the abnormal structure of pathological vasculatures, while patch-trained Fully Convolutional Neural Networks (FCNNs) struggle to learn the wide and extensive topology of the vessels because of their narrow receptive fields.

This paper proposes a novel Fully Convolutional Neural Network architecture capable of processing high resolution images through a large receptive field at a minimal memory and computational cost. First, a single branch CNN is trained on whole images at low resolution to learn large scale features. Then, this branch is incorporated into a standard encoder/decoder FCNN: its large scale features are concatenated to those computed by the central layer of the FCNN. Finally, the whole network architecture is trained on high-resolution patches. During this last phase, the FCNN benefits from the large scale features while the low resolution branch parameters are fine-tuned. This architecture was evaluated on the publicly available retinal fundus database DRIVE. The trained network achieves an accuracy of 96.1% in segmenting the full retinal vessels and improves by 5% the artery/vein classification compared to a basic U-Net.

Keywords: Retinal vessel segmentation · Retinal vessel classification
Convolutional neural network · Deep learning

1 Introduction

Early diagnosis is a key to reducing mortality rates in cardiovascular diseases, which caused 30.8% of deaths and were the top healthcare expenditure in the USA in 2013 [9]. Retinal fundus imaging allows the non-invasive observation of the retinal vascular system. This modality thereby offers a good overview of cardiovascular health: statistically, a patient suffering from retinopathy is twice as likely to have a stroke [13]. Clinicians already use fundus images to evaluate

© Springer Nature Switzerland AG 2018
D. Stoyanov et al. (Eds.): COMPAY 2018/OMIA 2018, LNCS 11039, pp. 201–209, 2018.
https://doi.org/10.1007/978-3-030-00949-6_24

cardiovascular risks, by analysing the retinal vasculature morphology through measures like the arterio-venous ratio (highly correlated with hypertension and diabetes risks). However, the accuracy of these measures relies on the robustness of the vessels segmentation and classification between arteries and veins.

In the last two decades, many algorithms were developed to handle those tasks. Most of them perform the vessel segmentation separately from the classification. For the former task, traditional computer vision algorithms such as the multi-scale line detector [10] have been out-performed by deep learning algorithms, using either convolution neural networks (*CNNs*) [8], CNNs combined with Conditional Random Fields (*CRFs*), e.g. the Deep Vessel architecture [3], or by adversarial architectures [6]. Meanwhile, for vessel classification, state of the art methods generally combine local features analysis by machine learning algorithms with prior knowledge of the vascular tree structure. Dashtbozorg *et al.* use linear discriminant analysis outputs in combination with the vascular graph corrected by rules derived from prior knowledge [1]. More recently, Estrada *et al.* proposed a graph-based algorithm to extract the vascular tree from a fundus image and classify each detected vessel using local features [2]. However, relying on rules derived from prior knowledge can impact the robustness of the algorithm, in particular for severe cases of retinopathy where the vasculature won't match the rules.

Recent progress in deep learning has made possible the training of larger and deeper Fully Convolutional Neural Networks (*FCNNs*). In particular, the U-Net achieves remarkably good performance in segmenting medical images thanks to its encoder/decoder architecture and to its skip-connexions [12]. However, to our knowledge, those architectures have never reached state of the art performance in artery/vein classification. Indeed, because of their narrow receptive fields, FCNNs struggle to learn the wide and extensive topology of retinal vessels.

This paper propose a novel Large Receptive Field Fully Convolutional Network architecture (*LRFFCN*), capable of segmenting very extensive shapes (i.e. vessels) in high-resolution images at a minimal memory and computational cost. This paper is organized as follows: (1) analysis of the U-Net's poor performances in classifying arteries and veins; (2) description of a novel FCNN architecture: the Large Receptive Field Fully Convolutional Network; (3) evaluation of the LRFFCN architecture experimentally in the semantic vessel segmentation task.

2   Methods

2.1   Large Receptive Field Fully Convolutional Network

Receptive Field Limitation in Convolutional Networks. The term *receptive field* is inherited from neurosciences and describes the region of the sensory space (e.g. the visual field) in which a stimulus will cause a neuron to be activated. In deep learning, the receptive field of a convolutional network is the region of the input analyzed by the network to produce the prediction for *one* pixel. Early CNN designers performing pixel-wise classification didn't care about this concept: because their models ended with fully connected layers, the receptive field was the whole input patch. However, this is not the case with FCNNs.

For example, the encoding branch of the U-Net architecture has a receptive field of 125 × 125 pixels even though its training patch size is 500 × 500 pixels.

There are several ways to increase the receptive field of an FCNN. Stacking more layers or extending their kernel size will theoretically increase the receptive field linearly, whereas sub-sampling the output features of a layer will increase it multiplicatively. In practice, Luo *et al.* have shown that the Effective Receptive Field (*ERF*) is always narrower than the theoretical one [7]. More precisely, the ERF follows a Gaussian distribution with a standard deviation depending on the model architecture and the weights initialization. Extending the kernel size will increase the ERF linearly; stacking n layers will only increase the ERF by a factor $\sqrt{n}$; sub-sampling will effectively increase the ERF quickly. On the contrary, skip-connections will shrink the ERF.

Focusing on the U-Net architecture, its decoding branch shouldn't have much impact on the ERF: the growth due to the convolutional layers is compensated by the upsampling and the skip-connections. Intuitively, this branch is only a complex upsampling interpolation of the deep features. In other words, the ERF of a U-Net is strictly lower than the theoretical receptive field of its encoding branch: 125 × 125 pixels. When processing high-resolution fundus images (2048 × 2048 pixels), such a receptive field is much too small to learn the topology of the vasculature, thus the network can only rely on local texture and color features. However, for small vessels far away from the optic disk, those features are not sufficient, even for clinicians, to efficiently discriminate arteries from veins.

As the U-Net architecture is already a large model, adding more layers to it would make the model too heavy. Namely adding a new pooling/up-conv stage would raise the forward-pass computation from 67 to 389 Mega Flops (mainly because the patch size is doubled).

LRFFCN Architecture. The Large Receptive Field Fully Convolutional Network is a novel network architecture which significantly increases the receptive field at a minimal memory and computational cost. The core of the architecture is a convolutional branch processing the full image at a low resolution (scaled down to a 128 × 128 pixels patch). This branch is structured as a repetition of fire/squeeze modules inspired by the SqueezeNet (a concise model with similar performance to AlexNet [4]). The theoretical receptive field of this branch is 21 × 21 pixels at $1/16^{th}$ of the full resolution, which corresponds to 336 × 336 pixels in the high resolution image. The large scale features learned by this branch are then incorporated in an encoder/decoder FCNN with skip-connections.

The encoding stage of this network contains 4 pooling layers (2 × 2 pooling), so the resolution of the deepest features F_e is $1/16^{th}$. Because the FCNN is trained on patches and not on the full image, a region corresponding to the patch is extracted from the large scale features and is concatenated to F_e. Thus, the decoding stage of the FCNN has access to textural information from high resolution features learned by the encoding stage, and to topological information from the large scale features at the cost of a forward-pass computation increase of only 6% (4 MFlops).

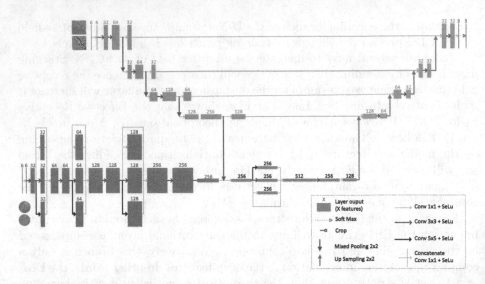

Fig. 1. The large receptive field fully convolution network architecture. Color code: *light blue*: encoding branch; *orange*: decoding branch; *dark blue*: low resolution branch. (Color figure online)

2.2 Model Specificities

The exact model trained for this paper is presented in Fig. 1. Its most interesting specificities are presented in this section.

Due to memory limitations, the maximum minibatch size during training is 8. Batch normalization layers would therefore be ineffective and have been replaced by a SeLu activation function [5]. This modified version of the ReLu activation function offers negative values which speed up learning by pushing mean activation towards 0, and a stable fixed point of the activation mean and variance so that exploding and vanishing gradients are impossible.

The model is trained to perform semantic segmentation: the vessels segmentation is performed simultaneously with the classification as each pixel is assigned a probability of being part of the background, an artery or a vein. Even if the LRFFCN architecture efficiently increases the ERF, performing this classification pixel-wise can cause local errors in the artery/vein classification. Indeed, because arteries and veins are visually similar and because of the local inconsistencies transmitted from the early layers to the later ones through the skip-connections, some pixels from a vessel can suddenly be predicted as veins even though they are surrounded by arteries (and vice versa). To correct such local inconsistencies, a CRF is added at the end of the model.

CRFs use the preprocessed version of the training patch as a reference and 2D Gaussian kernels to propagate the probabilities of being an artery or a vein along the vessel. We used the CRF as Recurrent Neural Network (RNN) implementation proposed by Zheng *et al.* [14] so its kernels are trained simultaneously with the LRFFCN parameters.

2.3 Training

The model was trained on 69 images: 30 images from the MESSIDOR training dataset, 19 images from the STARE dataset and 20 images from the DRIVE training dataset. We manually labelled the MESSIDOR and STARE images and asked an ophthalmologist to validate this labelling. Each image was preprocessed using a standard contrast enhancement technique. Both the enhanced and raw images were presented to the network since vessel detection strongly relies on the enhanced image whereas vessel classification relies on the vessels' true colors. Color and geometric data augmentation were used to double the training dataset size (contrast and gamma variations for color augmentation, horizontal mirroring and rotation for geometric augmentation).

The training process was split into 3 phases. The large scale branch was pre-trained first over 100 epochs on full images rescaled to 128×128 pixels. Then, the full LRFFCN was trained on 230×230 pixels patches; the large-scale features already learned by the low resolution branch allow a quick convergence of the network. This training phase was driven by the Adadelta gradient descent opti-mizer and lasted 30 epochs during which the learning rate was slowly decreased. Finally, the CRF as RNN was added to the classifier layer and the network was trained again for 10 epochs.

3 Experiments

3.1 Evaluating the LRFFCN Architecture

We used the 20 images from the DRIVE test dataset to evaluate the LRFFCN architecture. The architecture was tested with and without the CRF as RNN layer. Also, to quantify the contribution of the low resolution branch to the pre-diction, we evaluated the performance of the model when the scaled-down image input to this branch was replaced by a uniform noise. (We refer to this network as *LRFFCN w/o low branch*.) Finally, a classic U-Net was trained for the same number of epochs to estimate the performance gain. For each of these models, we measured the accuracy of the vessel segmentation and the accuracy, speci-ficity and sensitivity of the artery/vein classification. The results are presented in Table 1.

Table 1. Performance of LRFFCN architecture on DRIVE test dataset.

Architecture	Vessels	Arteries/Veins		
	Accuracy	Accuracy	True arteries	True veins
Basic U-Net	95.6 ± 0.4%	75.9 ± 3.0%	70.9%	80.6%
LRFFCN w/o low branch	96.0 ± 0.3%	63.1 ± 3.7%	57.7%	74.1%
LRFFCN	**96.1 ± 0.3%**	79.4 ± 3.5%	73.9%	**86.4%**
LRFFCN + CRF	95.9 ± 0.4%	**81.0 ± 3.8%**	**77.8%**	84.4%

The segmentation accuracy is consistent across architectures and is quite high, confirming the efficiency of convolutional networks in segmenting vessels. However, the LRFFCN architecture improves the classification accuracy by almost 5% compared to the basic U-Net. The large-scale features learned by the low resolution branch seems to be effectively used by the model to improve its generalizing capabilities. This is confirmed by the noise experiment: replacing the real data by noise make the classification accuracy drop by 16%. Thus, the large-scale feature branch contributes greatly to the network's predictions.

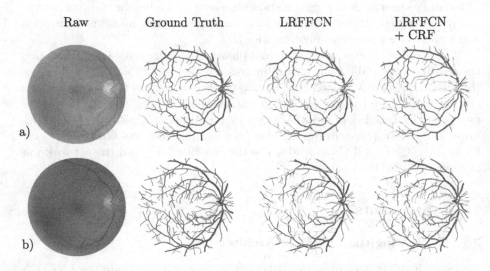

Fig. 2. Comparison of LRFFCN performance with and without CRF as RNN, on two images from DRIVE test dataset. Top row: 06_test; bottom row: 17_test. Color codes: *red*: true artery; *dark blue*: true vein; *light blue*: misclassified vein; *yellow*: misclassified artery. (Color figure online)

The CRF layer improves the classification accuracy by 2% and successfully propagates the vessel classes to correct local errors. Those corrections are visible in Fig. 2(b), where artery segments classified as veins by the LRFFCN are corrected by the CRF layer. However, the LRFFCN's predictions are sometimes not sufficiently accurate and the CRF layer propagates misclassifications (as visible for some veins in Fig. 2(a)). Overall, the CRF improves the topological plausibility of the predicted vessel network.

3.2   Model Segmentation and Classification Performance

In this section, we compare the performance of the LRFFCN to those of state of the art algorithms. For the segmentation task, the LRFFCN exceeds by 1.3% the Deep Vessel network in terms of accuracy. More precisely, our architecture is much more sensitive (80.8% against 72.7%) (Tables 2 and 3).

Table 2. Segmentation results on DRIVE test dataset.

Name	Accuracy	Specificity	Sensitivity
Mozzafarian *et al.* [9]	82.2%	–	–
Adversarial, Lahiri *et al.* [6]	94.%	–	–
Deep Vessel, Fu *et al.* [3]	94.6%	**97.7%**	72.7%
LRFFCN	**95.9 ± 0.4%**	97.3%	**80.8%**

Table 3. Classification results on DRIVE test dataset.

Name	Accuracy	True arteries	True veins
Niemeijer *et al.* [11]	80.0%	80.0%	80.0%
LRFFCN	81.0 ± 3.8%	77.8%	84.4%
Dashtbozorg *et al.* [1]	87.4%	90.0%	84.0%
Estrada *et al.* [2]	**91.7 ± 5%**	**91.7%**	**91.7%**

For the artery/vein classification, the performance of the LRFFCN architecture is still 10% below state of the art graph-based algorithms. Indeed, for the DRIVE dataset, the prior topological knowledge of retinal vessels provides a good enough estimation of the vascular tree. Graph-based algorithms can propagate the artery/vein probability through this tree, whereas our method didn't perfectly learn the vasculature topology and often misclassify the small vessels farther from the optic disk.

4 Discussion and Conclusion

The proposed LRFFCN architecture efficiently increases the receptive field of the FCNN by means of a low resolution branch and successfully takes advantage of large-scale features to better learn the retinal vessel topology. In particular, the LRFFCN architecture outperforms the U-Net in classifying retinal arteries and veins. It also does better than state of the art algorithms in vessel segmentation, yielding a higher sensitivity. For the classification task, the LRFFCN architecture does not reach state of the art performance. However, we believe that this particular result is due to the training conditions and not to the model design.

Nevertheless, these results are promising. Indeed, in contrast to static graph-based analysis, the performance of this model will improve as the training dataset grows. In particular, the bottleneck for this architecture is the training of the low resolution branch. Because this branch must be trained on whole images and not on patches, our current training dataset is too small for the LRFFCN architecture to show its full potential. But because the segmentation accuracy is high, the predictions from our method can quickly be fixed by a clinical expert. In other words, the LRFFCN architecture can be used to efficiently generate more ground truth images on which it can then be trained to improve its vessel classification performance.

References

1. Dashtbozorg, B., Mendonca, A.M., Campilho, A.: An automatic graph-based approach for artery/vein classification in retinal images. IEEE Trans. Image Process. **23**(3), 1073–1083 (2014)
2. Estrada, R., Allingham, M.J., Mettu, P.S., Cousins, S.W., Tomasi, C., Farsiu, S.: Retinal artery-vein classification via topology estimation. IEEE Trans. Image Process. **34**(12), 2518–2534 (2015)
3. Fu, H., Xu, Y., Lin, S., Kee Wong, D.W., Liu, J.: DeepVessel: retinal vessel segmentation via deep learning and conditional random field. In: Ourselin, S., Joskowicz, L., Sabuncu, M.R., Unal, G., Wells, W. (eds.) MICCAI 2016. LNCS, vol. 9901, pp. 132–139. Springer, Cham (2016). https://doi.org/10.1007/978-3-319-46723-8_16
4. Iandola, F.N., Moskewicz, M.W., Ashraf, K., Han, S., Dally, W.J., Keutzer, K.: SqueezeNet: AlexNet-level accuracy with 50x fewer parameters and <1 MB model size. CoRR abs/1602.07360 (2016). http://arxiv.org/abs/1602.07360
5. Klambauer, G., Unterthiner, T., Mayr, A., Hochreiter, S.: Self-normalizing neural networks. In: Advances in Neural Information Processing Systems, vol. 30, pp. 971–980. Curran Associates, Inc. (2017). http://papers.nips.cc/paper/6698-self-normalizing-neural-networks.pdf
6. Lahiri, A., Ayush, K., Biswas, P.K., Mitra, P.: Generative adversarial learning for reducing manual annotation in semantic segmentation on large scale miscroscopy images. In: 2017 IEEE Conference on Computer Vision and Pattern Recognition Workshops (CVPRW), pp. 794–800, July 2017
7. Luo, W., Li, Y., Urtasun, R., Zemel, R.: Understanding the effective receptive field in deep convolutional neural networks. In: Lee, D.D., Sugiyama, M., Luxburg, U.V., Guyon, I., Garnett, R. (eds.) Advances in Neural Information Processing Systems, vol. 29, pp. 4898–4906. Curran Associates, Inc. (2016). http://papers.nips.cc/paper/6203-understanding-the-effective-receptive-field-in-deep-convolutional-neural-networks.pdf
8. Maninis, K.-K., Pont-Tuset, J., Arbeláez, P., Van Gool, L.: Deep retinal image understanding. In: Ourselin, S., Joskowicz, L., Sabuncu, M.R., Unal, G., Wells, W. (eds.) MICCAI 2016. LNCS, vol. 9901, pp. 140–148. Springer, Cham (2016). https://doi.org/10.1007/978-3-319-46723-8_17
9. Mozaffarian, D., et al.: Heart disease and stroke statistics–2016 update. Circulation **133**(4), e38–e360 (2016). http://circ.ahajournals.org/content/133/4/e38
10. Nguyen, U.T., Bhuiyan, A., Park, L.A., Ramamohanarao, K.: An effective retinal blood vessel segmentation method using multi-scale line detection. Pattern Recognit. **46**(3), 703–715 (2013). http://www.sciencedirect.com/science/article/pii/S003132031200355X
11. Niemeijer, M., et al.: Automated measurement of the arteriolar-to-venular width ratio in digital color fundus photographs. IEEE Trans. Med. Imaging **30**, 1941–1950 (2011)
12. Ronneberger, O., Fischer, P., Brox, T.: U-Net: convolutional networks for biomedical image segmentation. In: Navab, N., Hornegger, J., Wells, W.M., Frangi, A.F. (eds.) MICCAI 2015. LNCS, vol. 9351, pp. 234–241. Springer, Cham (2015). https://doi.org/10.1007/978-3-319-24574-4_28. http://lmb.informatik.uni-freiburg.de/Publications/2015/RFB15a, (Available on arXiv:1505.04597 [cs.CV])

13. Wong, T.Y., Klein, R., Sharrett, A.R.: The prevalence and risk factors of retinal microvascular abnormalities in older persons: the cardiovascular health study. Ophthalmology **110**, 658–666 (2003)
14. Zheng, S., et al.: Conditional random fields as recurrent neural networks. In: International Conference on Computer Vision (ICCV) (2015)

Explaining Convolutional Neural Networks for Area Estimation of Choroidal Neovascularization via Genetic Programming

Yibiao Rong[1], Kai Yu[1], Dehui Xiang[1], Weifang Zhu[1], Zhun Fan[2],
and Xinjian Chen[1(✉)]

[1] School of Electrical and Information Engineering,
Soochow University, Suzhou 215006, China
`xjchen@suda.edu.cn`
[2] Key Laboratory of Digital Signal and Image Processing of Guangdong Provincial,
College of Engineering, Shantou University, Shantou 515063, China

Abstract. Choroidal neovascularization (CNV), which will cause deterioration of the vision, is characterized by the growth of abnormal blood vessels in the choroidal layer. Estimating the area of CNV is important for proper treatment and prognosis of the disease. As a noninvasive imaging modality, optical coherence tomography (OCT) has become an important modality for assisting the diagnosis. Due to the number of acquired OCT volumes increases, automating the OCT image analysis is becoming increasingly relevant. In this paper, we train a convolutional neural network (CNN) with the raw images to estimate the area of CNV directly. Experimental results show that the performance of such a simple way is very competitive with the segmentation based methods. To explain the reason why the CNN performs well, we try to find the function being approximated by the CNN. Thus, for each layer in the CNN, we propose using a surrogate model, which is desired to have the same input and output with the layer while its mathematical expression is explicit, to fit the function approximated by this layer. Genetic programming (GP), which can automatically evolve both the structure and the parameters of the mathematical model from the data, is employed to derive the model. Primary results show that using GP to derive the surrogate models is a potential way to find the function being approximated by the CNN.

Keywords: Choroidal neovascularization
Convolutional neural networks · Surrogate model
Genetic programming

1 Introduction

Choroidal neovascularization (CNV), which will cause deterioration of the vision, is characterized by the growth of abnormal blood vessels in the choroidal layer

© Springer Nature Switzerland AG 2018
D. Stoyanov et al. (Eds.): COMPAY 2018/OMIA 2018, LNCS 11039, pp. 210–218, 2018.
https://doi.org/10.1007/978-3-030-00949-6_25

[2]. One of the medical treatments for CNV is anti-vascular endothelial growth factor (anti-VEGF) injections, which may inhibit the growth of CNV. However, the treatment is very expensive and may be ineffective for some patients. Hence, accurate detection and quantification of CNV would be extremely useful to the clinicians in the diagnosis and evaluation of the therapeutic effect of the treatments [5].

As a non-invasion and non-contact imaging modality, Optical Coherence Tomography (OCT) [3] is becoming a powerful tool to monitor the change of CNV. Due to the number of acquired OCT volumes increases, estimating the area of CNV automatically is becoming increasingly relevant. Traditionally, segmentation based methods are popular for this process [1,5,12]. However, segmentation based methods are highly dependent on the intermediate steps, which inevitably induces cumulative errors.

Recently, some researchers proposed using the direct methods without segmentation for automated medical image analysis, whose effectiveness has been proved by many works [6,7,14]. In addition, with the revival of convolutional neural networks (CNNs) [4], the direct methods become more convenient since the CNNs can be trained directly by the raw data, and generally, achieve significant improvements compared with the traditional approaches.

In this paper, we train a CNN with raw images to estimate the area of CNV in retinal OCT images directly. The effectiveness of such a simple way is verified by the experimental results. However, the CNN is notorious due to the drawback that it is difficult to derive the function being approximated by it via studying the structure of the CNN. To deal with this problem, we try to find a surrogate model, whose mathematical expression is explicit, to fit the function of each layer in the CNN. To this end, genetic programming (GP) [9], which can automatically evolve both the structure and the parameters of the mathematical model from the data, is employed to derive the model. The contributions of this paper are concluded as follows.

- We demonstrate that using CNNs to estimate the area of CNV in retinal OCT images without segmentation is very promising.
- We propose using GP to derive mathematical models to fit the function being approximated by CNNs, which is a potential way to explain how do CNNs work.

The rest of this paper is organized as follows: In Sect. 2, the details of CNNs and GP are described. Section 3 presents the experimental results, including the performance of the CNN on direct area estimation and the functions obtained by GP. We have discussion and conclusion in Sect. 4.

2   Methods

2.1   Convolutional Neural Networks

Convolutional neural networks can be employed for different kinds of problems by constructing a proper network structure C_θ and cost function ϕ given the training set $\{(p_s, q_s)\}_{s=1}^S$, where p_s is the observed value and q_s is the desired output. This process can be formulated as

$$C_{learn} = \min_{C_\theta, \theta \in \Theta} \left(\sum_{s=1}^S \phi(q_s, C_\theta(p_s)) \right) \tag{1}$$

where Θ is the set of all possible parameters. Once the learning step is complete, C_{learn} can be used to the specific problem. In this work, we employ the CNNs to estimate the area of CNV in retinal OCT images directly. Since the area of CNV is continuous variable, using CNNs to estimate the area directly is a regression problem. The least squares is used as the loss function. Thus, the problem is formulated as

$$R_{learn} = \min_{R_\theta, \theta \in \Theta} \left(\sum_{s=1}^S \|y_s - R_\theta(x_s)\|^2 \right) \tag{2}$$

where x_s is an OCT image and y_s is the corresponding area value.

Particulary, the structure of a typical CNN includes several convolutional layers and pooling layers optionally followed by at least one fully connected layer. The structure of CNN applied on this work is shown in Table 1. The convolution kernels in all the convolutional layers are defined of rectangular size 7×3. The number of filters in each convolutional layer is $32 \times 2^{l-1}$, where $l = 1, 2, 3, 4$ is the lth convolutional layer. The receptive field size of pooling layer is 3×3 with stride 2 for downsampling. At the final layer, a 1×1 convolution is used to map each 256-component feature vector to the output. In total, there are four convolutional layers, three pooling layers and one fully connected layer in the CNN. It is noteworthy to point out that each convolutional layer is followed by a rectified linear units (ReLU) layer which is not displayed in the Table 1.

Table 1. The structure of the CNN applied on this work.

Layer	Input	Insize	K	S	Outsize
conv1	image	512×256	7×3	2	$255 \times 129 \times 32$
mpool1	conv1	$256 \times 129 \times 32$	3×3	2	$127 \times 64 \times 32$
conv2	mpool1	$127 \times 64 \times 32$	7×3	2	$63 \times 33 \times 64$
mpool2	conv2	$63 \times 33 \times 64$	3×3	2	$31 \times 16 \times 64$
conv3	mpool2	$31 \times 16 \times 64$	7×3	2	$15 \times 9 \times 128$
apool1	conv3	$15 \times 9 \times 128$	3×3	2	$7 \times 4 \times 128$
conv4	mpool2	$7 \times 4 \times 128$	7×3	2	$1 \times 1 \times 256$
fc1	conv4	$1 \times 1 \times 256$	1×1	1	1

K–kernel size, S–stride, conv–convolutional layer, mpool–max pooling layer, apool–average pooling layer, fc–fully connected layer.

2.2   Genetic Programming

Genetic programming (GP) is a biologically inspired machine learning method that evolves computer programs to perform a task. It does this by randomly generating a population of computer programs and then breeding together the best performing individuals to create a new population. Usually, the individual in a population is represented by tree structures, which consists of a set of functions, e.g. minus, square, as shown in Fig. 1(a). One of the popular applications of GP is symbolic data mining, which is a process of building an empirical mathematical model of data.

The general process of building an empirical mathematical model of data using GP is shown in Fig. 1(b). A population is first initialized. The fitness of each individual is then evaluated. The individual, which has the highest fitness value in the population, is selected if the evolution is completed. Otherwise, genetic operations are employed to breed new individual for the next generation. Generally, crossover, mutation and duplication are the common used genetic operations. The crossover operation is done by exchanging the subtrees at random positions in a pair of selected individuals, as shown in Fig. 1(c). Mutation is a process of replacing a subtree in an individual with a randomly generated tree, as shown in the upper part of Fig. 1(d). Duplication is a process which can result in a same individual, as shown in the lower part of Fig. 1(d). We refer reader to [9] for a comprehensive reading about the GP.

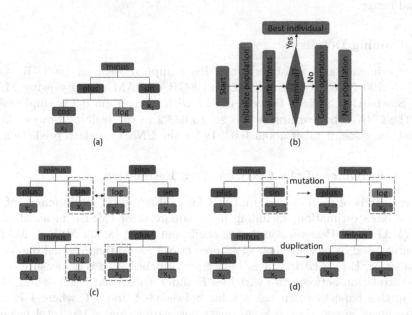

Fig. 1. (a) Example of a tree (gene) representing the model term $(cos(x_2) + log(x_3)) - sin(x_1)$. (b) The general process of GP. (c) Crossover. (d) Mutation and duplication.

3 Results

3.1 Data Collection and Augmentation

The OCT images used for the experiments are acquired from 7 patients. These images with $1024 \times 512 \times 128$ voxels ($11.72\,\mu m \times 5.86\,\mu m \times 15.6\,\mu m$), covering the volume of $6\,mm \times 6\,mm \times 2\,mm$, are obtained by ZEISS scanner. The number of B-scan images with CNV in these 3D OCT images is 4060, in which 60% of them are selected as training set randomly, 20% as validation set and 20% as test set. All the CNV in the B-scan images are delineated by a clinician. The ground truth used to train the CNN is calculated based on the delineated results and normalized to the range of $[0, 1]$ by the following equation:

$$\widehat{y} = \frac{y - y_{min}}{y_{max} - y_{min}} \tag{3}$$

where y and $\widehat{y}$ are the area values before and after normalization respectively. y_{max} and y_{min} are the maximal and minimal area values in the training set respectively.

Each image is cropped to a size of 512×256 to save memory. In addition, we can augment the data easily due to the cropping. A point is randomly selected from the CNV region. A sub image with size of 512×256 is then cropped center on this point. In this work, this cropping process is repeated ten times for each original image.

3.2 Training Details

The experiments are performed on a PC equipped with an Intel (R) Core (TM) i7-4790 M CPU at 3.60 GHZ and 8 GB of RAM capacity using MAT-LAB. Stochastic Gradient Descent (SGD) with momentum 0.9 is employed to train the CNN. The learning rate is set to 0.002 in the training process. Once the training done, it takes about 0.0494 s for the CNN to make a prediction.

3.3 Performance of the CNN on Area Estimation

Different kinds of metrics are employed for evaluating the performance of the CNN on area estimation, including mean square error (MSE), mean absolute error (MAE) and Pearson correlation coefficient (Pcc) [8,13]. MSE or MAE is a quantity used to measure how close predictions are to the eventual outcomes. A smaller MSE (MAE) indicates better performance. Pcc is a measure of the linear correlation between two variables F and Y (predicted results and ground truth in this paper), which has a value between $+1$ and -1, where 1 is total positive linear correlation, 0 is no linear correlation, and -1 is total negative linear correlation. A higher correlation coefficient indicates better performance.

Table 2 summarises the average results obtained by the trained CNN for estimating the area of CNV directly. It is observed that the values of MSE, MAE and Pcc obtained by the direct method are 0.0022, 0.0328 and 0.9763 respectively. To evaluate the performance of the direct method further, we also compare with the segmentation methods. In this work, we select the u-net [10], which is a popular network structure in biomedical image segmentation domain, for comparisons. Figure 2 shows several estimation results on different images obtained by these two methods respectively. The red curves in each sub-figure of Fig. 2 are the CNV boundaries delineated by hand and the white curves are the boundaries obtained by u-net. Table 2 aslo summarises the average results obtained by the segmentation based method. It is observed that the average results obtained by the direct method are better than the segmentation based method.

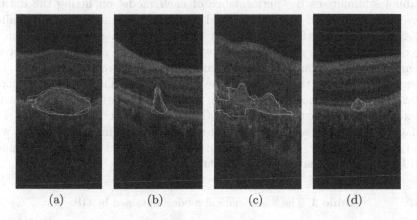

| | (a) | (b) | (c) | (d) |

Fig. 2. Estimation results obtained by different methods on the different images. Note that the red curves are the ground truth and the white curves are the results obtained by u-net. (a) gt = 0.2633, dir = 0.2653, seg = 0.2504; (b) gt = 0.0365, dir = 0.0337, seg = 0.0415; (c) gt = 0.1603, dir = 0.2503, seg = 0.2281; (d) gt = 0.0068, dir = 0.0453, seg = 0.0194; At each subfigure title, gt is the ground truth, dir is the value obtained by the direct method and seg segmentation based method. (Color figure online)

Table 2. The results obtained by the direct method and segmentation based method respectively.

	Direct estimation	Segmentation
MSE	0.0022	0.0033
MAE	0.0328	0.0332
Pcc	0.9763	0.9693

3.4   Deriving the Function Approximated by CNN

As the results shown above, the performance of CNN used for estimating the area of CNV directly is very competitive. However, the entire process looks like a black box. In addition, it is difficult to derive the function being approximated by the CNNs via studying the structure of the CNNs. To explain how does the CNN work, we try to find the function approximated by the CNN. To this end, we employ the GP, which can evolve an empirical mathematical model of data automatically, to derive the mathematical model in the last layer of the CNN, namely find a mathematical model to fit the input and output of the fc1 layer. The GPTIPS2 toolbox [11] with default settings is employed for building the mathematical models automatically. Table 3 lists several models obtained by GP. It is noteworthy to point out that x_i is the ith variable in the input of the fc1 layer, where $i = 1, 2, ..., 256$ since the input of the fc1 layer is a 256×1 vector.

Table 4 summarizes the performance of each model on fitting the data of fc1 layer. It is observed that the derived mathematical models only contain n ($n < 256$) variables, which means that the function in the fc1 layer can be approximated by the n dimensional curve. It is also observed that the derived mathematical model could be different at each genetic programming and their performance on fitting the data would be different also. The best model among the list is model 5, who achieves Pcc of 0.9426. Although the results are not very excellent, they illustrate that using GP to derive a mathematical model for each layer in the CNN, which is desired to have the same input and output with the corresponding layer, could be feasible since the derived mathematical model would be different at each genetic programming.

Table 3. The mathematical models obtained by GP.

Model ID	Model
1	$y = 0.323x_{18} + 0.742x_{33} + 0.689x_{45} + 0.323x_{107}$ $+ 0.323x_{155} + 0.29x_{161} + 0.267x_{231} + 0.0224$
2	$y = 0.585x_{10} + 0.585x_{33} + 0.0893x_{160} + 0.585x_{197} + 0.312tanh(x_{199})$ $+ (0.158exp(x_{85}))/log(x_{77}) + 0.081$
3	$y = 0.759x_{60} + 0.759x_{89} + 10.8x_{107} + 0.759x_{137} - 0.759x_{107}x_{209} + 1.62x_{148}^2 + 0.0863$
4	$y = 0.607x_{24} - 0.0735x_{53} + 0.158x_{164} - 0.607tanh(x_{83}) + 0.637tanh(x_{173})$ $+ 3.79x_{67}^3 - 0.607x_{238}^2 + 0.101$
5	$y = 0.442x_{33} + 0.752x_{125} + 0.752x_{223} + 0.752x_{253} + 0.468exp(x_{39} + x_{60} + x_{86})$ $+ 0.752x_{210}^3 - 0.444$

Table 4. The performance of each mathematical Model.

Model ID	Pcc	MSE	MAE
1	0.8784	0.0105	0.0792
2	0.9193	0.0076	0.0700
3	0.9228	0.0074	0.0682
4	0.9308	0.0065	0.0636
5	0.9426	0.0053	0.0563

4   Discussion and Conclusion

In this paper, we employ CNNs to estimate the area of CNV in retinal OCT images directly. The results show that the performance of the CNN trained by raw images is very competitive. To explain how does the trained CNN work, we try to find the function approximated by the CNN. To this end, we propose using GP to derive a surrogate model for each layer in the CNN to demonstrate the function being approximated by CNN. In this work, we derive such a model for the fc1 layer in the CNN. The experimental results show that the function in fc1 layer can be approximated by a high dimensional curve. However, so far, the GP is only suitable for the case that the input is multiple variables while the output is single variable. If we search such a surrogate model for each single variable in the CNN, the amount of calculation would be very huge. Could we derive a mathematical model to fit the layer well where both of the input and output are multiple variables? It is the issue being explored in our future work.

Acknowledgment. This work has been supported by the National Basic Research Program of China (973 Program) under Grant 2014CB748600.

References

1. Abdelmoula, W.M., Shah, S.M., Fahmy, A.S.: Segmentation of choroidal neovascularization in fundus fluorescein angiograms. IEEE Trans. Biomed. Eng. **60**(5), 1439–1445 (2013)
2. Donoso, L.A., Kim, D., Frost, A., Callahan, A., Hageman, G.: The role of inflammation in the pathogenesis of age-related macular degeneration. Surv. Ophthalmol. **51**(2), 137–152 (2006)
3. Huang, D., et al.: Optical coherence tomography. Science **254**(5035), 1178–1181 (1991)
4. Lecun, Y., Bottou, L., Bengio, Y., Haffner, P.: Gradient-based learning applied to document recognition. Proc. IEEE **86**(11), 2278–2324 (1998)
5. Liu, L., Gao, S.S., Bailey, S.T., Huang, D., Li, D., Jia, Y.: Automated choroidal neovascularization detection algorithm for optical coherence tomography angiography. Biomed. Opt. Express **6**(9), 3564–3576 (2015)
6. Luo, G., Dong, S., Wang, K., Zuo, W., Cao, S., Zhang, H.: Multi-views fusion CNN for left ventricular volumes estimation on cardiac MR images. IEEE Trans. Biomed. Eng. **PP**(99), 1924–1934 (2018)
7. Manit, J., Schweikard, A., Ernst, F.: Deep convolutional neural network approach for forehead tissue thickness estimation. Curr. Dir. Biomed. Eng. **3**(2), 103–107 (2017)
8. Miao, F., et al.: A novel continuous blood pressure estimation approach based on data mining techniques. IEEE J. Biomed. Health Inform. **21**(6), 1730–1740 (2017)
9. Poli, R., Langdon, W.B., McPhee, N.F., Koza, J.R.: A field guide to genetic programming, pp. 1–93. Lulu.com (2008). http://www.gp-field-guide.org.uk
10. Ronneberger, O., Fischer, P., Brox, T.: U-Net: convolutional networks for biomedical image segmentation. In: Navab, N., Hornegger, J., Wells, W.M., Frangi, A.F. (eds.) MICCAI 2015. LNCS, vol. 9351, pp. 234–241. Springer, Cham (2015). https://doi.org/10.1007/978-3-319-24574-4_28

11. Searson, D.P.: GPTIPS 2: an open-source software platform for symbolic data mining. In: Gandomi, A.H., Alavi, A.H., Ryan, C. (eds.) Handbook of Genetic Programming Applications, pp. 551–573. Springer, Cham (2015). https://doi.org/10.1007/978-3-319-20883-1_22

12. Tsai, C.L., Yang, Y.L., Chen, S.J., Chan, C.H., Lin, W.Y.: Automatic characterization of classic choroidal neovascularization by using AdaBoost for supervised learning. Investig. Ophthalmol. Vis. Sci. **52**(5), 2767–2774 (2011)

13. Wang, Z., Bovik, A.C., Sheikh, H.R., Simoncelli, E.P.: Image quality assessment: from error visibility to structural similarity. IEEE Trans. Image Process. **13**(4), 600–612 (2004)

14. Zhen, X., Islam, A., Bhaduri, M., Chan, I., Li, S.: Direct and simultaneous four-chamber volume estimation by multi-output regression. In: Navab, N., Hornegger, J., Wells, W.M., Frangi, A.F. (eds.) MICCAI 2015. LNCS, vol. 9349, pp. 669–676. Springer, Cham (2015). https://doi.org/10.1007/978-3-319-24553-9_82

Joint Segmentation and Uncertainty Visualization of Retinal Layers in Optical Coherence Tomography Images Using Bayesian Deep Learning

Suman Sedai[✉], Bhavna Antony, Dwarikanath Mahapatra, and Rahil Garnavi

IBM Research - Australia, Melbourne, VIC, Australia
ssedai@au1.ibm.com

Abstract. Optical coherence tomography (OCT) is commonly used to analyze retinal layers for assessment of ocular diseases. In this paper, we propose a method for retinal layer segmentation and quantification of uncertainty based on Bayesian deep learning. Our method not only performs end-to-end segmentation of retinal layers, but also gives the pixel wise uncertainty measure of the segmentation output. The generated uncertainty map can be used to identify erroneously segmented image regions which is useful in downstream analysis. We have validated our method on a dataset of 1487 images obtained from 15 subjects (OCT volumes) and compared it against the state-of-the-art segmentation algorithms that does not take uncertainty into account. The proposed uncertainty based segmentation method results in comparable or improved performance, and most importantly is more robust against noise.

Keywords: Retinal imaging · Bayesian segmentation · OCT
Fully convolution neural networks · Uncertainty maps

1 Introduction

Optical Coherence Tomography (OCT) is a popular non-invasive imaging modality for retinal imaging. OCT provides volumetric scans of retinal layers for the diagnosis and evaluation of different diseases such as Glaucoma and Age regated macular degeneration (AMD). For example, [1] have shown the correlation between outer retinal layer thickness and visual acuity in early AMD patients. It has also been shown that retinal layer features can be used to predict vision loss and progression [6].

The segmentation of retinal layers in OCT has been tackled in a number of ways, such as dynamic programming [13], graph-based shortest path algorithms [4], graph-based minimum s-t cut formulations [8] and level sets [3,14]. Machine-learning based approaches have also been proposed, where the retinal layer and boundary probability maps are detected using a trained classifier. The final segmentation is then obtained by imposing a model such as active contours [20] or minimum s-t cut framework [12] on the soft labels.

D. Stoyanov et al. (Eds.): COMPAY 2018/OMIA 2018, LNCS 11039, pp. 219–227, 2018.
https://doi.org/10.1007/978-3-030-00949-6_26

In the past few years, Convolutional Neural Networks (CNNs) based methods such as Unet [15,17] and fully convolutional Densenet (FC-DN) [10] have achieved remarkable performance gain in medical image and natural image segmentation. The networks are trained end-to-end, pixels-to-pixels on semantic segmentation exceeded the most state-of-the-art methods without further machinery. [2,16] used Unet like network to perform pixelwise semantic segmentation of retinal layers. In another approach, [5], used CNN and graph search method for layer boundary classification. Once trained, these methods acts as a black box where one has to assume that the segmentation output is accurate which is not always the case. For example, the model will produce incorrect segmentation when the test image is different from the distribution of images used to train the model. This may happen when the model is trained using limited number of training images. In other scenario, the model will produce inaccurate segmentation when trained using normal images, yet pathologies are observed in the test image or the test image is noisy. Quantification of uncertainties associated with the segmentation output is therefore important to determine the region of incorrect segmentation, e.g., region associated with higher uncertainty can either be excluded from subsequent analysis or highlighted for manual attention. In another scenario, when the retinal layer segmentation map is used to diagnose the diseases such as AMD and Glaucoma, the uncertainty map can be used to determine the confidence of final automatic or clinical diagnosis.

Previous works have explored the uncertainty quantification in biomedical segmentation [9], however, these approaches do not utilize the representative power of deep learning. Recent research has shown that Bayesian probability theory offers a mathematically grounded technique to reason about uncertainty in deep learning models [7,11]. In this paper, we explore Bayesian fully convolutional neural network for segmentation and uncertainty quantification of retinal layers in OCT images. We experimentally demonstrate that in addition to the uncertainty based confidence measure, our method provides improved layer segmentation accuracy and robustness towards noise in the test images.

2   Methodology

We model two types of uncertainties for retinal layer segmentation; epistemic uncertainty and aleatoric uncertainty. The epistemic uncertainty captures the uncertainty related to the model parameters, e.g., when the model does not take into account certain aspect of the training data. Therefore, the epistemic uncertainty can be reduced by training the model using more images. Aleoretic uncertainty, on the other hand, captures the noise inherent in the images, therefore, it cannot be reduced with additional training images. We model the aleatoric uncertainty as an additional output variance for both deep learning networks.

We enhance fully convolutional Densenet (FC-DN) [10] for segmentation and uncertainty quantification of retinal layers. FC-DN is a fully convolutional neural network with several dense-blocks connected in encoder-decoder architecture with skip connections across them which effectively combines coarse semantic features with fine image details for pixel-wise semantic segmentation. Each layer in the *dense block* is connected to all the preceding layers by iterative concatenation of previous *feature maps*. This allows all layers to access *feature maps* from their preceding layers which encourages heavy feature reuse. As a result, FC-DN uses less parameter and is less prone to over-fitting. The networks is then trained using the proposed class weighted Bayesian loss function by taking into account the output variance which is described in Sect. 2.1. Once the networks are trained, in the test phase, we use *dropout variational inference* technique [7] to compute the epistemic uncertainty which we describe in Sect. 2.2.

Let $F_{\mathbf{W}}(X)$ be a FC-DN model parameterized by $\mathbf{W}$ which takes input image X and produces the *logit* vector $\mathbf{z}$ for each pixel as $\mathbf{z} = F_{\mathbf{W}}(X)$. The logit vector $\mathbf{z}$ consists of *logits* for each class as $\mathbf{z} = (z_1,...z_C)$ where C is the number of classes i.e., number of retinal layers for segmentation. The final probability vector for a pixel $\mathbf{y} = (y_1,...y_C)$ can be computed by applying the *softmax* function over the *logits* as $\mathbf{y} = \text{Softmax}(\mathbf{z})$. The *softmax* function gives the relative probabilities between classes, but fails to measure the model's uncertainty.

2.1 Bayesian Fully Convolution Network

Here we present a method to convert FC-DN to output the pixelwise uncertainty map in addition to the pixel-wise segmentation map. We name the proposed method Bayesian FC-DN (BFC-DN). In BFC-DN, we apply 1x1 convolution to the feature maps of last layers followed by *softplus* activation to output the variance $\mathbf{v}$ for each pixel in addition to the *logit* vector $\mathbf{z}$ i.e., $(\mathbf{z}, \mathbf{v}) = F_{\mathbf{W}}(X)$. This variance gives aleatoric uncertainty of the model which the network learns to predict during the training. In addition, we include the *dropout layer* before every convolution layer which allows us to compute *epistemic uncertainty* which will be described in Sect. 2.2.

The output of the model is the Gaussian distribution $\mathcal{N}(\mathbf{z}, \mathbf{v})$. Computing the categorical cross entropy loss over this distribution is not feasible. Therefore, we approximated it using the monte-carlo integration. Given a set of training images and corresponding ground truth segmentation mask, $D = \{X_n, Y_n\}_{n=1}^{N}$, output *logit* for each sample in the mini-batch is perturbed T times with a Gaussian noise $\epsilon_t \sim \mathcal{N}(0, \mathbf{v})$ as $\hat{\mathbf{z}}_{\mathbf{t}} = \mathbf{z} + \epsilon_t$ and the final pixel-wise bayesian loss is computed as:

$$L(W) = -\frac{1}{T} \sum_{t=1}^{T} \sum_{c=1}^{C} \beta_c \sum_{\forall Y_c} \log y_c^t \tag{1}$$

where y_c^t is obtained by applying *softmax* to the *logit* vector $\hat{\mathbf{z}}_{\mathbf{t}}$; Y_c denotes the pixels region of the cth class in the ground truth Y and the scale factor $\beta_c = 1/|Y_c|$ weights the contribution of each class to mitigate the class imbalances of different OCT layers and the background by increasing the weight of under represented classes while decreasing the effect of over represented classes. The proposed Bayesian loss function encourage the network to minimize the larger losses by increasing the variance, therefore is more robust towards noise.

We train the proposed BFC-DN using the *bayesian loss* given by Eq. 1 for 40000 iterations. We have used mini-batch gradient descent and the Adam optimizer with momentum and a batch size of 2. The learning rate is set to 10^{-5} which is decreased by one tenth after 10000 iterations of the training. Data augmentation is an important step in training deep networks. We augment the training images and corresponding label map masks through a mirror-image reflection and random rotation within the range of $[-15, 15]$ degrees.

2.2   Segmentation and Uncertainty Quantification

Epistemic uncertainty is generally computed by assuming distribution over the network weights which allows the computation of distribution of class probabilities rather than point estimate [18]. Such methods require optimization over weights distribution and therefore is computationally expensive [7]. We adopt more practical approach introduced by [7] which is based on the *dropout variational inference*. We train the BFC-DN with a *dropout* layer before every convolution layer and use the *dropout* in test phase as well. Specifically, segmentation samples from the output predictive distribution are obtained by performing T stochastic forward passes through the network, i.e., $(\mathbf{z}^t, \mathbf{v}^t) = F_{\hat{\mathbf{W}}_t}(X), t = 1, \cdots, T$ where $\hat{\mathbf{W}}_t$ is an effective network weight after the *dropout*. In each forward pass, the fraction of network weights (denoted by *dropout rate*) are disabled and the segmentation score is computed using only the remaining weights. The segmentation score vector $\bar{\mathbf{y}}$ and the aleatoric variance $\bar{\mathbf{v}}$ is obtained by averaging the T samples, via *monte carlo integration*:

$$\bar{\mathbf{y}} = \frac{1}{T} \sum_{t=1}^{T} \mathrm{Softmax}(\mathbf{z}^t) \tag{2}$$

$$\bar{\mathbf{v}} = \frac{1}{T} \sum_{t=1}^{T} \mathbf{v}^t \tag{3}$$

The average score vector contains the probability score for each retinal layers class, i.e. $\bar{\mathbf{y}} = [\bar{y}_1, \cdots, \bar{y}_C]$. The overall segmentation uncertainty for each pixel can then be obtained as:

$$U(\bar{\mathbf{y}}) = - \sum_{c=1}^{C} \bar{y}_c \log \bar{y}_c + \bar{\mathbf{v}} \tag{4}$$

where the first term denotes *epistemic uncertainty* of the score computed as the entropy of the average score vector obtained by averaging T stochastic predictions (Eq. 2) and the second term is the uncertainty output produced by the network itself (Eq. 3). We set the *dropout* rate $= 0.4$ and $T = 50$ to allow sufficient sampling of network weights for final prediction.

For uncertain predictions, network assigns higher probabilities to different classes for different forward passes, resulting in higher epistemic uncertainty given by Eq. 4. For the certain predictions, network assigns higher probability to the true class for different forward passes, resulting in lower epistemic uncertainty. Since epistemic uncertainty is related to the model parameters weights, it can be reduced by observing more data. This is because, the network becomes robust towards weight dropout in test phase as it observes more data.

3   Experiments

The dataset [19] consists of 1487 images from 15 spectral-domain optical coherence tomography (OCT) volumes from unique normal subjects acquired on a Spectralis scanner. The size of each volume is $512 \times 496 \times N_{slices}$ where N_{slices} is different for each volume and ranges from 49–100. All scans have axial resolution of $3.87\,\mu m$. The ground truth has been obtained by manual annotation of the nine boundaries from eight retinal layers [12]. To facilitate the pixel-wise semantic segmentation, we convert the layer boundaries to the probability map for the eight layers regions and the background region. Therefore, the number of classes is $C = 9$.

Out of 1487 images, we select 1116 images from 12 volumes to create a training set and remaining 291 images from 3 volumes for validation. We compare our method with the baseline FC-DN [10] which do not take into account uncertainty, i.e., the networks do not output aleatoric variance and segmentation is performed in a single forward pass by disabling the *dropout* is the test phase. To train these networks, we use non-bayesian class weighted cross entropy loss function which can be derived by setting $T = 1$ and $v = 0$ in Eq. 1.

Table 1. Performance of our proposed retinal layer segmentation method compared with the state-of-the-art Jégou et al. [10] and Lang et al. [12] segmentation methods.

Layer	Dice Coefficient		Boundary	Absolute error μm	
	FC-DN [10]	**BFC-DN**		Lang et al. [12]	**BFC-DN**
RNFL	0.94 ± 0.01	0.95 ± 0.01	ILM	2.6 ± 3.89	1.81 ± 4.12
GCL+IPL	0.96 ± 0.01	0.97 ± 0.01	RNFL-GCL	4.0 ± 6.11	3.6 ± 6.3
INL	0.91 ± 0.01	0.93 ± 0.01	IPL-INL	3.78 ± 4.41	2.6 ± 2.73
OPL	0.90 ± 0.01	0.91 ± 0.01	INL-OPL	3.66 ± 3.84	2.9 ± 2.8
ONL	0.96 ± 0.008	0.96 ± 0.005	OPL-ONL	3.4 ± 4.24	2.64 ± 2.54
IS	90 ± 0.01	0.91 ± 0.01	ELM	2.79 ± 2.68	2.44 ± 2.4
OS	0.91 ± 0.01	0.92 ± 0.01	IS-OS	2.38 ± 2.49	2.1 ± 2.3
RPE	0.95 ± 0.01	0.96 ± 0.008	OS-RPE	4.16 ± 4.13	3.8 ± 3.3
–	–	–	BrM	3.87 ± 3.69	3.14 ± 2.81

Table 1 compares the average Dice coefficient (DC) between the ground truth and predicted segmentation of the 8 layers using the proposed Bayesian method (BFC-DN) and non-Bayesian method (FC-DN [10]). The proposed method BFC-DN resulted in highest DC of 0.97 for GCL+IPL layer and lowest DC of 0.91 for OPL and IS layer. Moreover, BFC-DN resulted in improved segmentation for most of the layers in comparison to FC-DN. Table 1 also compares the average absolute error for 9 boundaries of our method with [12]. We observe that BFC-DN resulted in lower error than [12] which indicates proposed uncertainty based method is effective in segmenting retinal layers.

Figure 1 shows the examples of segmentation and uncertainty maps produced by our proposed method on few images from the validation set. It can be seen that our method produces pixel-wise uncertainty associated with the segmentation output where high uncertainty correlates with the inaccurate segmentation in the corresponding region. In order to validate the robustness of our proposed method against noise, we evaluate the performance by adding random block noise to the test images as shown in the last row image of Fig. 1. We observe that BFC-DN performs much better than FC-DN in presence of large noise levels as shown in Fig. 2. This demonstrates that BFC-DN is more robust towards the noisy images than FC-DN.

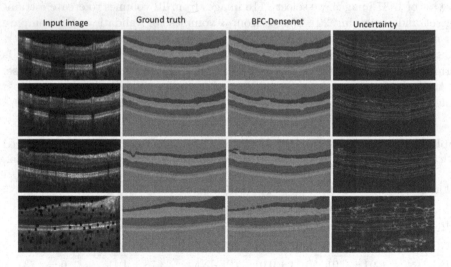

Fig. 1. Examples of retinal layer segmentation and uncertainty quantification using proposed BFCN-Densenet. (a) test images, (b) ground truth, (c) predicted segmentation map from FBC-Densenet (d) uncertainty map (warmer color denotes regions with higher uncertainty). The last row shows an example of layer segmentation in test images with added random block noise.

The average execution time for the retinal layer segmentation for BFC-DN is 2.5 s per image Tesla-K40 GPU which is somewhat slower than that of FC-DN

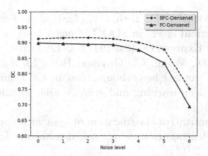

Fig. 2. Comparison of average segmentation performance of proposed BFC-DN with FC-DN [10] for different noise levels. The number of random block noise components at a given noise level is double than that at previous level.

which took 300 ms. This is because our model requires T forward passes in the test phase in contrast to FC-DN which requires one forward pass.

4   Conclusion

In this paper, we proposed a Bayesian deep learning based method for retinal layer segmentation in OCT images. Our method produces layer segmentation and corresponding uncertainty maps depicting the pixel-wise confidence measure of the segmentation output. Experimental results demonstrate that our method compares favorably with non-bayesian DL methods, particularly in the presence of noise and outperforms sate of the art boundary based segmentation method. We have shown qualitatively that the resulting uncertainty maps correlates with the inaccuracies in segmentation output. The proposed method is applicable in determining the confidence of image analysis modules that utilizes the segmentation output for downstream analysis. Such uncertainty visualization can also be useful in computer-assisted diagnostic systems where clinician have additional insight about various measurements generated by the system to make necessarily adjustments and make more informed decisions Also, the resulting uncertainty map can be integrated within active learning systems to correct the segmentation output.

References

1. Acton, J.H., Smith, R.T., Hood, D.C., Greenstein, V.C.: Relationship between retinal layer thickness and the visual field in early age-related macular degeneration. Investig. Ophthalmol. Vis. Sci. **53**(12), 7618–7624 (2012)
2. Apostolopoulos, S., Zanet, S.D., Ciller, C., Wolf, S., Sznitman, R.: Pathological OCT retinal layer segmentation using branch residual u-shape networks. CoRR abs/1707.04931 (2017)
3. Carass, A., Lang, A., Hauser, M., Calabresi, P.A., Ying, H.S., Prince, J.L.: Multiple-object geometric deformable model for segmentation of macular OCT. Biomed. Opt. Express **5**(4), 1062 (2014)

4. Chiu, S.J., Li, X.T., Nicholas, P., Toth, C.A., Izatt, J.A., Farsiu, S.: Automatic segmentation of seven retinal layers in SDOCT images congruent with expert manual segmentation. Optics Express **18**(18), 19413–19428 (2010)
5. Fang, L., Cunefare, D., Wang, C., Guymer, R.H., Li, S., Farsiu, S.: Automatic segmentation of nine retinal layer boundariès in OCT images of non-exudative AMD patients using deep learning and graph search. Biomed. Opt. Express **8**(5), 2732 (2017)
6. Farsiu, S., et al.: Quantitative classification of eyes with and without intermediate age-related macular degeneration using optical coherence tomography. Opthamalogy **121**(1), 162–172 (2014)
7. Gal, Y., Ghahramani, Z.: Dropout as a Bayesian approximation: representing model uncertainty in deep learning. In: International Conference on International Conference on Machine Learning, pp. 1050–1059 (2016)
8. Garvin, M.K., Abràmoff, M.D., Wu, X., Russell, S.R., Burns, T.L., Sonka, M.: Automated 3D intraretinal layer segmentation of macular spectral-domain optical coherence tomography images. IEEE Trans. Med. Imaging **28**(9), 1436–1447 (2009)
9. Iglesias, J.E., Sabuncu, M.R., Leemput, K.V.: Improved inference in bayesian segmentation using monte carlo sampling: application to hippocampal subfield volumetry. Med. Image Anal. **17**(7), 766–778 (2013)
10. Jégou, S., Drozdzal, M., Vázquez, D., Romero, A., Bengio, Y.: The one hundred layers tiramisu: fully convolutional densenets for semantic segmentation. In: 2017 IEEE Conference on Computer Vision and Pattern Recognition Workshops, CVPR Workshops, Honolulu, HI, USA, 21–26 July 2017, pp. 1175–1183 (2017)
11. Kendall, A., Gal, Y.: What uncertainties do we need in bayesian deep learning for computer vision? In: Neural Information Processing Systems (NIPS) (2017)
12. Lang, A., et al.: Retinal layer segmentation of macular OCT images using boundary classification. Biomed. Opt. Express **4**(7), 1133–1152 (2013)
13. Mishra, A., Wong, A., Bizheva, K., Clausi, D.A.: Intra-retinal layer segmentation in optical coherence tomography images. Opt. Express **17**(26), 23719–28 (2009)
14. Novosel, J., Thepass, G., Lemij, H.G., de Boer, J.F., Vermeer, K.A., van Vliet, L.J.: Loosely coupled level sets for simultaneous 3D retinal layer segmentation in optical coherence tomography. Med. Image Anal. **26**(1), 146–158 (2015)
15. Ronneberger, O., Fischer, P., Brox, T.: U-Net: convolutional networks for biomedical image segmentation. In: Navab, N., Hornegger, J., Wells, W.M., Frangi, A.F. (eds.) MICCAI 2015. LNCS, vol. 9351, pp. 234–241. Springer, Cham (2015). https://doi.org/10.1007/978-3-319-24574-4_28
16. Roy, A.G., et al.: Relaynet: retinal layer and fluid segmentation of macular optical coherence tomography using fully convolutional network. CoRR abs/1704.02161 (2017)
17. Sedai, S., Tennakoon, R., Roy, P., Cao, K., Garnavi, R.: Multi-stage segmentation of the fovea in retinal fundus images using fully convolutional neural networks. In: ISBI, pp. 1083–1086, April 2017
18. Springenberg, J.T., Klein, A., Falkner, S., Hutter, F.: Bayesian optimization with robust Bayesian neural networks. In: Lee, D.D., Sugiyama, M., Luxburg, U.V., Guyon, I., Garnett, R. (eds.) Advances in Neural Information Processing Systems 29, pp. 4134–4142. Curran Associates, Inc. (2016)

19. Srinivasan, P.P., et al.: Fully automated detection of diabetic macular edema and dry age-related macular degeneration from optical coherence tomography images. Biomed. Opt. Express **5**(10), 3568–3577 (2014)
20. Vermeer, K.A., van der Schoot, J., Lemij, H.G., de Boer, J.F.: Automated segmentation by pixel classification of retinal layers in ophthalmic OCT images. Biomed. Opt. Express **2**(6), 1743–1756 (2011)

cGAN-Based Lacquer Cracks Segmentation in ICGA Image

Hongjiu Jiang[1], Yuhui Ma[1], Weifang Zhu[1], Ying Fan[2], Yihong Hua[2], Qiuying Chen[2], and Xinjian Chen[1(✉)]

[1] School of Electronic and Information Engineering, Soochow University, Suzhou 215006, Jiangsu, China
xjchen@suda.edu.cn
[2] Shanghai General Hospital, Shanghai 200080, China

Abstract. The increasing prevalence of high myopia has raised concern worldwide. In high myopia, myopia macular degeneration (MMD) is a major cause of vision impairment and lacquer crack (LC) is one of the main signs of MMD. Since the development of LC can reflect the severity of MMD, it is important and meaningful to segment LCs. Indocyanine green angiography (ICGA) has been used for visualizing LCs and is considered to be superior to fluorescein angiography (FA). However, LCs segmentation is difficult due to the image blurring and the confusion between LCs and the background. In this paper, we propose an automatic LCs segmentation method based on the improved conditional generative adversarial nets (cGAN). To apply the advanced cGAN on ICGA images, Dice loss function is added to improve the accuracy of segmentation. Experiments on the ICGA images of high myopia denoted that the proposed method can successfully segment LCs with the trained model and achieve better performance than other popular nets.

Keywords: Lacquer crack · Conditional generative adversarial networks Dice loss function

1 Introduction

The prevalence of myopia and high myopia are increasing globally at an alarming rate, with significant increases in the risk for vision impairment from pathologic conditions associated with high myopia, including retinal damage, cataract and glaucoma [1]. Myopia macular degeneration (MMD) is a major cause of vision impairment in high myopia. Lacquer cracks (LCs), signs of MMD, typically present as yellowish to white lines in the posterior segment in eyes with high myopia and are believed to be breaks in the choroid/retinal pigment epithelium (RPE)/Bruch membrane [2]. The prevalence of LC is 4.3%–9.2% in high myopia eyes [3]. Patients with LCs are at high risk of visual impairment because LCs may lead to further adverse changes in the fundus, such as patchy chorioretinal atrophy or myopic choroidal neovascularization [4]. Thus, the segmentation of LCs are quite important in clinical ophthalmology, which helps doctors diagnose and analyze the development of MMD.

© Springer Nature Switzerland AG 2018
D. Stoyanov et al. (Eds.): COMPAY 2018/OMIA 2018, LNCS 11039, pp. 228–235, 2018.
https://doi.org/10.1007/978-3-030-00949-6_27

Indocyanine green angiography (ICGA) is considered to be the ground truth for LCs detection. It provides details of choroidal vasculature in high myopia eyes and allows observation of the location and extent of LCs much more clearly than fundus photography and typical fluorescein angiography (FA) [2, 5, 6].

There is little studies on LCs segmentation recently. In order to achieve the accurate segmentation of LCs in ICGA images, we propose a novel method based on conditional generative adversarial networks (cGAN) [7]. As generative adversarial networks (GANs) [8] learn a generative model of data, cGANs learn a condition generative model. This makes cGANs much more suitable for image segmentation tasks, where we condition on an input image and generate the corresponding output segmentation image. Previous cGANs have tackled inpainting [9], image prediction from a normal map [10], image manipulation guided by user constraint [11], future frame prediction [12], etc. We are the first to apply cGAN on LCs segmentation. According to the features of ICGA images with LCs, Dice loss function [13] is added in the cGAN model to deal with the situation where there is a strong imbalance between the number of object the background pixels so that the generator can finally achieve better segmentation.

2 Method

2.1 Conditional Generative Adversarial Networks and Improvements

Image-conditional generative adversarial nets consists of two adversarial models: a generative model G that extracts the image features and generates fake images, and a discriminator D that estimates the probability that the image came from the training data rather than the generator.

It is easier to understand the network through the diagram below (see Fig. 1). cGANs learn a mapping from original image x and random noise vector z to the ground truth y. The generator G is trained to produce outputs that cannot be distinguished by real images, while the discriminator D is trained to detect the fake images produced by generator. This training process is diagrammed in Fig. 1.

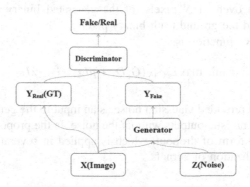

Fig. 1. Diagram of conditional GAN.

The objective function of a cGAN can be expressed as following [7]:

$$L_{cGAN}(G,D) = E_{x,y \sim p_{data}(x,y)}[\log D(x,y)] + E_{x \sim p_{data}(x), z \sim p_z(z)}[\log(1 - D(x, G(x,z)))]$$

(1)

Since the generator tries to minimize the objective function against the adversarial discriminator that tries to maximize it, the final objective function is:

$$F = \arg \min_G \max_D L_{cGAN}(G,D)$$

(2)

Previous approaches to cGANs have found it beneficial to mix the GAN objective with a traditional loss, such as L1 and L2 loss functions [9]. By adding L1 or L2 loss, discriminator's job remains unchanged, but the generator is tasked to produce not only undistinguishable fake images but also images much more similar to the ground truth. According to previous work [7], L1 loss, which encourages less blurring than L2 loss, is adopted in this paper.

To apply cGANs on LCs segmentation, improvement is made based on the objective function. In ICGA images, LCs usually occupy a relatively small part of whole image. This data imbalance problem often causes the learning process to get trapped in a local minimum of the loss function and finally achieves predictions which are mainly biased to backgrounds. Also, it may just mistake the vessel, choroidal hemorrhage and the shade at the edge of images for LCs. To solve this problem, Dice loss function is added in the objective function. Dice loss function can effectively deal with the imbalance between the number of object and background pixels and finally make the proposed segmentation much more accurate.

Thus, both L1 loss function and Dice loss function [13] as follows are adopted:

$$L_{L1}(G) = E_{x,y \sim p_{data}(x,y), z \sim p_z(z)}[\|y - G(x,z)\|_1]$$

(3)

$$L_{Dice}(G) = E_{x,y \sim p_{data}(x,y), z \sim p_z(z)}[1 - \frac{2\sum_i^N y_i G(x,z)_i}{\sum_i^N y_i^2 + \sum_i^N G(x,z)_i^2}]$$

(4)

where the sums run over all N pixels, of the generated binary segmentation pixel $G(x,z)_i \in G(x,z)$ and the ground truth binary pixel $y_i \in y$.

The final objective function is:

$$F = \arg \min_G \max_D L_{cGAN}(G,D) + \mu L_{L1}(G) + \lambda L_{Dice}(G)$$

(5)

Past cGANs [10] provided Gaussian noise as an input to the generator since the net would produce deterministic outputs without the noise. In the proposed net, we provide random noise in the form of dropout, which is applied in several layers of our generator, and the initialization of kernels.

2.2   Network Architectures

The architecture of our network including generator and discriminator is illustrated in Figs. 2 and 3. The generator in Fig. 2 is similar to the traditional encoder-decoder architecture. Each encoding layer is a convolution layer with batch normalization and ReLU activation function. Each layer of decoder consists of a deconvolution, batch normalization and ReLU activation function. Dropout with a rate of 50% is adopted in the first three decoding layers to efficiently prevent the overfitting during training. In practise, leaky ReLU function, a variant of ReLU function, is adopted with the slope of 0.2. Leaky ReLU function can be used to mitigate the vanishing gradient problem and makes the network converge much faster during training.

All convolutions and deconvolutions are 4×4 spatial filters applied with stride 2. Compared to other traditional deep convolution networks, our network adopt filters in convolution layer applied with stride 2 instead of pooling layer to reduce the spatial size of the representation, since discarding the pooling layer performs better in training good generative models [14].

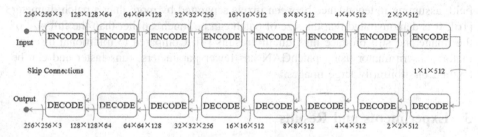

Fig. 2. Architecture of generator.

We adopt the recent popular U-Net [15] as our main framework in generator. For medical image segmentation tasks, predicted segmentation shares the structure information with original images. Skip connections can constrain the output to be aligned with the input and make segmentation result more reasonable and accurate. In the proposed generator, high resolution features from the contracting path are combined with the upsampled output by skip connections. Thus, a successive convolution layer can then learn to assemble a more precise output based on this information. Since LCs in ICGA are mostly tiny and irregular, skip connections, which helps generator produce images with more details that look similar to the real LCs, can improve the accuracy of our segmentation.

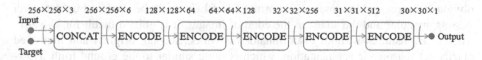

Fig. 3. Architecture of discriminator.

PatchGAN is adopted in the discriminator which is shown in Fig. 3. Traditional discriminator in GANs for image processing estimates the probability that the image is real or fake and outputs a single scalar to represent. In contrast, patchGAN tries to classify if each $N \times N$ patch in an image is real or fake. We run this discriminator convolutionally across the image, averaging all responses to provide the ultimate output possibility.

We create 2 copies of discriminator with the same underlying variables, one for real pairs and one for fake pairs. For real pairs, we concatenate the input and the ground truth first. For fake pairs, we concatenate the input and the output first. They all run over 5 encoding layers right after the concatenation. The convolutions are 4×4 spatial filters with stride 2 except for those in last two layers with stride 1. These two layers applied zero-padding before convolution to change the size of representation in every channel from 32×32 to 30×30 and make the size of reception field, also the N in patchGAN, to be 70, which has better image quality than the full size image [7]. In the final 30×30 image, each pixel represents the probability of a 70×70 patch in the original image.

Discriminator with patchGAN effectively models the image as a Markov random field, assuming independence between pixels separated by more than a patch diameter [16]. It is demonstrated that the size of patch can be much smaller than the full size of the image and still produce high quality results [7]. Compared to traditional discriminators, discriminator using patchGAN has fewer parameters, runs faster and can be applied on arbitrarily large images.

3 Experiments and Results

The proposed network is evaluated on the ICGA data set of patients with LCs. The training data set consists of 22 annotated ICGA images of size 768×768. Because of the lack of training data, we use excessive data augmentation by flipping images vertically and horizontally. During the experiments, 6 images were randomly chosen as the testing images. Data augmentation is applied to the rest 16 images and finally the training set consists of 64 images.

In order to evaluate the performance of segmentation, our method is compared with U-Net, a recently popular network that specializes on biomedical image segmentation, DenseNet [17], a new network which is good at feature extraction, and also the original cGAN without Dice loss function. U-Net with 5 layers and DenseNet with 7 dense blocks are adopted in the comparison, which have the best performance on the ICGA data set. The segmentation results are shown in Fig. 4.

As shown in Fig. 4, original cGAN, improved cGAN and DenseNet perform better than U-Net on segmenting LCs. Since ICGA images with LCs does not include many obvious features and the intensity information of LCs might be confused by nets with vessels, macular and the shade at the edge of images, it is quite difficult for U-Net to extract key features without discriminator networks. Second, original and improved cGAN get reasonable segmentation, which is quite similar to the ground truth. However, original cGAN and DenseNet segment some part of choroidal hemorrhage and

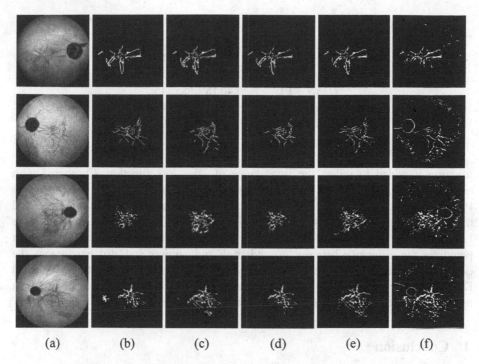

Fig. 4. Segmentation results of different networks. (a) Original ICGA image. (b) The ground truth. (c) Results of original cGAN. (d) Results of the proposed net. (e) Results of DenseNet. (f) Results of U-Net.

retinal vessels as false positives. This situation is drastically suppressed in our proposed net due to the addition of Dice loss function.

To make the comparison more quantitatively, we adopt intersection-over-union (IoU) and pixel accuracy (PA) to evaluate the segmentation results in Table 1. IoU is the standard metric for segmentation purposes. It computes a ratio between the intersection and the union of the ground truth and the predicted segmentation. The ratio can be reformulated as the number of true positives (TP) over the sum of true positives, false positives (FP) and false negatives (FN). Pixel accuracy is another metric, simply computing a ratio between the amount of properly segmented pixels and the total number of them [18].

$$IoU = \frac{TP}{TP + FP + FN} \tag{6}$$

$$PA = \frac{TP}{TP + FP} \tag{7}$$

As we have seen in the Table 1, all three cGANs and DenseNet achieve better segmentation according to both IoU and PA. Results from DenseNet is overall better than original cGAN but worse than the improved cGAN. cGAN with only Dice loss

Table 1. Quantitative segmentation results of different networks.

	IoU	PA
U-Net	22.92%	27.51%
DenseNet	43.44%	59.11%
cGAN + L1	40.15%	55.54%
cGAN + Dice	31.01%	61.84%
cGAN + L1 + Dice	47.38%	69.59%

function is added to reflect the importance of Dice loss function. Compared with original cGAN, segmentation of cGAN with only Dice loss function can achieve higher PA but lower IoU. In theory, Dice loss function adds more strict constraint to net, and the segmentation just contains the most obvious part of LCs and losses some part of the LCs, which is not obvious to the net. L1 loss is also important in cGAN since it penalizes the difference between ground truth and outputs and also encourages the output to be aligned with the input. Finally, the proposed net with both L1 loss and Dice loss achieves better IoU and better PA than other nets. It seems to be the most appropriate method to solve the LCs segmentation problem.

4   Conclusion

We propose an improved conditional GAN to segment LCs in ICGA images. Compared with original cGAN, U-Net and DenseNet, adding Dice loss function can solve the data imbalance problem and optimize the segmentation results. According to the results of experiments on the data set, the segmentation in the proposed network is overall better than other nets.

Acknowledgement. This study was supported in part by the National Basic Research Program of China (973 Program) under Grant 2014CB748600, in part by the National Nature Science Foundation of China for Excellent Young Scholars under Grant 61622114, in part by the National Nature Science Foundation of China under Grants 81371629, 81401472, 61401294, 61401293 and 61601317, and in part by the International Cooperation Project of Ministry of Science and Technology (2016YFE010770).

References

1. The Impact of Myopia and High Myopia: Report of the Joint World Health Organization–Brien Holden Vision Institute Global Science Meeting on Myopia. World Health Organization, Geneva (2015)
2. Ikuno, Y., Sayanagi, K., Soga, K., Sawa, M., Gomi, F., et al.: Lacquer crack formation and choroidal neovascularization in pathologic myopia. Retina **28**, 1124–1131 (2008)
3. Wang, N.K., Lai, C.C., Chou, C.L., Chen, Y.P., Chuang, L.H., Chao, A.N., et al.: Choroidal thickness and biometric markers for the screening of lacquer cracks in patients with high myopia. PLoS One **8**, e53660 (2013)

4. Ohno-Matsui, K., Tokoro, T.: The progression of lacquer cracks in pathologic myopia. Retina **16**, 29–37 (1996)
5. Ohno-Matsui, K., Morishima, N., Ito, M., Tokoro, T.: Indocyanine green angiographic findings of lacquer cracks in pathologic myopia. Jpn. J. Ophthalmol. **42**, 293–299 (1998)
6. Quaranta, M., Arnold, J., Coscas, G., Francais, C., Quentel, G., et al.: Indocyanine green angiographic features of pathologic myopia. Am. J. Ophthalmol. **122**, 663–671 (1996)
7. Isola, P., Zhu, J.-Y., Zhou, T., Efros, A.A.: Image to image translation with conditional adversarial networks. In: CVPR (2017)
8. Goodfellow, I., et al.: Generative adversarial nets. In: NIPS (2014)
9. Pathak, D., Krahenbuhl, P., Donahue, J., Darrell, T., Efros, A.A.: Context encoders: feature learning by inpainting. In: CVPR (2016)
10. Wang, X., Gupta, A.: Generative image modeling using style and structure adversarial networks. In: Leibe, B., Matas, J., Sebe, N., Welling, M. (eds.) ECCV 2016. LNCS, vol. 9908, pp. 318–335. Springer, Cham (2016). https://doi.org/10.1007/978-3-319-46493-0_20
11. Zhu, J.-Y., Krähenbühl, P., Shechtman, E., Efros, A.A.: Generative visual manipulation on the natural image manifold. In: Leibe, B., Matas, J., Sebe, N., Welling, M. (eds.) ECCV 2016. LNCS, vol. 9909, pp. 597–613. Springer, Cham (2016). https://doi.org/10.1007/978-3-319-46454-1_36
12. Mathieu, M., Couprie, C., LeCun, Y.: Deep multi-scale video prediction beyond mean square error. In: ICLR (2016)
13. Milletari, F., Navab, N., Ahmadi, S.-A.: V-Net: fully convolutional neural networks for volumetric medical image segmentation. In: Proceeding of 3D Vision, pp. 565–571. IEEE (2016)
14. Radford, A., Metz, L., Chintala, S.: Unsupervised representation learning with deep convolutional generative adversarial networks. arXiv preprint arXiv:1511.06434 (2015)
15. Ronneberger, O., Fischer, P., Brox, T.: U-Net: convolutional networks for biomedical image segmentation. In: Navab, N., Hornegger, J., Wells, W.M., Frangi, A.F. (eds.) MICCAI 2015. LNCS, vol. 9351, pp. 234–241. Springer, Cham (2015). https://doi.org/10.1007/978-3-319-24574-4_28
16. Li, C., Wand, M.: Precomputed real-time texture synthesis with Markovian generative adversarial networks. In: Leibe, B., Matas, J., Sebe, N., Welling, M. (eds.) ECCV 2016. LNCS, vol. 9907, pp. 702–716. Springer, Cham (2016). https://doi.org/10.1007/978-3-319-46487-9_43
17. Huang, G., Liu, Z., Weinberger, K.Q.: Densely connected convolutional networks. In: CVPR (2017)
18. Garcia-Garcia, A., Orts-Escolano, S., Oprea, S., Villena-Martinez, V., Garcia-Rodriguez, J.: A review on deep learning techniques applied to semantic segmentation. arXiv preprint arXiv:1704.06857 (2017)

Localizing Optic Disc and Cup for Glaucoma Screening via Deep Object Detection Networks

Xu Sun[1], Yanwu Xu[1,4(✉)], Mingkui Tan[2], Huazhu Fu[3], Wei Zhao[1],
Tianyuan You[1], and Jiang Liu[4]

[1] Guangzhou Shiyuan Electronic Technology Company Limited, Guangzhou, China
ywxu@ieee.org
[2] Institute for Infocomm Research, A*STAR, Singapore, Singapore
[3] South China University of Technology, Guangzhou, China
[4] Cixi Institute of Biomedical Engineering, Chinese Academy of Sciences, Cixi, China

Abstract. Segmentation of the optic disc (OD) and optic cup (OC) from a retinal fundus image plays an important role for glaucoma screening and diagnosis. However, most existing methods only focus on pixel-level representations, and ignore the high level representations. In this work, we consider the high level concept, *i.e.*, objectness constraint, for fundus structure analysis. Specifically, we introduce a deep object detection network to localize OD and OC simultaneously. The end-to-end architecture guarantees to learn more discriminative representations. Moreover, data from a similar domain can further contributes to our algorithm through transfer learning techniques. Experimental results show that our method achieves state-of-the-art OD and OC segmentation/localization results on ORIGA dataset. Moreover, the proposed method also obtains satisfactory glaucoma screening performance with the calculated vertical cup-to-disc ratio (CDR).

1 Introduction

As the second leading cause of blindness, glaucoma is predicted to affect about 80 million people by 2020 [7]. Since damage to optic nerves cannot be reversed, early detection of glaucoma is critical in preventing further deterioration. The vertical cup-to-disc ratio (CDR) is a commonly-used metric for glaucoma screening. Thus, accurate segmentation of the optic disc (OD) and optic cup (OC) is essential for developing practical automated glaucoma screening systems. Most existing methods tackle this challenging problem by using traditional segmentation techniques like thresholding, edge-based and region-based methods [5,13]. While these solutions work well with images of healthy retina, they tend to be misleading in illness cases where retinas suffer from different types of retinal lesions (*e.g.*, drusen, exudates, hemorrhage, *etc.*). Alternatively, some methods based on conventional machine learning pipelines have been proposed [6]. However, since these approaches rely too heavily on handcrafted features, their applicability is limited. A promising way to improve the performance is to employ

© Springer Nature Switzerland AG 2018
D. Stoyanov et al. (Eds.): COMPAY 2018/OMIA 2018, LNCS 11039, pp. 236–244, 2018.
https://doi.org/10.1007/978-3-030-00949-6_28

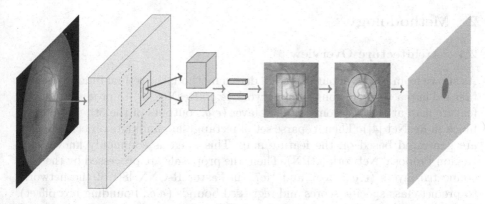

Fig. 1. Architecture of the proposed method for OD and OC segmentation/localization, where purple and magenta regions denote OD and OC respectively. (Color figure online)

the deep neural networks (DNN) architectures as they are capable of learning more discriminating features.

The effectiveness of DNN structure, indeed, has been well demonstrated in a recent state-of-the-art work termed M-Net [3]. Nevertheless, like most of the existing algorithms, M-Net is still a pixel-wise classification based approach, which first classifies each pixel as one of the three classes, *i.e.*, OD, OC and non-target, and then uses ellipse-fitting to approximate the smooth boundaries of OD and OC. In fact, the ellipse fitting step can be easily bypassed if OD and OC are assumed to be in a non-rotated ellipse shape. Therefore, the OD and OC can be treated as a whole object instead of a bunch of pixels without objectness constraint, which enables tackling the segmentation task from an object detection perspective. Follow this basic idea, two typical methods are presented in literature [12,14]. Unfortunately, these two methods are initially developed for OC localization only and not easy to adapt to OD localization.

In this paper, we formulate the OD and OC segmentation as a multiple object detection problem, with the introduction of the objectness constraints to improve the accuracy. Different from tradition pixel-wise based two-step approaches, we propose a simple yet effective method to jointly localize/segment OD and OC in a retinal fundus image based on deep object detection networks. The proposed method inherently holds four desirable features: (1) the multi-object network involves the OD and OC relationship and localizes them simultaneously; (2) the object detection network contains the objectness property, which presents the high-level discriminate representation; (3) the end-to-end architecture guarantees learning image features automatically, and also allows for transfer learning to address the challenging of small scale data; (4) by simply using Faster R-CNN [8] as the deep object detector, our method outperforms state-of-the-art OC and/or OD segmentation/localization methods on ORIGA dataset, and obtains satisfactory glaucoma screening performances with calculated CDR on ORIGA and SCES datasets.

2   Methodology

2.1   Architecture Overview

As shown from Fig. 1, in our detection driven method, the retinal fundus image is first fed into a deep convolutional network (*e.g.*, ResNet [4]) to produce a shared feature map at the last convolutional layer (*e.g.*, outputs of the 5th convolution block in ResNet [4]). Then a sparse set of rectangular candidate object locations are generated based on the feature map. This stage is commonly known as a Region Proposal Network (RPN). Then, the proposals are processed by the fully connected layers (*e.g.*, "fc6" and "fc7" in Faster R-CNN [8]) of the networks to predict class-specific scores and regressed bounds (*e.g.*, bounding box offset). For each foreground class (*i.e.*, OD and OC), we keep the bounding box with the highest confidence score as the final output of the detector.

Provided these two detected bounding boxes, the next stage is how to generate satisfactory OD and OC boundaries. It is widely accepted by many ophthalmologists and researchers that the shape of OD and OC can be well approximated by a vertical ellipse. Inspired by this concept, we propose to obtain the OD and OC boundaries by simply redrawing the predicted bounding boxes as vertical ellipses. The fundus image segmentation problem thus reduces to a relatively more straightforward localization task in our setting.

2.2   Implementation

In this paper, we adopt *Faster R-CNN* [8] as the object detector due to its flexibility and robustness comparing to many follow-up architectures. Faster R-CNN consists of two stages. During training, the loss for the first stage RPN is defined as

$$L(\{p_i, t_i\}) = \beta \sum_i L_{\text{cls}}(p_i, p_i^*) + \gamma \sum_i p_i^* L_{\text{reg}}(b_i, b_i^*) \tag{1}$$

where β, γ are weights balancing localization and classification losses. i is the index of an anchor in a training mini-batch. p_i is the predicted probability of the ith anchor being OD/OC. The ground truth label p_i^* indicates if the overlapping ratio between the anchor and the manual OD/OC mask is either larger than an given threshold (*e.g.*, 0.3) or the largest among all anchors. b_i is a vector standing for the 4 coordinates of the predicted bounding box, and b_i^* is that of the ground-truth box associated with a positive anchor (*i.e.*, with $p_i^* = 1$). The classification loss L_{cls} is the log loss over target and non-target classes, and the regression loss is a robust loss function (*e.g.*, the smooth L_1 loss). We refer readers [8] to for the more details of these entries. Meanwhile, the loss function for the second stage box classifier also takes a similar form of (1) using proposals produced from the RPN as anchors.

Data Augmentation: We employ two distinct forms of data augmentation in our experiment. The first form is to rotate fundus images from the training set using a set of angles over $-10(2)10$ degrees, where the notation $N_1(\Delta)N_2$

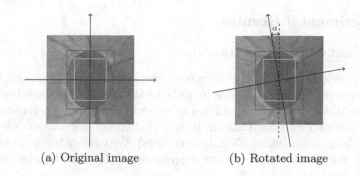

(a) Original image   (b) Rotated image

Fig. 2. The generated "ground truth" OD and OC bounding boxes for the augmented image (*right*) from the original manual segmentation masks (*left*), where purple and magenta regions denote OD and OC respectively. The right image is obtained by rotating the whole original fundus image with an angle α regarding to its center. (Color figure online)

represents a list ranging from N_1 to N_2 with an increment of Δ. We limit the degree of rotation into such a small interval because of the assumption that the OD and OC are in a vertical ellipse shape. The second form is to generate image horizontal reflections on both the original training set and its rotated counterparts. With this transformation operation, a left eye image is artificially turned into the "right eye" image, and vice versa. This is desirable as we now have a balanced training set that consists of equal number of images from the left eye and the right eye. These two augmentation schemes increase the amount of our training set by a factor of 20.

Training Details: To enable training of the deep object detector, we first need to transform the manual segmentation masks into the "ground truth" bounding boxes. As illustrated in Fig. 2, this can be simply achieved by finding a vertical rectangle whose bounds lie exactly on the edge of the provided mask for each type of targets. Faster R-CNN [8] is implemented using `Tensorflow` based on a publicly available code [1].

We train the detection networks on a single-scale image using a single model. Before feeding images to the detector, we rescale their shorter side to 600 pixels. A 101-layer *ResNet* [4] is used as the backbone of Faster R-CNN. For anchors, we use 5 naive scales with box areas of $32^2, 64^2, 128^2, 256^2$, and 512^2 pixels, and 3 naive aspect ratios of $1:1, 1:2$, and $2:1$. Instead of training all parameters from scratch, we fine-tune the network end-to-end from an ImageNet pre-trained model on a single NVIDIA TITAN XP GPU. We use a weight decay of 0.0001 and momentum of 0.9 for optimization. We start with a learning rate of 0.001, divide it by 10 at 100k iterations, and terminate training at 200k iterations.

3   Experimental Results

3.1   OD and OC Segmentation

Following previous work in the literature, we evaluate and compare the OD and OC segmentation performance on ORIGA dataset [14]. In each image, OD and OC are labelled as vertical ellipses by experienced ophthalmologists. These images are divided into 325 training images (including 73 glaucoma cases) and 325 testing images (including 95 glaucoma cases). We employ two measurements to evaluate the performance, the overlapping error (E) and the absolute CDR error (δ) defined as:

$$E = 1 - \frac{A_{\mathrm{GT}} \cap A_{\mathrm{SR}}}{A_{\mathrm{GT}} \cup A_{\mathrm{SR}}}, \text{ and } \delta = |d_{\mathrm{GT}} - d_{\mathrm{SR}}| \tag{2}$$

where A_{GT} and A_{SR} denote the areas of the ground truth and segmented mask, respectively. d_{GT} denotes the manual CDR provided by ophthalmologists, and d_{SR} denotes the CDR that is calculated by the ratio of vertical cup diameter to vertical disc diameter from the segmentation results.

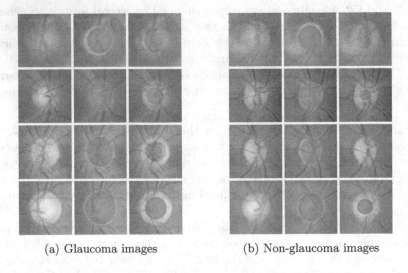

(a) Glaucoma images   (b) Non-glaucoma images

Fig. 3. The segmentation results of the proposed method, where the purple, cyan and blue regions denote the manual masks, the segmentation outputs and their overlapping regions, respectively. From top to bottom rows are images with highest disc overlapping error, lowest disc overlapping error, highest cup overlapping error and lowest cup overlapping error, for cases with and without glaucoma, respectively. The overlapping errors from top to bottom rows, left to right are 0.219, 0.021, 0.096, 0.044, 0.247, 0.119, 0.471, 0.038, 0.264, 0.008, 0.045, 0.062, 0.293, 0.175, 0.752, and 0.035, respectively. (Color figure online)

We compare the proposed method to the state-of-the-art methods in OD and OC segmentation, including the relevant-vessel bends method (R-bend) [5],

Table 1. OD and OC Segmentation Performance Comparison of Different Methods on ORIGA Dataset.

Method	E_{disc}	E_{cup}	δ
R-Bend [5]	0.129	0.395	0.154
ASM [15]	0.148	0.313	0.107
SP [2]	0.102	0.264	0.077
LRR [11]	-	0.244	0.078
SW [14]	-	0.284	0.096
Reconstruction [12]	-	0.225	0.071
U-Net [9]	0.115	0.287	0.102
M-Net [3]	0.083	0.256	0.078
M-Net + PT [3]	0.071	0.230	0.071
Proposed	**0.069**	**0.213**	**0.067**

active shape model (ASM) [15], superpixel-based classification method (SP) [2], low-rank superpixel representation method (LRR) [11], sliding-window based method (SW) [14], reconstruction based method (Reconstruction) [12] and three deep learning based methods, *i.e.*, U-Net [9], M-Net [3] and M-Net with polar transformation (M-Net + PT). As shown in Table 1, our proposed deep object detection based method outperforms all state-of-the-art OD and OC segmentation algorithms on ORIGA dataset in terms of all aforementioned three evaluation criteria. Figure 3 shows some visual outputs of our method.

3.2 Glaucoma Screening/Classification Based on CDR

Following clinical convention, we evaluate the proposed method for glaucoma screening by using the calculated CDR value. Generally, the larger CDR value indicates the higher risk of glaucoma. We train our model using 7,150 images augmented from ORIGA training set, and then test it on ORIGA testing set and the whole SCES dataset [3] individually. We evaluate glaucoma screening/classification performance using the area under Receiver Operating Characteristic curve (AUC). As illustrated in Fig. 4, the AUC values of our method on ORIGA and SCES are 0.845 and 0.898, respectively, which are slightly lower than M-Net. Here we justify that: 1) the major objective of this work is to minimize OD and OC segmentation errors, which are not directly associated to glaucoma classification accuracy; 2) the state-of-the-art method M-Net [3] has no significant difference from our proposed method ($p >> 0.05$ on ORIGA and $p >> 0.05$ on SCES using DeLong's test [10]); 3) on the independent test dataset SCES, our proposed object detection method with objectness constraint achieves consistent higher sensitivity (*i.e.*, true positive rate) than other two competitive methods when false positive rate (*i.e.*, 1-Specificity) is lower than 0.2, which indicates that our approach is promising for practical glaucoma screening.

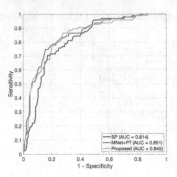 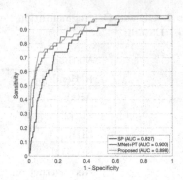

Fig. 4. Glaucoma screening performance on the ORIGA (*left*) and SCES (*right*) datasets.

4   Discussion

To illustrate why the proposed method is more preferable, below we highlight its main features by comparing it with two most related work in literature. The first one is the sliding-window based method [14], which first introduces the concept to segment OC via object detection technique. However, it is only developed for detecting OC after obtaining OD from another individual procedure. Our method, instead, incorporates these two separate tasks into a joint framework. Additionally, the sliding window method relies on handcrafted features. In contrast, our method learns deep representation directly from data. It should be pointed out that a fairly large amount of annotated data is usually required for training a highly accurate deep model, while in practice, such annotated data are expensive to acquire, especially in the field of medical imaging. One typical way of addressing a lack of data problem is by using a technique known as transfer learning and fortunately, this can be easily performed in DNN-based frameworks including our method. We also highlight that the training takes much longer time to converge and can hardly get satisfactory results, when the pre-trained model on ImgaeNet is not used to initialize the networks.

The second work to be compared is M-Net [3], which also trains a DNN for extracting image features and shares some aforementioned advantages of our method. To deploy M-Net, besides the end-to-end U-shape segmentation network, we also require an OD detector for detecting the disc center, a polar transformation method for mapping the disc image from the Cartesian coordinate system to polar coordinate system, an inverse polar transformation operation for recovering the segmentation result back to the Cartesian coordinate system, and an ellipse-fitting for generating smooth boundaries of OD and OC. In contrast, our method requires only a deep object detector.

5   Conclusion

In this paper, we tackle the fundus image segmentation problem from an object detection perspective, based on the circumstance that OD/OC can be well approximated with vertical ellipse shape. The proposed method is not only conceptually simpler but also easier to deploy comparing to other multi-step based approaches such as M-Net [3]. Evaluated on the ORIGA dataset, our method outperforms all existing methods, achieving state-of-the-art segmentation results. Moreover, the proposed method also obtains satisfactory glaucoma screening performance with CDR calculated on the ORIGA and SCES datasets. In the future, we plan to investigate other deep object detectors and to explore more diagnostic indicators for glaucoma screening.

References

1. Chen, X., Gupta, A.: An implementation of faster RCNN with study for region sampling. arXiv preprint arXiv:1702.02138 (2017)
2. Cheng, J., et al.: Superpixel classification based optic disc and optic cup segmentation for glaucoma screening. IEEE Trans. Med. Imaging **32**(6), 1019–1032 (2013)
3. Fu, H., Cheng, J., Xu, Y., Wong, D.W.K., Liu, J., Cao, X.: Joint optic disc and cup segmentation based on multi-label deep network and polar transformation. IEEE Trans. Med. Imaging **37**(7), 1597–1605 (2018)
4. He, K., Zhang, X., Ren, S., Sun, J.: Deep residual learning for image recognition. In: CVPR, pp. 770–778. IEEE (2016)
5. Joshi, G.D., Sivaswamy, J., Krishnadas, S.: Optic disk and cup segmentation from monocular color retinal images for glaucoma assessment. IEEE Trans. Med. Imaging **30**(6), 1192–1205 (2011)
6. Li, A., et al.: Learning supervised descent directions for optic disc segmentation. Neurocomputing **275**, 350–357 (2018)
7. Quigley, H., Broman, A.: The number of people with glaucoma worldwide in 2010 and 2020. Br. J. Ophthalmol. **90**(3), 262–267 (2006)
8. Ren, S., He, K., Girshick, R., Sun, J.: Faster R-CNN: towards real-time object detection with region proposal networks. In: NIPS, pp. 91–99 (2015)
9. Ronneberger, O., Fischer, P., Brox, T.: U-Net: convolutional networks for biomedical image segmentation. In: Navab, N., Hornegger, J., Wells, W.M., Frangi, A.F. (eds.) MICCAI 2015. LNCS, vol. 9351, pp. 234–241. Springer, Cham (2015). https://doi.org/10.1007/978-3-319-24574-4_28
10. Sun, X., Xu, W.: Fast implementation of DeLong's algorithm for comparing the areas under correlated receiver operating characteristic curves. IEEE Sig. Process. Lett. **21**(11), 1389–1393 (2014)
11. Xu, Y., et al.: Optic cup segmentation for glaucoma detection using low-rank superpixel representation. In: Golland, P., Hata, N., Barillot, C., Hornegger, J., Howe, R. (eds.) MICCAI 2014. LNCS, vol. 8673, pp. 788–795. Springer, Cham (2014). https://doi.org/10.1007/978-3-319-10404-1_98
12. Xu, Y., Lin, S., Wong, D.W.K., Liu, J., Xu, D.: Efficient reconstruction-based optic cup localization for glaucoma screening. In: Mori, K., Sakuma, I., Sato, Y., Barillot, C., Navab, N. (eds.) MICCAI 2013. LNCS, vol. 8151, pp. 445–452. Springer, Heidelberg (2013). https://doi.org/10.1007/978-3-642-40760-4_56
13. Xu, Y., et al.: Efficient optic cup detection from intra-image learning with retinal structure priors. In: Ayache, N., Delingette, H., Golland, P., Mori, K. (eds.) MICCAI 2012. LNCS, vol. 7510, pp. 58–65. Springer, Heidelberg (2012). https://doi.org/10.1007/978-3-642-33415-3_8

14. Xu, Y., et al.: Sliding window and regression based cup detection in digital fundus images for glaucoma diagnosis. In: Fichtinger, G., Martel, A., Peters, T. (eds.) MICCAI 2011. LNCS, vol. 6893, pp. 1–8. Springer, Heidelberg (2011). https://doi.org/10.1007/978-3-642-23626-6_1
15. Yin, F., et al.: Model-based optic nerve head segmentation on retinal fundus images. In: EMBC, pp. 2626–2629. IEEE (2011)

Fundus Image Quality-Guided Diabetic Retinopathy Grading

Kang Zhou[1,2(✉)], Zaiwang Gu[2,3], Annan Li[4], Jun Cheng[2], Shenghua Gao[1], and Jiang Liu[2]

[1] School of Information Science and Technology, ShanghaiTech University,
Shanghai, China
zhoukang@shanghaitech.edu.cn
[2] Cixi Institute of Biomedical Engineering,
Ningbo Institute of Materials Technology and Engineering, Ningbo, China
[3] School of Mechatronic Engineering and Automation, Shanghai University,
Shanghai, China
[4] School of Computer Science and Engineering,
Beijing University of Aeronautics and Astronautics, Beijing, China

Abstract. With the increasing use of fundus cameras, we can get a large number of retinal images. However there are quite a number of images in poor quality because of uneven illumination, occlusion and so on. The quality of images significantly affects the performance of automated diabetic retinopathy (DR) screening systems. Unlike the previous methods that did not face the unbalanced distribution, we propose weighted softmax with center loss to solve the unbalanced data distribution in medical images. Furthermore, we propose Fundus Image Quality (FIQ)-guided DR grading method based on multi-task deep learning, which is the first work using fundus image quality to help grade DR. Experimental results on the Kaggle dataset show that fundus image quality greatly impact DR grading. By considering the influence of quality, the experimental results validate the effectiveness of our propose method. All codes and fundus image quality label on Kaggle DR dataset are released in https://github.com/ClancyZhou/kaggle_DR_image_quality_miccai2018_workshop.

Keywords: Fundus image quality classification · DR screening
Multi-task · Deep learning

1 Introduction

The fundus image quality has a significant effect on the performance of automated ocular disease screening, such as diabetic retinopathy (DR), glaucoma and age-related macular degeneration (AMD). The symptoms of the above diseases are well defined and visible in fundus images. Research communities have put great effort towards the automation of a computer screening system which is able to promptly detect DR in fundus images. The evaluation of fundus image

© Springer Nature Switzerland AG 2018
D. Stoyanov et al. (Eds.): COMPAY 2018/OMIA 2018, LNCS 11039, pp. 245–252, 2018.
https://doi.org/10.1007/978-3-030-00949-6_29

Table 1. In our Kaggle DR image quality dataset (Sect. 3.1), the number of good and poor quality images are shown as follows. The ratio is extremely unbalanced.

Data set	Total	Good	Poor	Ratio (poor/good)
Training	35126	33841	1285	0.038
Validation	10906	10680	226	0.021
Testing	42670	41797	873	0.021

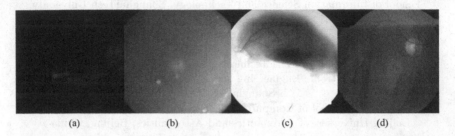

(a)   (b)   (c)   (d)

Fig. 1. Four instances of poor quality images in Kaggle DR dataset, and the quality of these images are too poor to identify the lesion.

quality involves a computer-aided retinal image analysis system that is designed to assist ophthalmologists to detect eye diseases. Consequently, automated evaluations of ophthalmopathy can be performed to support the diagnosis of doctors. However, the success of these automatic diagnostic systems heavily relies on the image quality. In reality, due to some inevitable disturbances in the image acquisition, e.g. the operator's expertise, the type of image acquisition equipment, the situation of different individuals, the images are often blurred, which affects the follow-up diagnosis. Therefore, the image quality plays an extremely important role in the computer-aided screening system (Fig. 1).

In the context of retinal image analysis, image quality classification is used to determine whether an image is useful or the quality of a retinal image is sufficient for the subsequent automated diagnosis. Many methods based on hand-crafted features have been proposed for fundus image quality assessment for disease screening. Lee *et al.* [6] use a quality index Q which is calculated by the convolution of a template intensity histogram to measure the retinal image quality. Lalonde *et al.* [5] adopt the features which are based on the edge amplitude distribution and the pixel gray value to automatically assess the quality of retinal images. Traditional feature extraction methods with low computational complexity only can obtain some characteristic that represents image quality rather than always acquiring diversity factors that affect image quality.

With the development of convolution neural network (CNN) in image and video processing [4], automatic feature learning algorithms using deep learning have emerged as feasible approaches and are applied to handle the medical image analysis. Recently, some methods based on deep learning have been proposed for fundus images [2,3]. Specially, methods to handle the fundus image quality

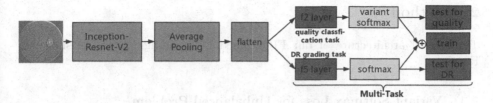

Fig. 2. The overall architecture of our method.

assessment problem also have been proposed. Yu *et al.* [9] first introduced CNN and treated it as a fixed high-level feature extractor, replacing low-level features such as hand-crafted geometric and structural features. Then, SVM algorithm was adopted to automatically classify high quality and poor quality retinal fundus images. Sun *et al.* [7] directly used four CNN architectures to assess fundus images quality. However, in these two papers the authors randomly selecting training set and testing set in Kaggle DR dataset [1], which make it difficult for other to reproduce and compare. In addition, in these two papers the amount of training set and testing set are equal, but it dose not reflect the real data distribution, in which the amount of good quality fundus images is much more than that of poor quality. For example, as Table 1 shown, in Kaggle DR dataset the amount of good quality fundus images and poor quality fundus images are extremely unbalanced. Both of the work avoided the unbalanced data distribution, which is a very common but complex problem in the field of medical image analysis. In this paper, we propose weighted softmax with center loss to handle the problem of unbalanced data distribution.

In the realistic process of computer-aided screening system, fundus image quality assessment is important for subsequent disease diagnosis, such as DR grading. To the best of our knowledge, there is no work using fundus image quality information to help grade DR. In this paper, we propose Fundus Image Quality (FIQ)-guided DR grading method based on multi-task deep learning.

The contributions of our work are summarized as follows:

1. We propose weighted softmax with center loss to solve the unbalanced data distribution in medical images.
2. We propose FIQ-guided DR grading method based on multi-task deep learning, which is the first work using fundus image quality information to help grade DR.
3. Experimental results on the Kaggle dataset show that fundus image quality greatly impact DR grading. By considering the influence of quality, the experimental results validate the effectiveness of our propose method.

The rest of the paper is organized as follows. In Sect. 2, we introduce our method in detail. Section 3 introduce kaggle image quality dataset, as well as the experimental results and quantitative analysis. In the last section, the conclusion is presented.

2   Method

The overall architecture of our FIQ-guided DR grading method is shown in Fig. 2.

2.1   Variant Softmax Loss for Unbalanced Problem

A commonly used loss function for classification in machine learning is softmax loss function, which is shown in Eq. (1):

$$L_{q0} = -\frac{1}{m}\left[\sum_{i=1}^{m}\sum_{j=1}^{k}1\{y^{(i)} = j\}\log(\text{Prob}_{ij})\right] \tag{1}$$

where m denotes the number of input instances, k denotes the number of classes, $1\{\cdot\}$ denotes the indicator function, $y^{(i)}$ denotes the label of i-th instance and Prob_{ij} denotes the probabilities output by softmax activation. However, this loss function is not appropriate for unbalanced problem because the loss dosen't consider the unbalanced distribution.

The image quality data distribution of Kaggle DR dataset is shown in Table 1, which is extremely unbalanced. To solve the unbalanced problem, there are two popular variant softmax loss called weighted softmax loss (i.e. Eq. 2) and center loss (i.e. Eq. 4).

Weighted Softmax Loss. The weighted softmax loss is shown as follow, where each class is weighted inversely proportional to the number of its samples.

$$L_{q1} = -\frac{1}{\sum_{i=1}^{m}w_i}\left[\sum_{i=1}^{m}w_i\sum_{j=1}^{k}1\{y^{(i)} = j\}\log(\text{Prob}_{ij})\right] \tag{2}$$

where

$$w_i = \begin{cases} \beta, & y^{(i)} = 0 \\ 1, & y^{(i)} = 1 \end{cases} \tag{3}$$

and scalar β is a hyperparameter.

Center Loss. In order to enhance the discriminative power of the deeply learned features, Wen *et al.* [8] proposed a new supervision signal, called center loss. Specifically, the center loss simultaneously learns a center for deep features of each class and penalizes the distances between the deep features and their corresponding class centers.

$$L_{q2} = -\frac{1}{\sum_{i=1}^{m}w_i}\left[\sum_{i=1}^{m}\sum_{j=1}^{k}1\{y^{(i)} = j\}\log(\text{Prob}_j) + \lambda L_c\right] \tag{4}$$

where

$$L_c = \frac{1}{2} \sum_i^m \|x_i - c_{y_i}\|_2^2 \tag{5}$$

and scalar λ is a hyperparameter, which is used for balancing the two loss functions.

Weighted Softmax with Center Loss. In order to make full use of weighted softmax loss and center loss, we propose weighted softmax with center loss:

$$L_{q3} = -\frac{1}{\sum_{i=1}^m w_i} \left[\sum_{i=1}^m \sum_{j=1}^k 1\{y^{(i)} = j\} \log(\text{Prob}_j) w_i + \lambda L_c \right] \tag{6}$$

The conventional softmax loss can be considered as a special case of this joint supervision, if λ is set to 0 and β is set to 1.

2.2 Multi-task Learning

To use fundus image quality information for improving DR grading, we propose multi-task learning that train quality classification task and DR grading task at the same time. As shown in Fig. 2, the propose loss function in training stage is defined as follow:

$$L = L_{dr} + L_q + L_{reg} \tag{7}$$

where L_{dr} denotes the softmax loss of DR grading task, L_q denotes the loss of image quality classification task and L_{reg} denotes the regularization loss (weight decay term) used to avoid overfitting. In testing period, we can simultaneously predict image quality class and DR grade.

3 Experiment

3.1 Datasets

To validate the propose multi-task method and analysis the influence of image quality, we use two dataset as follows:

Kaggle DR Dataset. Kaggle organized a comprehensive competition in order to design an automated retinal image diagnosis system for DR screening in 2015 [1]. The retinal images were provided by EyePACS, which is a free platform for retinopathy screening. The dataset consists of 35126 training images, 10906 validate images and 42670 testing images. Each image is labeled as $\{0, 1, 2, 3, 4\}$ and the number represents the level of DR. We will use this dataset to evaluate the performance of DR grading.

Kaggle DR Image Quality Dataset. To verify the effectiveness of variant softmax loss methods for unbalanced medical images and analysis the influence of image quality qualitatively, we label Kaggle DR Dataset as Image Quality Dataset, which is shown in Table 1. All images are tagged by the professionals to identify the quality of the dataset, in which label 1 represents the image of good quality and label 0 stands for the poor quality images.

3.2 Evaluation Protocols

DR Grading. To evaluate the performance of DR grading, we use the quadratic weighted kappa (shown as Eq. 8) to evaluate our methods, which is used in Kaggle DR Challenge [1]. The quadratic weighted kappa not only measures the agreement between two ratings but also considers the distance between the prediction and the ground truth.

$$k = 1 - \frac{\sum_{i,j} w_{i,j} O_{i,j}}{\sum_{i,j} w_{i,j} E_{i,j}} \tag{8}$$

where $w_{i,j} = \frac{(i-j)^2}{(N-1)^2}$ and O, E are N-by-N histogram matrix.

Image Quality Classification. On the one hand, since this is a binary classification problem, we use the popular metrics: specificity, sensitivity, precision. On the other hand, this is an unbalanced binary classification problem and these negative samples are few but important, so we use mean_acc and specificity as the mainly metrics:

$$\text{mean\_acc} = \frac{\text{acc\_0} + \text{acc\_1}}{2} = \frac{\text{specificity} + \text{sensitivity}}{2} \tag{9}$$

where acc_0, acc_1 denoted the accuracy of class 0, class 1 respectively. Futhermore, specificity = acc_0, sensitivity = acc_1.

3.3 Hyper-parameters

During the training stage, the learning rate in our network is empirically set as 0.001, $\beta = 27$ in weighted softmax loss, $\lambda = 0.1$ in center loss.

3.4 Experiments

A. Image Quality Classification

To evaluate each softmax loss and its variant, we conduct ablation experiments and the results are shown in Tables 2 and 3. All of these results are evaluated on Kaggle Image Quality Dataset.

Performance on **validation set** is shown in Table 2. Results about mean_acc and specificity in row 1 (i.e. L_{q0} with Adadelta) and row 2 (i.e. L_{q1} with Adadelta) show that weighted softmax loss is more appropriate for unbalanced

Table 2. Performance on **validation set**. L_{q0} denotes naive softmax loss, L_{q1} denotes weighted softmax loss, L_{q3} denotes weighted softmax with center loss. For the unbalanced binary classification problem and the negative samples are few, mean_acc and specificity metrics are important.

Loss	Optimizer	*mean_acc*	*Specificity*	acc	Sensitivity	Precision
L_{q0}	Adadelta	0.845	0.704	0.980	0.986	0.994
L_{q1}	Adadelta	0.897	0.827	0.965	0.968	0.996
L_{q1}	Momentum	0.961	0.947	0.974	0.974	0.999
L_{q3}	Momentum	**0.962**	**0.969**	0.955	0.954	0.999

Table 3. Performance on **testing set**, on which is similar with validate set.

Loss	Optimizer	*mean_acc*	*Specificity*	acc	Sensitivity	Precision
L_{q0}	Adadelta	0.850	0.711	0.983	0.989	0.994
L_{q1}	Adadelta	0.905	0.838	0.969	0.971	0.997
L_{q1}	Momentum	0.966	0.954	0.977	0.978	0.980
L_{q3}	Momentum	**0.965**	**0.976**	0.955	0.954	0.999

quality dataset. Results in row 3 (i.e. L_{q1} with Momentum) and row 4 (i.e. L_{q3} with Momentum) show that our weighted softmax with center loss is effective. Performance on **testing set** is shown in Table 3, which is similar in Table 2.

B. DR Grading and Quantitative Analysis

The performance of our method and quantitative experimental results are shown in Table 4, and these results show: (i) $b > a > c$: Fundus image quality greatly impact DR grading; (ii) $d > a$: Our proposed FIQ-guided DR grading method is effective; (iii) $e > b, f < c$ and the raise of ratio: Explain why our proposed method is effective.

Table 4. Quantitative analysis on Kaggle DR dataset. **Single-task** denotes single naive DR grading task, **multi-task** denotes our FIQ-guided DR grading method, **good** denotes kappa on good quality images set while $poor_k$ denotes kappa on the opposite set, **true** denotes the number of true prediction while $poor_n$ denotes the number of poor quality image in true set.

Date set	Methods	Kappa			Num		
		Overall	Good	$poor_k$	True	$poor_n$	Ratio
Validation	Single-task	0.718_a	0.721_b	0.629_c	8854	164	18.52‰
	Multi-task	0.745_d	0.750_e	0.616_f	9095	167	18.36‰
Testing	Single-task	0.710	0.715	0.589	34298	633	18.46‰
	Multi-task	0.724	0.730	0.549	34908	623	17.85‰

4   Conclusion

In this paper we propose weighted softmax with center loss to solve the unbalanced data distribution in medical images. Futhermore, we propose FIQ-guided DR grading method based on multi-task deep learning, which is the first work using fundus image quality information to help grade DR. Experimental results on the Kaggle dataset show that fundus image quality greatly impact DR grading. By considering the influence of quality, the experimental results validate the effectiveness of our propose method.

References

1. EyePACS: Diabetic retinopathy detection. https://www.kaggle.com/c/diabetic-retinopathy-detection/data
2. Fu, H., Cheng, J., Xu, Y., Wong, D.W.K., Liu, J., Cao, X.: Joint optic disc and cup segmentation based on multi-label deep network and polar transformation. IEEE Trans. Med. Imaging (2018)
3. Fu, H., et al.: Disc-aware ensemble network for glaucoma screening from fundus image. IEEE Trans. Med. Imaging (2018)
4. Krizhevsky, A., Sutskever, I., Hinton, G.E.: ImageNet classification with deep convolutional neural networks. In: Advances in Neural Information Processing Systems, pp. 1097–1105 (2012)
5. Lalonde, M., Gagnon, L., Boucher, M.C.: Automatic visual quality assessment in optical fundus images. In: Proceedings of Vision Interface, Ottawa, vol. 32, pp. 259–264 (2001)
6. Lee, S.C., Wang, Y.: Automatic retinal image quality assessment and enhancement. In: Medical Imaging 1999: Image Processing, vol. 3661, pp. 1581–1591. International Society for Optics and Photonics (1999)
7. Sun, J., Wan, C., Cheng, J., Yu, F., Liu, J.: Retinal image quality classification using fine-tuned CNN. In: Cardoso, M. (ed.) FIFI/OMIA-2017. LNCS, vol. 10554, pp. 126–133. Springer, Cham (2017). https://doi.org/10.1007/978-3-319-67561-9_14
8. Wen, Y., Zhang, K., Li, Z., Qiao, Y.: A discriminative feature learning approach for deep face recognition. In: Leibe, B., Matas, J., Sebe, N., Welling, M. (eds.) ECCV 2016. LNCS, vol. 9911, pp. 499–515. Springer, Cham (2016). https://doi.org/10.1007/978-3-319-46478-7_31
9. Yu, F., Sun, J., Li, A., Cheng, J., Wan, C., Liu, J.: Image quality classification for DR screening using deep learning. In: 2017 39th Annual International Conference of the IEEE Engineering in Medicine and Biology Society (EMBC), pp. 664–667. IEEE (2017)

DeepDisc: Optic Disc Segmentation Based on Atrous Convolution and Spatial Pyramid Pooling

Zaiwang Gu[1,2(✉)], Peng Liu[1,3], Kang Zhou[4], Yuming Jiang[1,3], Haoyu Mao[1], Jun Cheng[1], and Jiang Liu[1]

[1] Cixi Institute of Biomedical Engineering,
Ningbo Institute of Materials Technology and Engineering,
Chinese Academy of Sciences, Beijing, China
guzaiwang@nimte.ac.cn
[2] School of Mechatronic Engineering and Automation, Shanghai University,
Shanghai, China
[3] University of Electronic Science and Technology of China, Chengdu, China
[4] School of Information Science and Technology,
ShanghaiTech University, Shanghai, China

Abstract. The optic disc (OD) segmentation is an important step for fundus image base disease diagnosis. In this paper, we propose a novel and effective method called *DeepDisc* to segment the OD. It mainly contains two components: atrous convolution and spatial pyramid pooling. The atrous convolution adjusts filter's field-of-view and controls the resolution of features. In addition, the spatial pyramid pooling module probes convolutional features at multiple scales and encodes global context information. Both of them are used to further boost OD segmentation performance. Finally, we demonstrate that our DeepDisc system achieves state-of-the-art disc segmentation performance on the ORIGA and Messidor datasets without any post-processing strategies, such as dense conditional random field.

Keywords: Disc segmentation · Atrous convolution
Spatial pyramid pooling

1 Introduction

The optic disc (OD) contains lots of information and is an important anatomical structure in the retina. Locating and segmenting OD is an essential step in many retinal image analysis for disease detection and monitoring. For example, in age-related macular degeneration detection, localization of the macula is often conducted by finding the OD first. In retinal image analysis for coronary heart disease detection, OD is segmented for retinal vessel caliber measurement [10]. In glaucoma diagnosis, OD segmentation is often conducted before the cup to

© Springer Nature Switzerland AG 2018
D. Stoyanov et al. (Eds.): COMPAY 2018/OMIA 2018, LNCS 11039, pp. 253–260, 2018.
https://doi.org/10.1007/978-3-030-00949-6_30

disc ratio (CDR) is computed [2,5]. Therefore, a reliable OD segmentation is necessary in automatic retinal image analysis for different diseases detection.

Many methods [2,3,7,9,15] have been proposed to segment the OD. Cheng *et al.* proposed a method based on peripapillary atrophy elimination [1]. Then, they improved the OD segmentation performance by adopting superpixel classification method [2]. Li *et al.* [9] built a shape-appearance model to segment the OD, which could learn a sequence for supervised decent directions between the coordinates of OD boundary and their surrounding visual appearances for OD segmentation.

With the development of convolution neural network (CNN) in image and video processing [8], automatic feature learning algorithms using deep learning have emerged as feasible approaches and are applied to handle the retinal image analysis. Recently, some OD segmentation algorithms [5,14] based on the fully convolutional network (FCN) [11] have been proposed. In FCN-like structure, the deep CNNs are originally designed for image classification tasks. The consecutive pooling and strided convolutional operations reduce the feature resolution and extract increasingly abstract feature representations. Meanwhile, the segmentation is a dense prediction task, which is a pixel-wise classification. The detailed spatial information is often neglected in the pooling and strided convolutional operations, however, such information is important for segmentation.

To overcome this limitation, Fu *et al.* [5] proposed M-Net, which is modified from U-Net [13]. The architecture contains multi-scale inputs and deep supervision. The methods attempted to solve the problem of losing some detailed information with pooling and strided convolution operations. The multi-scale inputs introduced the resized images into each middle training stage to ensure that the architecture could learn more from the full-size image. The deep supervision was adopted to optimize the deep network more easily. However, the multi-scale inputs and deep supervision cannot prevent losing some spatial information inevitably. More generally, the FCN-like structures (such as U-Net) can be considered as a Encoder-Decoder architecture. The Encoder aims to reduce the spatial dimension of feature maps gradually and capture more high-level semantic features. The Decoder aims to recover the object details and spatial dimension. FCN employs deconvolution operation to learn the upsampling of low resolution features, while U-Net mainly adds skip connections from the encoder features to the corresponding decoder features. Preserving more spatial information in the convolution is critical to improve the performance of the OD segmentation. Intuitively, maintaining high-resolution feature maps at the middle stage can boost segmentation performance. However, to accelerate training and ease the difficulty of optimization, the size of feature map should be small. Therefore, there is a trade-off between accelerating the training and maintaining the high resolution.

In this paper, different from the FCN-like structures, we propose a novel architecture for OD segmentation. It combines atrous convolution [16] and spatial pyramid pooling. In particular, the atrous convolution allows us to efficiently enlarge the field of view of filters to incorporate multi-scale context. It learns

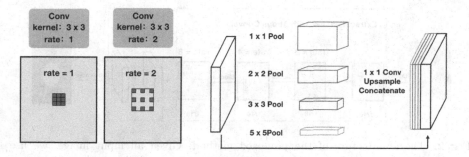

(a) The standard convolution can be seen as atrous convolution with rate = 1

(b) Spatial pyramid pooling employs four-level pooling kernels to extract both local and global context information.

Fig. 1. The illustrations of atrous convolution and spatial pyramid pooling

high-level semantic features in high resolution and preserve more spatial details. The spatial pyramid pooling strategy is adopted to ensure the pooling operation at multiple kernel sizes and effective fields of view. Our proposed OD segmentation architecture is validated on ORIGA and Messidor datasets. It outperforms state-of-the-art methods and achieves an overlapping error of 0.069 in the ORIGA dataset and 0.064 in the Messidor dataset.

The rest of the paper is organized as follows. In Sect. 2, we introduce our proposed method in detail. Section 3 contains our adopted datasets including ORIGA and Messidor, as well as the experimental results. In the last section, the conclusion is presented.

2 Method

The OD segmentation can be considered as a pixel classification problem. It assigns each pixel a label, indicating whether this pixel belongs to the OD or not. To achieve high performance in this dense prediction task, it should not only maintain high-resolution features containing much spatial information, but also extract high-level semantic features. Therefore, we propose to use atrous convolution operation and spatial pyramid pooling strategy for OD segmentation.

2.1 Atrous Convolution

The atrous convolution is originally developed for the efficient computation of the wavelet transform. Mathematically, the atrous convolution under two-dimensional signals is expressed as the following formula:

$$y[i] = \sum_k x[i + rk]w[k], \tag{1}$$

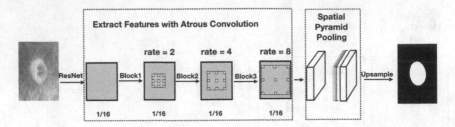

Fig. 2. The architecture of the proposed method. Given an input image, we use pretrained-ResNet to extract features. Then, atrous convolution and spatial pyramid pooling are applied to extract high-level semantic features under high resolution. Finally, upsampling operation is adopted to get the final per-pixel prediction map.

where the convolution of the input feature map x and a filter w yields the output y, where the atrous rate r corresponds to the stride with which we sample the input signal. It is equivalent to convolute the input x with upsampled filters produced by inserting $r-1$ zeros between two consecutive filter values along each spatial dimension (hence the name atrous convolution where the French word atrous means holes in English). Standard convolution is a special case for rate $r = 1$, and atrous convolution allows us to adaptively modify filter's field-of-view by changing the rate value. See Fig. 1(a) for illustration.

2.2 Spatial Pyramid Pooling

A challenge in segmentation is the large variation of OD sizes. In this paper, we adopt spatial pyramid pooling to address the problem, which mainly relies on multiple effective fields of view to detect objects at different sizes.

In the deep neural network, the size of receptive field can roughly indicates how much we can use the context information. The adopted pyramid pooling encodes global context information with four different-size receptive fields, $1 \times 1, 2 \times 2, 3 \times 3$ and 5×5 respectively. The four-level outputs contain the feature maps with varied sizes. To reduce the dimension of weights and computational cost, we use a 1×1 convolution after each level of pooling. It reduces the dimension of feature maps to the $1/N$ of original dimension, where N represents number of channels in original feature maps. Then we upsample the low-dimension feature map to get the same size features as the origin feature map via bilinear interpolation. Finally, we concatenate the original pyramid pooling features with upsampled feature maps.

2.3 Network Architecture

With the atrous convolution and spatial pyramid pooling, we propose our OD segmentation network as illustrated in Fig. 2. Given an input image, we adopted the ImageNet-pretrained ResNet [6] as our main body. Different from the ResNet under classification task, we use atrous convolution to extract the dense high-level semantic features. Atrous convolution allows us to explicitly control the denseness of feature responses in neural convolution networks. Here, all subsequent convolutional layers are replaced with atrous convolution with rate $r = 2$. This allows us to extract denser feature responses without requiring learning extra parameters.

The pretrained-ResNet is used to extract the high-level feature maps and the image is $1/16$ of the original input image. Then we continue with atrous convolution layers with rate $r = 2, 4, 8$. After that, we use spatial pyramid pooling module to gather the context information. The four-level pooling kernels are adopted to cover most multi-scale size objects. Finally, we bilinearly upsample and deconvolve the feature maps to the desired spatial dimension.

To summarize, we use the ImageNet-pretrained ResNet as the feature extractor, avoiding the training difficulty with lacking of sufficient training dataset. With the atrous convolution and spatial pyramid pooling, the proposed method ensures the spatial resolution. At the same time, it can also detect discs with different sizes.

Compared to the background in a fundus image, OD just occupied a smaller area. Therefore, common cross-entropy loss function could not achieve high performance on facing the problem of imbalanced data distribution. In this paper, we use dice-coefficient loss function [12].

3 Experiments and Results

3.1 Datasets and Evaluation Protocols

We validated the effectiveness of the proposed method using two datasets, the ORIGA [17] and Messidor datasets. The ORIGA dataset contains $325/325$ images for training/testing respectively [4], and the OD boundaries have been manually demarcated. The Messidor dataset is a public dataset provided by the Messidor program partners. It consists of 1200 images with three different sizes: $1440 \times 960, 2240 \times 1488, 2340 \times 1536$. The Messidor dataset is originally collected for Diabetic Retinopathy (DR) grading. Later, disc boundary for each image has also been provided[1].

To evaluate the performance, we adopt the overlapping error [2], E given as:

$$E = 1 - \frac{Area(S \cap G)}{Area(S \cup G)}, \tag{2}$$

where S and G denote the segmented and the manual ground truth OD respectively.

---

[1] http://www.uhu.es/retinopathy/.

3.2 Implementation Details

The OD only occupies a small area in full-size image, which causes difficulty in the training. Therefore, we crop a 800×800 region containing the OD of ORIGA dataset and a 448×448 region of Messidor dataset. Data augmentation is adopted on the training set by flipping images, which simulates the relationship between left-right eyes and also simulates the image rotation caused in capturing the fundus images. Our architecture is implemented with PyTorch. During the training process of whole architecture, we trained the parameters of atrous convolution and spatial pyramid pooling modules with 0.001 learning rate. In addition, we fine-tuned the parameters of ResNet with one tenth of origin learning rate. Additionally, the ℓ_2 norm regularization with factor 0.00004 is applied in the final loss function. The whole code will be released at github after acceptance of the paper.

Table 1. Performance comparison of the different methods on ORIGA dataset

Method	E_{disc}
R-Bend [7]	0.129
ASM [15]	0.148
Superpixel [2]	0.102
QDSVM [3]	0.110
U-Net [13]	0.089
M-Net [5]	0.083
Our	0.069

Table 2. Performance comparison of the different methods on Messidor dataset.

Method	E_{disc}
Li's [9]	0.087
Superpixel [2]	0.125
U-Net [13]	0.069
DeepDisc	0.064

3.3 Experiment Results

Experiment Results on the ORIGA Dataset. Limited by the memory of our computational devices, we choose the ImageNet-pretrained ResNet-34 as main body of architecture. As can be seen in Table 1, we show the performances of most OD segmentation algorithms. Compared with some state-of-the-art OD segmentation methods, our DeepDisc achieves an overlapping error of 0.069 in the ORIGA dataset. It outperforms the other algorithms based on deep learning or traditional image processing method. From the comparison between M-Net and our DeepDisc, we observe that there is a clear improvement of overlapping error by 16.9% from 0.083 to 0.069.

Experiment Results on the Messidor Dataset. To further validate the performance of OD segmentation, the Messidor dataset is also tested. The image of Messidor only has rough boundary of the OD, however it is enough for our validation. As three different sizes in the Messidor, we first resize all images to a fix size 1440×960, then crop a 448×448 size image. For a fair comparison, we follow the same setting as in [9]: 1000 images for training and 200 images for testing. From the Table 2, our proposed method achieves an overlapping error of 0.064 and it outperforms the other algorithms. Different from the FCN-like structures, our DeepDisc system based on atrous convolution and spatial pyramid pooling can save more detailed information and improve performance on OD segmentation, as illustrated in Fig. 3.

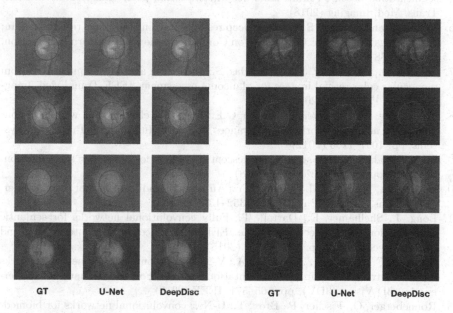

GT   U-Net   DeepDisc   GT   U-Net   DeepDisc

Fig. 3. The visual examples of optic disc segmentation. The left is examples from ORIGA dataset and the right is from Messidor dataset.

4   Conclusion

We propose an effective architecture to segment the OD, which employs atrous convolution to extract the dense image-level feature maps with high resolution and the spatial pyramid pooling module to collect contextual information with multiple effective field-of-views. The proposed method achieves an overlapping error of 0.069 in the ORIGA dataset and 0.064 in the Messidor dataset, better than other methods.

References

1. Cheng, J., et al.: Automatic optic disc segmentation with peripapillary atrophy elimination. In: 2011 Annual International Conference of the IEEE Engineering in Medicine and Biology Society (EMBC), pp. 6224–6227. IEEE (2011)
2. Cheng, J., et al.: Superpixel classification based optic disc and optic cup segmentation for glaucoma screening. IEEE Trans. Med. Imaging **32**(6), 1019–1032 (2013)
3. Cheng, J., Tao, D., Wong, D.W.K., Liu, J.: Quadratic divergence regularized SVM for optic disc segmentation. Biomed. Opt. Express **8**(5), 2687–2696 (2017)
4. Cheng, J., et al.: Similarity regularized sparse group lasso for cup to disc ratio computation. Biomed. Opt. Express **8**(8), 3763–3777 (2017)
5. Fu, H., Cheng, J., Xu, Y., Wong, D.W.K., Liu, J., Cao, X.: Joint optic disc and cup segmentation based on multi-label deep network and polar transformation. IEEE Trans. Med. Imaging (2018)
6. He, K., Zhang, X., Ren, S., Sun, J.: Deep residual learning for image recognition. In: Proceedings of the IEEE Conference on Computer Vision and Pattern Recognition, pp. 770–778 (2016)
7. Joshi, G.D., Sivaswamy, J., Krishnadas, S.: Optic disk and cup segmentation from monocular color retinal images for glaucoma assessment. IEEE Trans. Med. Imaging **30**(6), 1192–1205 (2011)
8. Krizhevsky, A., Sutskever, I., Hinton, G.E.: ImageNet classification with deep convolutional neural networks. In: Advances in Neural Information Processing Systems, pp. 1097–1105 (2012)
9. Li, A., et al.: Learning supervised descent directions for optic disc segmentation. Neurocomputing **275**, 350–357 (2018)
10. Li, H., Hsu, W., Lee, M.L., Wong, T.Y.: Automatic grading of retinal vessel caliber. IEEE Trans. Biomed. Eng. **52**(7), 1352–1355 (2005)
11. Long, J., Shelhamer, E., Darrell, T.: Fully convolutional networks for semantic segmentation. In: Proceedings of the IEEE Conference on Computer Vision and Pattern Recognition, pp. 3431–3440 (2015)
12. Milletari, F., Navab, N., Ahmadi, S.A.: V-Net: fully convolutional neural networks for volumetric medical image segmentation. In: 2016 Fourth International Conference on 3D Vision (3DV), pp. 565–571. IEEE (2016)
13. Ronneberger, O., Fischer, P., Brox, T.: U-Net: convolutional networks for biomedical image segmentation. In: Navab, N., Hornegger, J., Wells, W.M., Frangi, A.F. (eds.) MICCAI 2015. LNCS, vol. 9351, pp. 234–241. Springer, Cham (2015). https://doi.org/10.1007/978-3-319-24574-4_28
14. Sevastopolsky, A.: Optic disc and cup segmentation methods for glaucoma detection with modification of U-Net convolutional neural network. arXiv preprint arXiv:1704.00979 (2017)
15. Yin, F., et al.: Model-based optic nerve head segmentation on retinal fundus images. In: 2011 Annual International Conference of the IEEE Engineering in Medicine and Biology Society (EMBC), pp. 2626–2629. IEEE (2011)
16. Yu, F., Koltun, V.: Multi-scale context aggregation by dilated convolutions. arXiv preprint arXiv:1511.07122 (2015)
17. Zhang, Z., et al.: ORIGA-Light: an online retinal fundus image database for glaucoma analysis and research. In: 2010 Annual International Conference of the IEEE Engineering in Medicine and Biology Society (EMBC), pp. 3065–3068. IEEE (2010)

Large-Scale Left and Right Eye
Classification in Retinal Images

Peng Liu[1,2(✉)], Zaiwang Gu[2,3], Fan Liu[1], Yuming Jiang[1], Shanshan Jiang[2],
Haoyu Mao[2], Jun Cheng[2], Lixin Duan[1], and Jiang Liu[2]

[1] Big Data Research Center,
University of Electronic Science and Technology of China, Chengdu, China
liupengimed@nimte.ac.cn
[2] Cixi Institute of Biomedical Engineering,
Ningbo Institute of Materials Technology and Engineering, Ningbo, China
[3] School of Mechatronic Engineering and Automation, Shanghai University,
Shanghai, China

Abstract. Left and right eye information is an important priori for automatic retinal fundus image analysis. However, such information is often not available or even wrongly provided in many datasets. In this work, we spend a considerable amount of efforts in manually annotating the left and right eyes from the large-scale Kaggle Diabetic Retinopathy dataset consisting of 88,702 fundus images, based on our developed online labeling system. With the newly annotated large-scale dataset, we also train classification models based on convolutional neural networks to discriminate left and right eyes in fundus images. As experimentally evaluated on the Kaggle and Origa dataset, our trained deep learning models achieve 99.90% and 99.23% in term of classification accuracy, respectively, which can be considered for practical use.

Keywords: Left and right eye classification
Convolutional neural networks

1 Introduction

Retinal fundus images contain rich information for ophthalmic disease diagnosis. Ophthalmologists often determine the health conditions of eyes by examining blood vessels, optic nerve head, vitreous and macula on the corresponding retinal fundus images. Among those, information on left and right eyes is often used in ophthalmic disease diagnosis. For instance, such information is used to determine the nasal and temporal side of an eye. Also, left and right eye information is also considered in the glaucoma diagnosis, when comparing asymmetric cup to disc ratios of different eyes. Moreover, for the diagnosis of age-related macula degeneration, eye side information is also used to determine the location of the macula.

© Springer Nature Switzerland AG 2018
D. Stoyanov et al. (Eds.): COMPAY 2018/OMIA 2018, LNCS 11039, pp. 261–268, 2018.
https://doi.org/10.1007/978-3-030-00949-6_31

Although some fundus cameras automatically record left and right eye information when taking the retinal images, many others still do not record such information. In our practical study, we often find that clinicians put all the retinal fundus images from one subject into one folder without having the labeling information of either left or right eyes. Moreover, we have found that some images are inverted (rotated for 180°), which can affect the visual appearance of left vs. right. According to the description of Kaggle Diabetic Retinopathy dataset (Kaggle DR dataset) [4], the retinal images provided are a mix of images shown as standard retina anatomy and images taken through a microscope condensing lens (which are inverted). Figure 1 shows two pairs of retinal images from the dataset. The above cases can actually pose big problems when later eye side information gets necessary needs, which requires immediate actions.

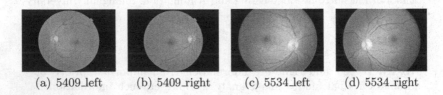

(a) 5409_left   (b) 5409_right   (c) 5534_left   (d) 5534_right

Fig. 1. Examples of retinal images of a pair of eyes not inverted and an pair of eyes inverted in Kaggle DR dataset. (a) and (b) A pair of retinal fundus images from a patient not inverted; (c) and (d) A pair of inverted retinal fundus images from a patient.

In recent years, methods have been proposed to classify left and right eyes. In [9], Tan et al. proposed a classification method by examining the intensity changes across the optic disc. In [8], support vector machine was used to train a more robust classification model. In [7], the vessel distribution within the optic disc was further used to distinguish the left and right eyes in retinal fundus images and tested on the Origa dataset [10]. Those methods have two major limitations. On one hand, they are all built on holistic features (e.g., intensity changes or vessel changes within the disc) and shallow models (e.g., SVM), and are thus sensitive to images of different quality (e.g., taken from different machines) as well as not fully capable of capturing the high semantic meaning based on the image content. On the other hand, the datasets used in their experiments are of small scales. To solve their limitations, in this work we propose to employ deep learning based methods to train better classification models for left and right eye classification. And also, we relabel the large-scale Kaggle DR dataset, which consists of 88,702 fundus images, by providing the information of left and right eyes. To the best of our knowledge, this dataset is so far the largest one for left and right eye classification.

The rest of the paper is organized as follows. Section 2 introduces our online system for left and right eye annotation. Section 3 details the procedures to train deep learning models for left and right eye classification. Experimental results are provided in Sect. 4. Finally, conclusive remarks are presented in Sect. 5.

2   Labeling Protocol

In this paper, we spend a considerable amount of efforts in providing the left and right eye information on the Kaggle DR dataset which is public available[1]. This large-scale dataset contains 88,702 fundus images in total: 35,126 training images and 53,576 test images which are already split and provided. Base on the filename, each image is named by a patient id followed by left or right eye information, for example, "5409_left" denotes left eye of patient 5409. However, we found that many images were inverted. According to our statistics, more than 36% images are inverted or contain wrong labels. This surprising number actually motivates us to develop an online system for left and right eye labeling.

2.1   Manual Labeling of Left and Right Eyes and Inverted Images

The left and right eye information can be determined by comparing the location of optic disc with that of macula. If the optic disc on the left side of the macula, the retinal image is from a left eye, otherwise, right eye. However, in some images, the macula is not captured or its location cannot be easily identified due to pathological changes. Therefore, the above method cannot be used. A second method is to examine the intensity changes within the optic disc. Typically, the temporal side of the optic disc is brighter than the nasal side. A third method is to use the blood vessels. Very often, the main vessels bend toward the macula. Simultaneously, we label whether the retinal image is inverted via examining if there is a notch on the side of the image.

In our manual labeling process, we combine the above rules to determine the left and right eye information.

2.2   Online Labeling System

For the ease of manual labeling, we have developed a labeling system to relabel all the 88,702 fundus images from the Kaggle DR dataset. A group of six researchers has been trained to identify left and right eyes and whether the retinal image is inverted. In our relabeling, we label each image as left, right or unable to tell. We label each image as inverted or not inverted as while. In the relabeling, each image will be labeled by two researchers independently. When the label result by the two researchers are different, the image will be examined and discussed by a group of at least three researchers to reach a consensus. For images where nobody can tell the left and right eye information, we retain the original label.

3   Left and Right Eye Classification Based on Deep Learning

Convolutional neural networks (CNNs) [3,5] have achieved superior performance in object classification and detection. We use three typical CNN architectures,

---

[1] https://www.kaggle.com/c/diabetic-retinopathy-detection.

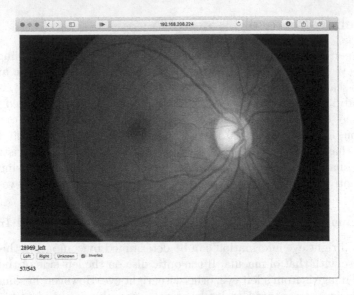

Fig. 2. Examples of the collaboration labeling system. Researchers need to determine left and right eye and whether the image is inverted, then choose "Right" or "Left" to label the retinal fundus image or "Unknown" to indicate a pair of poor quality fundus images.

including the VGG-16 [6], the 50-layer and 101-layer ResNet [3], to automatically determine the left and right eye information (Fig. 2).

3.1 Image Normalization

Since the images in the Kaggle dataset are obtained under different conditions including the age of subjects, the models of the fundus camera and the settings, dilated or un-dilated eyes, illumination changes, these images show a variety change of colors. Very often, image normalization to reduce the changes from one to another image is beneficial for subsequent retinal image analysis.

In this paper, we first preprocessing the images from the Kaggle and Origa datasets. In our processing, we first extract the effective image region by applying a thresholding to the full image to remove the unnecessary black edge, where the threshold is set as 20 of gray value in our implementation.

Next, we normalize the image following the min-pooling's solution [2]. Mathematically, each image is computed as

$$I_c = \alpha \cdot I + \beta \cdot G(\rho) * I + \gamma, \tag{1}$$

where I represents the input image, $G(\rho)$ denotes the Gaussian filter with a standard deviation of ρ, $*$ means the convolution operation, and α, β, γ are predefined parameters (we use $\alpha = 4, \beta = -4, \rho = 10, \gamma = 128$ in the experiments).

In the last step, we resize all the images to 224 × 224. For inverted images according to our relabeling, we inverted back the images as retina anatomically(macula on the left, optic nerve on the right for the right eye). Figure 3 shows two sample images before and after our normalization. After normalization, the color difference between the two images is reduced.

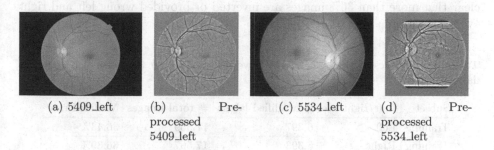

(a) 5409_left   (b)   Pre-   (c) 5534_left   (d)   Pre-
                    processed   processed
                    5409_left   5534_left

Fig. 3. Two sample images before and after our pre-processing.

3.2 Deep Learning Models

We train the deep learning models from pre-trained networks. Three different CNNs architectures, including VGG-16, 50-layer and 101-layer ResNet, are used in this paper. Since we need to classify left and right eye instead of 1000 classes of objects, we add a 2-d fully connected layer before the softmax layer in the three original architectures.

In the training, we use 35,126 images in the training set of Kaggle DR dataset after our pre-processing with relabeled left and right eye information. We fine-tune the network from the pre-trained using the ImageNet dataset [1]. All the parameters were involved into the finetune. We use Adam optimizer with a learning rate of 0.0001. The models are optimized for 40 epochs and the mini-batch size is 128.

4 Experiments

4.1 Datasets

Kaggle Diabetic Retinopathy Dataset: we utilize the relabeled Kaggle DR dataset which has in total 88,702 retinal fundus images. It was divided into a training set with 35,126 images and a test set with 53,576 images, which is exactly the same as the original partition. According to the our relabeled left and right eye, there are 17,559 left eyes and 17,567 right eyes in the training set, 26,742 left eyes and 26,834 right eyes in the test set.

Origa Dataset: Origa dataset, consists 336 left eyes and 314 right eyes retinal images. In this paper, we applied the finetuned models to classify left and right eyes on all the 650 retinal fundus images.

4.2   Results

Labeling Results. We have labeled left and right eye and inverted information for all the 88,702 fundus images from the Kaggle dataset. Based on our statistics, a total number of 32,199 images have been inverted or contain wrong labels in the original dataset. The detailed statistics are provided in Table 1. It is quite clear that more than 36% images are inverted or provided wrong left and right eye information.

Table 1. Statistics of the left and right eye information on the original Kaggle DR dataset.

Subset	Left/right eye	# Modified labels	# total Images	Modified rate
Training	Left	6,397	17,559	36.43%
Training	Right	6,393	17,567	36.39%
Test	Left	9,754	26,742	36.47%
Test	Right	9,708	26,834	36.18%

Left and Right Eye Classification. We evaluate the finetuned VGG-16, 50-layer and 101-layer ResNet on our newly labeled Kaggle DR dataset and the Origa datasets. We summarize the classification accuracies on the two datasets in Tables 2 and 3, respectively. We can see from the results that image normalization using the Gaussian filter improve the classification performance in all the settings. The best result obtained by ResNet with image normalization indicates its effectiveness and potential in practical use.

Table 2. Classification accuracies of different models on the Kaggle DR dataset.

Model	Accuracy without Gaussian	Accuracy with Gaussian
ResNet-50	99.53%	99.90%
ResNet-101	99.37%	99.69%
VGG-16	99.20%	99.49%

We also present sample images that are both correctly and incorrectly predicted by ReNet-50 in Fig. 4. It is worth noting that the incorrectly predicted images are quite hard to classify even for human experts.

Table 3. Classification accuracies of different models on the Origa dataset.

Model	Accuracy without Gaussian	Accuracy with Gaussian
Tan et al. [7]	98.00%	–
ResNet-50	98.62%	99.23%
ResNet-101	98.31%	99.23%
VGG-16	98.15%	98.92%

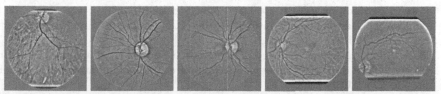

(a) Normalized fundus images containing left eyes, correctly classified.

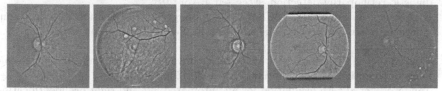

(b) Normalized fundus images containing right eyes, correctly classified.

(c) Normalized fundus images containing left eyes, incorrectly classified.

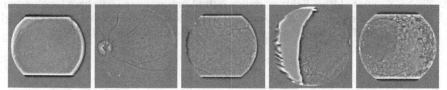

(d) Normalized fundus images containing right eyes, incorrectly classified.

Fig. 4. Sample images categorized by the classification results of ResNet-50.

5   Conclusion

In this work, we newly annotate the left and right eye and inverted information for all the 88,702 fundus images from the Kaggle Diabetic Retinopathy dataset. Based on such newly annotated large-scale dataset, we train three CNN models for left and right eye classification from the Kaggle dataset, and evaluate them on the additional Origa dataset. Extensive experiments clearly show the good generalization ability of the deep learning models as well as their great potential in applying those models for practical use.

References

1. Deng, J., Dong, W., Socher, R., Li, L.J., Li, K., Fei-Fei, L.: ImageNet: a large-scale hierarchical image database. In: IEEE Conference on Computer Vision and Pattern Recognition, CVPR 2009, pp. 248–255. IEEE (2009)
2. Graham, B.: Kaggle diabetic retinopathy detection competition report. University of Warwick (2015)
3. He, K., Zhang, X., Ren, S., Sun, J.: Deep residual learning for image recognition. In: Proceedings of the IEEE Conference on Computer Vision and Pattern Recognition, pp. 770–778 (2016)
4. Kaggle: Diabetic retinopathy detection (2015). https://www.kaggle.com/c/diabetic-retinopathy-detection. Accessed 4 Apr 2016
5. Krizhevsky, A., Sutskever, I., Hinton, G.E.: ImageNet classification with deep convolutional neural networks. In: Advances in Neural Information Processing Systems, pp. 1097–1105 (2012)
6. Simonyan, K., Zisserman, A.: Very deep convolutional networks for large-scale image recognition. arXiv preprint arXiv:1409.1556 (2014)
7. Tan, N.M., et al.: An evaluation for an automated left and right eye identification system for digital fundus images for glaucoma diagnosis. Invest. Ophthalmol. Vis. Sci. **53**(14), 649–649 (2012)
8. Tan, N.M., et al.: Classification of left and right eye retinal images. Proc. SPIE **7624**, 762438 (2010)
9. Tan, N.M., et al.: Automatic detection of left and right eye in retinal fundus images. In: Lim, C.T., Goh, J.C.H. (eds.) 13th International Conference on Biomedical Engineering, pp. 610–614. Springer, Heidelberg (2009). https://doi.org/10.1007/978-3-540-92841-6_150
10. Zhang, Z., et al.: Origa-light: an online retinal fundus image database for glaucoma analysis and research. In: 2010 Annual International Conference of the IEEE Engineering in Medicine and Biology Society (EMBC), pp. 3065–3068. IEEE (2010)

Automatic Segmentation of Cortex and Nucleus in Anterior Segment OCT Images

Pengshuai Yin[1], Mingkui Tan[1], Huaqing Min[1], Yanwu Xu[2,4](✉),
Guanghui Xu[1], Qingyao Wu[1], Yunfei Tong[2], Higashita Risa[3], and Jiang Liu[4]

[1] South China University of Technology, Guangzhou, China
[2] Guangzhou Shiyuan Electronic Technology Company Limited, Guangzhou, China
ywxu@ieee.org
[3] Tommy Corporation, Nagoya, Japan
[4] Cixi Institute of Biomedical Engineering, Chinese Academiy of Sciences,
Cixi, China

Abstract. We propose a pipeline for automatically segmenting cortex
and nucleus in a 360-degree anterior segment optical coherence tomography (AS-OCT) image. The proposed pipeline consists of a U-shaped
network followed by a shape template. The U-shaped network predicts
a mask for cortex and nucleus. However, the boundary between cortex
and nucleus is weak, so that the boundary of the prediction is an irregular shape and does not satisfy the physiological structure of nucleus.
To address this problem, in the second step, we design a shape template according to the physiological structure of nucleus to refine the
boundary. Our method integrates both appearance and structure information. The accuracy is measured by the normalized mean squared error
(NMSE) between ground truth line and predicted line. We achieve NMSE
7.09/7.94 for nucleus top/bottom boundary and 2.49/2.43 for cortex
top/bottom boundary.

Keywords: SS-OCT · AS-OCT · Image segmentation

1 Introduction

Anterior segment optical coherence tomography (AS-OCT) can assist the diagnosis of many eye diseases, such as glaucoma and cataract [2]. The measurement
is made without contact and with low risk of infection. The role of AS-OCT in
research and clinical care continues to accelerate [3].

AS-OCT nuclear density measurement is a repeatable and reliable objective
cataract grading method, It is correlated with the Lens Opacity Classification
System Version III (LOCS III) grading [5,8]. The nuclear density is got by delin-
eates the lens nucleus and calculates the total average pixel intensity. If nucleus

---

P. Yin and M. Tan—Equally contribution to this work.

© Springer Nature Switzerland AG 2018
D. Stoyanov et al. (Eds.): COMPAY 2018/OMIA 2018, LNCS 11039, pp. 269–276, 2018.
https://doi.org/10.1007/978-3-030-00949-6_32

can be automatically segmented, cataract grading can be automatically acquired. To the best of our knowledge, no previous work focus on automatically segmenting the nucleus in AS-OCT images.

In this paper, we propose a pipeline to automatically segment cortex and nucleus in AS-OCT images. The proposed pipeline consists of a U-shaped network followed by a shape template. The U-shaped network predicts a preliminary mask for cortex and nucleus. However, the predicted boundary of nucleus is arbitrary because the boundary between cortex and nucleus is weak. To solve this problem, we design a shape template based on the physiological structure of nucleus to refine the boundary of nucleus. The basic idea of the refinement is to find a template in the training set to replace the boundary of the prediction. After the refinement, the boundary of nucleus satisfies the physiological structure of nucleus.

We summarize the contributions of this work as follows:

- We propose a simple and effective pipeline to segment cortex and nucleus by using a U-shaped network and a shape template. This method integrates both structure information and appearance information.
- We design a shape template that imitates the intrinsic concentric layers structure of nucleus. By using the template to refine the boundary of nucleus, the final prediction satisfies the physiological structure of nucleus.

2   Proposed Method

The proposed pipeline is shown in Fig. 1. First, we find the lens region using the Canny edge detector and divide the lens region into three sub-regions: capsule region, cortex region and nucleus region. We train a U-shaped network to predict a mask for each region. However, the output of the U-shaped network has no regular shape, especially for the nucleus which has a weak boundary. So we design a shape template to model the structure of nucleus and use the template to refine the boundary of nucleus.

2.1   Network Architecture

Motivated by [6]. We design a U-shaped network to predict a preliminary mask for capsule, cortex and nucleus. The U-shaped network can obtain a high-resolution mask with a clear boundary by using skip connections to restore the information loss caused by pooling layers.

Our network mainly includes two modules: encoding module and decoding module. The encoding module consists of six blocks. Each block contains two or three convolutional layers, and each convolutional layer is followed by a rectified linear unit (ReLU) and a 2×2 max pooling operation with stride 2. The decoding module is also composed of six blocks. Each block consists of a concatenation with the corresponding feature maps from the contracting path, a spatial upsampling of the feature maps with a factor of 2 followed by two convolutions and

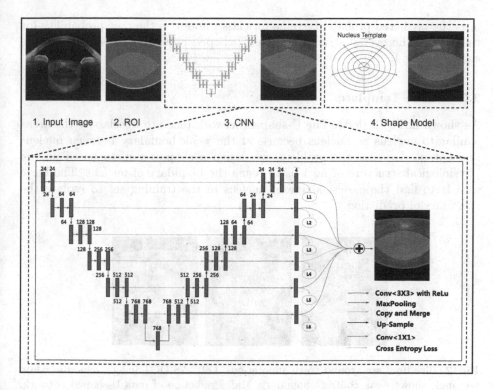

Fig. 1. The proposed pipeline: 1, AS-OCT image. 2, Region of interest (ROI). We locate the lens and divide the lens into three sub-regions according to the ground truth annotations (blue lines), pink: cortex region, yellow: nucleus region, cyan: capsule region. 3, U-shaped network with side-output predicts a mask for each region. 4, Shape template refines the boundary of nucleus. (Color figure online)

a ReLU layer. At the final layer, a 1×1 convolution maps each 24-dim feature vector to the desired number of classes (here 4).

Motivated by [1], we add a side-output layer which acts as a classifier that produces a companion local output for early layers and also integrates different level information. Cross-entropy loss is used for each side-output layer. There are M side-outputs in the network. The loss function of the side-output layer is given as:

$$L_{side-output} = \frac{1}{M} \sum_{m=1}^{M} L_{cross-entropy}(y, y'),$$
(1)

$L_{cross-entropy}$ is the cross entropy loss:

$$L_{cross-entropy} = -\sum_{i} (y'_i \log(y_i)),$$
(2)

y_i is the predicted probability value for class i and y_i' is the true probability for that class. The overall structure of the U-shaped network with side-outputs is shown in Fig. 1.

2.2 Shape Template

As shown in Fig. 2 (left). The U-shaped network tends to misclassify the areas similar to nucleus as nucleus because of the weak boundary between nucleus and cortex. To solve the problem, we design a shape template based on the physiological structure of nucleus to refine the boundary of nucleus. The basic idea is to find the closest shape of nucleus in the training set to replace the boundary of prediction.

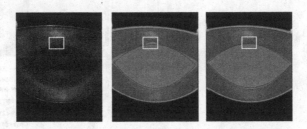

Fig. 2. Blue lines are ground truth annotations. Left: AS-OCT image, region in white rectangle shows weak contrast boundary. Mid: Prediction of ours U-shaped network, the U-shaped network misclassify the regions similar to nucleus as nucleus, shown in white rectangle. Right: Shape template refine the boundary of nucleus. (Color figure online)

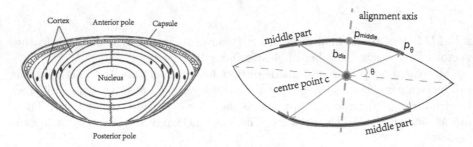

Fig. 3. Left: The structure of the lens, the lens fibres are arranged in concentric layers. Right: The shape template of nucleus, p_{middle} is the intersection between boundary and alignment axis, b_{dis} is the distance from center point c to point p_{middle}, the middle part of the template is shown in green. (Color figure online)

Lens fibers form the bulk of the lens. The lens fibers stretch lengthwise from the posterior to the anterior poles and are arranged in concentric layers rather like the layers of an onion when cutting horizontally [4], as shown in Fig. 3

(left). Motivated by the ray feature [7], the structure of concentric layers can be represented by a center point and the distance between the center point and the nearest point on the boundary, as shown in Fig. 3 (right). Different layers share the same center point and different distances to the center point.

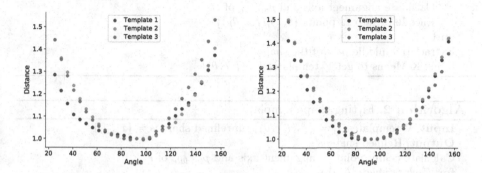

Fig. 4. Three nucleus templates of the top part (left) and the bottom part (right). Different templates are shown in different colors.

The boundary of nucleus is represented by n points: $S_i = \{x_i, y_i\}$. The center point of nucleus is defined as $c = (\sum \frac{x_i}{n}, \sum \frac{y_i}{n})$. Then the boundary of nucleus is encoded by c and the distance from c to the nearest point p_θ on the boundary of nucleus in direction θ:

$$f(I, c, \theta) = \frac{\|c - p_\theta\|}{z}, \tag{3}$$

where $z = \|c - p_{middle}\| = b_{dis}$ is a normalization factor. p_{middle} is the intersection of alignment axis and the boundary of nucleus as shown in Fig. 3 (right). p_{middle} is different in each image due to the eye movement between AS-OCT images mentioned in [9]. Normalization not only ease the rotational misalignment problem but also eliminates scale effects. We sample θ every 5 degrees. In this way all the shapes are encoded by K points $\{\theta_k, f(I, c, \theta_k)\}$.

The training procedure is shown in Algorithm 1. The purpose of training is to learn all possible shapes of nucleus. For each shape in the training set, we encode it into K points $\{\theta_k, f(I, c, \theta_k)\}$ and then cluster them into v templates. In the experiments, we only learn the middle part of shape and use quadratic curve fitting to get the other part because the shape of middle part is relatively stable. The middle part is divided into a top part and a bottom part as shown in Fig. 3 (right). For each part, we cluster the corresponding shapes into v' templates using K-means. Figure 4 shows three normalized templates of the top part and the bottom part.

The refinement procedure is shown in Algorithm 2, s is the boundary of nucleus predicted by the U-shaped network. We calculate the center point c', alignment axis and p'_{middle} of s. The next step is to find a template closest to s. For each template T_v, we finetune the template by multiply T_v by z, where

---

Algorithm 1. Training shape template

Input: Training shape $S_i = \{x_i, y_i\}$
Output: V templates $T_v = \{f_v(I, c, \theta)\}$
for *Each shape S_i in training set* **do**
    Center point: $c = (\sum \frac{x_i}{n}, \sum \frac{y_i}{n})$
    Calculate alignment axis and p_{middle} of S_i
    Encode S_i into K points $\{\theta_k, f_i(I, c, \theta_k)\}$
end
Extract the middle part of the shape
Using K-Means to get V templates $\{f_v(I, c, \theta)\}$

Algorithm 2. Fitting shape template

Input: V templates $T_v = \{f_v(I, c, \theta)\}$, unrefined shape $s = \{x_j, y_j\}$
Output: Refined shape s'
Calculate center point c', alignment axis and p'_{middle} of s
for *Each template T_v* **do**
    $T'_v = T_v * z$, where $z = \|c' - p'_{middle}\|$
    for *offset* $\leftarrow \{-10, -9, ..., 9, 10\}$ **do**
        $B = T'_v + \text{offset}$
        Calculate the similarity between s and B
    end
end
Find the template $B' = B$ with maximum similarity
The refined shape $s' = B'$

$z = \|c' - p'_{middle}\|$. After the multiplication, p_{middle} of T_v is coincide with p'_{middle} of s. Then we add an offset to T_v. The positive offset means the template move to outer layer. For each transformed template B, we calculate the similarity between B and s. We find a most similar template B' to replace s. B and s can be represent by K points: $\{\theta_k, f_B(I, c, \theta_k)\}$ and $\{\theta_k, f_s(I, c, \theta_k)\}$. The similarity between B and s is defined as:

$$Similarity(B, s) = \Sigma_{k=1}^{K} |f_B(I, c, \theta_k) - f_s(I, c, \theta_k)|. \tag{4}$$

3   Performance Evaluation and Discussion

We acquire the data from CASIA-2000 anterior SS-OCT produced by Tomey Co. Ltd. The dataset contains 20 eyes from 10 people, 8 images per eye. We select 7 people (120 images) for training and 3 people (40 images) for testing. All the images are annotated by one experienced ophthalmologist. The accuracy is measured by the normalized mean squared error (NMSE) between the predicted shape $S_p = \{x_i, y_i\}$ and the ground truth $S_g = \{x_j, y_j\}$, it is defined as

$$NMSE = \frac{\sum_{n_g} \sqrt{(x_i - x_j)^2 + (y_i - y_j)^2}}{n_g}, \tag{5}$$

where n_g is the number of annotation points. The results are shown in Table 1. Nucleus top is the top boundary of nucleus. Nucleus bottom is the bottom boundary of nucleus. The same thing is conducted for the cortex.

The U-shaped network is trained from scratch. The initial learning rate is 0.001. The network is trained for 70 epochs. The origin image size is 2130×1864, we resize the image into 1024×1024. For shape template, we learn 40 templates for nucleus top and 40 templates for nucleus bottom from the training set. The offset range of template is set to $[-10, 10]$. As shown in Table 1, the U-shaped network is better than U-Net [6] and M-Net [1] because of the larger reception field. Side-output layer adds the supervision to mid layers and eases the difficulty to train the network. Side-output layer also integrates different scale information and obviously improve the results. However, multi-scale structure such as M-Net shows no improvement. The shape template refines the boundary of nucleus predicted by the U-shaped network with side-output. The entire refinement process can be seen as finding the most similar template in the training set to replace the boundary of nucleus. Using shape template to refine the nucleus boundary not only improves the performance but also makes the prediction consistent with the physiological structure of nucleus. More results are shown in Fig. 5.

Table 1. Accuracy of the segmentation

Method	Nucleus top	Nucleus bottom	Cortex top	Cortex bottom
U-Net [6]	12.4	12.8	3.19	2.73
M-Net [1]	15.75	13.04	3.03	2.82
U-shaped Net (ours)	8.57	9.83	**2.45**	2.74
Side output (ours)	8.28	8.96	2.49	**2.43**
Shape template (ours)	**7.09**	**7.94**		

Note: Shape template only refines the boundary of nucleus

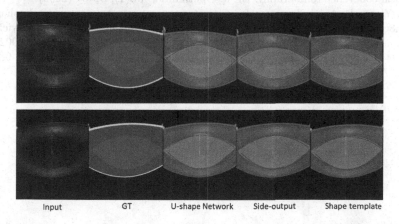

Input   GT   U-shape Network   Side-output   Shape template

Fig. 5. Example results. From left to right: Input image, ground truth, prediction of the U-shape Network, prediction after side-output layer, prediction after the refinement. Blue lines are ground truth annotations. (Color figure online)

4   Conclusions

In this paper, we use a U-shaped network to produce a preliminary mask and we design a shape template to refine the mask. The experiments show the effectiveness of our method. After the refinement, the result satisfies the physiological structure of the lens.

References

1. Fu, H., Cheng, J., Xu, Y., Wong, D.W.K., Liu, J., Cao, X.: Joint optic disc and cup segmentation based on multi-label deep network and polar transformation. IEEE Trans. Med. Imag. **37**(7), 1597–1605 (2018)
2. Fu, H., et al.: Segmentation and quantification for angle-closure glaucoma assessment in anterior segment OCT. IEEE Trans. Med. Imag. **36**(9), 1930–1938 (2017)
3. Fujimoto, J., Huang, D.: Foreword: 25 years of optical coherence tomography. Invest. Ophthalmol. Vis. Sci. **57**(9), OCTi–OCTii (2016)
4. Kaufman, P.L., Levin, L.A., Adler, F.H., Alm, A.: Adler's Physiology of the Eye. Elsevier Health Sciences, New York (2011)
5. Panthier, C., Burgos, J., Rouger, H., Saad, A., Gatinel, D.: New objective lens density quantification method using swept-source optical coherence tomography technology: comparison with existing methods. J. Cataract Refract. Surg. **43**(12), 1575–1581 (2017)
6. Ronneberger, O., Fischer, P., Brox, T.: U-Net: convolutional networks for biomedical image segmentation. In: Navab, N., Hornegger, J., Wells, W.M., Frangi, A.F. (eds.) MICCAI 2015. LNCS, vol. 9351, pp. 234–241. Springer, Cham (2015). https://doi.org/10.1007/978-3-319-24574-4_28
7. Smith, K., Carleton, A., Lepetit, V.: Fast ray features for learning irregular shapes. In: IEEE 12th International Conference on Computer Vision, pp. 397–404. IEEE (2009)
8. Wong, A.L., et al.: Quantitative assessment of lens opacities with anterior segment optical coherence tomography. Br. J. Ophthalmol. **93**(1), 61–65 (2009)
9. Xu, Y., et al.: Axial alignment for anterior segment swept source optical coherence tomography via robust low-rank tensor recovery. In: Ourselin, S., Joskowicz, L., Sabuncu, M.R., Unal, G., Wells, W. (eds.) MICCAI 2016. LNCS, vol. 9902, pp. 441–449. Springer, Cham (2016). https://doi.org/10.1007/978-3-319-46726-9_51

Local Estimation of the Degree of Optic Disc Swelling from Color Fundus Photography

Samuel S. Johnson[1(✉)], Jui-Kai Wang[1,2], Mohammad Shafkat Islam[1],
Matthew J. Thurtell[3], Randy H. Kardon[2,3], and Mona K. Garvin[1,2]

[1] Department of Electrical and Computer Engineering, The University of Iowa,
Iowa City, IA, USA
{sam-johnson,mona-garvin}@uiowa.edu
[2] Iowa City VA Health Care System, Iowa City, IA, USA
[3] Department of Ophthalmology and Visual Sciences, The University of Iowa,
Iowa City, IA, USA

Abstract. Swelling of the optic nerve head (ONH) is most accurately quantitatively assessed via volumetric measures using 3D spectral-domain optical coherence tomography (SD-OCT). However, SD-OCT is not always available as its use is primarily limited to specialized eye clinics rather than in primary care or telemedical settings. Thus, there is still a need for severity assessment using more widely available 2D fundus photographs. In this work, we propose a machine-learning approach to locally estimate the degree of the optic disc swelling at each pixel location from only a 2D fundus photograph as the input. For training purposes, a thickness map of the swelling (reflecting the distance between the top and bottom surfaces of the ONH and surrounding retina) as measured from SD-OCT at each pixel location was used as the ground truth. A random-forest classifier was trained to output each thickness value from local fundus features pertaining to textural and color information. Eighty-eight image pairs of ONH-centered SD-OCT and registered fundus photographs from different subjects with optic disc swelling were used for training and evaluating the model in a leave-one-subject-out fashion. Comparing the thickness map from the proposed method to the ground truth via SD-OCT, a root-mean-square (RMS) error of 1.66 mm^3 for the entire ONH region was achieved, and Spearman's correlation coefficient was $R = 0.73$. Regional volumes for the nasal, temporal, inferior, superior, and peripapillary regions had RMS errors of 0.64 mm^3, 0.61 mm^3, 0.74 mm^3, 0.71 mm^3, and 1.30 mm^3, respectively, suggesting that there is enough evidence in a singular color fundus photograph to estimate local swelling information.

1 Introduction

For many years, color fundus photographs have been a common imaging modality for ophthalmologists to examine the back of the eye in cases of optic

© Springer Nature Switzerland AG 2018
D. Stoyanov et al. (Eds.): COMPAY 2018/OMIA 2018, LNCS 11039, pp. 277–284, 2018.
https://doi.org/10.1007/978-3-030-00949-6_33

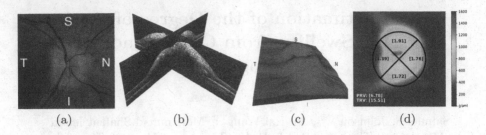

(a)   (b)   (c)   (d)

Fig. 1. (a) An example registered and cropped fundus photograph at the optic-nerve-head (ONH) region with the nasal (N), temporal (T), superior (S), and inferior (I) sides marked. (b) A 3D rendering of the corresponding OCT image depicting a swollen optic disc. (c) A surface rendering of the internal limiting membrane (ILM) shown in red and the retinal pigment epithelium complex (RPE) shown in blue. (d) The ONH thickness map with a grid showing regional volumes (in mm^3) displayed within the grid, peripapillary volume (PRV), and total retinal volume (TRV) both shown in the bottom left-hand corner (also in mm^3). (Color figure online)

disc swelling Fig. 1(a) [5]. Traditionally, assessment of swelling via fundus photographs has been a very challenging task due to the specialized expertise required and the inability to compute volumetric measures of the swelling [8]. With the introduction of spectral-domain optical coherence tomography (SD-OCT) Fig. 1(b) and development of associated retinal-layer-segmentation algorithms [3,10] Fig. 1(c), SD-OCT-based volumetric measures have recently shown great promise in overcoming many of the limitations of a traditional fundus-based assessment.

However, SD-OCT is not always available as its use is primarily limited to specialized eye clinics rather than in primary care or telemedical settings. Thus, the need for better approaches for the assessment of optic disc swelling via fundus photographs alone still exists. In this area, Echegaray et al. [2] proposed a decision-tree system that automatically grades/stages the optic disc swelling using image features directly extracted from monocular fundus photographs. However, this approach only provides ordinal severity grades rather than continuous volumetric measures. Tang et al. [9] demonstrated that the stereoscopic color fundus photographs have the potential to reconstruct the depth information and allow the volumetric estimation for the optic disc swelling, but the requirement of carefully acquired stereo images rather than monocular images limits its applicability. More recently, Agne et al. [1] proposed a regression approach, which can directly estimate the total retinal volume (TRV) at the optic-nerve-head (ONH) region by only inputting a single fundus photograph; however this approach doesn't predict local thickness values as may be needed to compute regional volumes.

Thus, to overcome the limitations above, we propose a machine-learning method that estimates the local volumetric information by only requiring a single monocular color fundus photograph Fig. 1(a) as the input. The proposed method outputs a thickness map with 200×200 pixels covering $6 \times 6\,mm^2$ at the ONH region. Based on the resulting thickness map, the volumes of the peripapillary region, the nasal, temporal, inferior, and superior quadrants, as well as the TRV can be computed shown in Fig. 1(d). Results are quantitatively assessed using the root-mean-square errors between the model's outputs and the OCT ground truths, as well as Spearman's rank correlation coefficients. Visualizations of the predicted thickness maps are also provided for qualitative assessment.

2   Methods

2.1   Overview

For the purpose of a fair comparison between the two image modalities, the input fundus photographs were registered to the SD-OCT image domain and centered/cropped at the ONH Fig. 1(a). Next, the blood vessels were inpainted and the resulting fundus image had features pertaining to textural and color information extracted. Based on these selected features, random forest classifiers, which will be discussed more thoroughly in Sect. 3, were trained to estimate the depth information of the retina at the pixel level and be able to output an ONH thickness map that makes regional volumetric measurements computable Fig. 1(d).

2.2   Preprocessing

We first registered the input fundus photograph with the ONH-centered SD-OCT en-face image. In particular, the SD-OCT images were segmented using 2D/3D graph-theoretical algorithms [3,10], and the en-face image was created by averaging the pixel intensities along each A-scan within the retinal pigment epithelium (RPE) complex. After that, we applied blood vessel inpainting on the ONH-centered/registered fundus photograph Fig. 2(a) to suppress the negative effects from blood vessels on the predicted thickness map. During the processes of vessel inpainting, a blood vessel probability map was computed using a deep learning based approach using U-Net [6] Fig. 2(b). Then, this vessel probability map was thresholded ($p = 0.5$) into a blood vessel mask. Next, a binary morphology dilation (spherical filter size: $r = 1$ pixel) was used to ensure that the vessels were completely encompassed by the mask. By overlapping the cropped fundus photograph with the dilated blood vessel mask Fig. 2(c), a blood vessel inpainted image using second order interpolation was created Fig. 2(d).

2.3   Feature Extraction

In the processes of assessing optic disc swelling via fundus photographs, neuro-ophthalmologists grade the swelling severity by inspecting key observable features on the image. Similarly, in this work, several feature sets (categories include: image intensity, color representations, gradient, and texture information) were extracted from the inpainted image to help the proposed classifier to

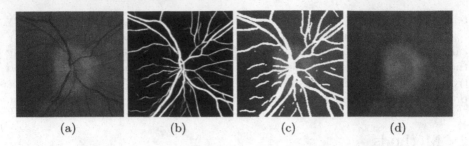

(a)   (b)   (c)   (d)

Fig. 2. (a) An example cropped ONH-centered fundus image. (b) The corresponding blood vessel probability map obtained from a deep-learning neural network. (c) Overlapped dilated vessel mask on the cropped fundus photograph. (d) The blood vessel inpainted image.

Table 1. Complete list of features for regression analysis.

Feature list			
Number	Description	Number	Description
1–5	GLCM properties[†] (10 × 10)	30–49	Gabor responses
6–10	GLCM properties[†] (15 × 15)	50–52	RGB intensity values
11–15	GLCM properties[†] (20 × 20)	53–55	HSV intensity values
16–27	Entropy	56	Histogram equalized image
28	Gradient magnitude	57	Saturated image
29	Gradient direction	58	Grayscale intensity

[†]GLCM properties include: variance, contrast, entropy, homogeneity, and inertia

estimate the thickness information at pixel-level; a total of 58 features were used as listed in Table 1.

To quantify the textural information, Gabor filters [7] are commonly used to analyze image objects with specific combinations of frequencies, directions, and regions of interest. Here, Gabor magnitude responses were computed at $0°$, $45°$, $90°$, and $135°$ with wavelengths of two and four pixels at each orientation Fig. 3.

Textural features were also obtained via use of gray-level co-occurrence matrices (GLCM) which involves statistically considering the spatial relationship of pixels [4]. The GLCMs were computed for each pixel in the inpainted image at an offset of one pixel at the right using three different neighborhood sizes: $10 × 10$, $15 × 15$, and $20 × 20$. For each GLCM, statistical properties, including variance, contrast, entropy Fig. 4(a), homogeneity Fig. 4(b), and inertia, were used to create different feature images.

In addition, entropy can also be used to evaluate the image information in a particular region of interest. In this work, both small and large sliding window sets were applied on the inpainted fundus photographs to compute image entropy

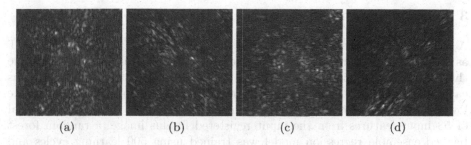

(a)   (b)   (c)   (d)

Fig. 3. Example Gabor responses with a wavelength of two pixels with directions (a) 0°, (b) 45°, (c) 90°, (d) 135°.

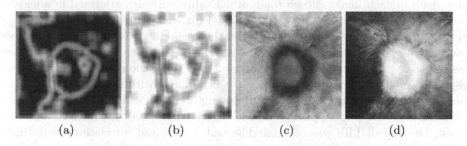

(a)   (b)   (c)   (d)

Fig. 4. Example feature images. (a) GLCM entropy (15 × 15). (b) GLCM homogeneity (15 × 15). (c) Saturation image. (d) Histogram equalized fundus image.

in regions[1]. For the smaller windows, these computed entropy values were interpreted as quantitative indicators of the presence of homogeneous textures; for the larger windows, the computed entropy values indicated the presence of heterogeneous textures. The image gradient direction and magnitude were also included as features as well to enhance the borders among the regions with texture and/or color changes. Along with the red, green, and blue intensity values, the inpainted fundus image was also converted to the hue-saturation-value (HSV) color space, and the separate channel values were used as features as well. Differences in color contrast can be accentuated in the hue and saturation Fig. 4(c) channels, which are highly indicative of texture since an opaque texture is more associated with swelling. In addition, the histogram-equalized image Fig. 4(d) as well as an intensity mapped image with the top and bottom one percent of pixels saturated were used as features. Both images work to emphasize contrast in color between regions of differing colors or differing intensities.

---

[1] The small sliding window sizes include: 5 × 5, 7 × 7, 11 × 11, 13 × 13, 15 × 15, 17 × 17, 21 × 21, 25 × 25; the large sliding window sizes include: 37 × 37, 49 × 49, 73 × 73, 101 × 101.

3   Experimental Methods

A total of 88 subjects with optic disc swelling having both volumetric SD-OCT as well as color fundus images were used for experimental analysis. The true thickness information at the ONH (i.e., the ground truth) for each subject was calculated based on the segmented internal limiting membrane (ILM) and the lower bounding surface of the RPE complex in the SD-OCT image. With a total of 58 input features from the input registered fundus image, a random forest bagged ensemble regression model was trained using 500 learning cycles and feature importance was calculated as part of the training process. To reduce computational complexity, the model was trained on two and a half percent of the pixels in each image chosen randomly. Evaluation was performed in a leave-one-subject-out approach so that the model for each subject was obtained by training the classifier on the images from the remaining 87 subjects. After predictions were made for individual pixel locations, volumes were calculated for the peripapillary, nasal, temporal, inferior, and superior regions. The peripapillary region was defined as the region inside a central circle with radius 1.73 mm. The nasal, temporal, inferior, and superior regions were defined as the four interior quadrants of the peripappilary circle using the 135° and 45° lines as boundaries. The overall TRV was calculated as well. Errors and correlations were then calculated for each individual region.

4   Results

When comparing the total retinal volume (TRV) calculated from the retinal thickness predictions generated from the described model and the ground truth from OCT images, a root-mean-square-error of $1.66\,\text{mm}^3$ was achieved. Spearman's correlation coefficient was $R = 0.73$. When comparing regional volumes, the nasal, temporal, inferior, superior, and peripapillary regions had root-mean-square-errors of $0.64\,\text{mm}^3$, $0.61\,\text{mm}^3$, $0.74\,\text{mm}^3$, $0.71\,\text{mm}^3$, and $1.30\,\text{mm}^3$, respectively. The correlations (R) were 0.71 (nasal), 0.72 (temporal), 0.61 (inferior), 0.65 (superior), and 0.75 (peripapillary). Examples of comparisons between the total retinal thickness maps from the SD-OCTs (i.e., ground truths) and from the monocular fundus photographs are shown in Fig. 5.

Average feature importance across all models was calculated by permuting the features and looking for change in the model error. Top features were found to be entropy in large neighborhoods as well as features that accentuate color change, such as hue, saturation Fig. 4(c), or the histogram equalized image Fig. 4(d). All features that had distinctly different values for the optic disc compared to the peripheral area were helpful in distinguishing swollen regions from non-swollen regions.

5   Discussion and Conclusion

In this preliminary study, we have shown that the proposed method demonstrates the monocular fundus photographs as a potentially lower cost but more available

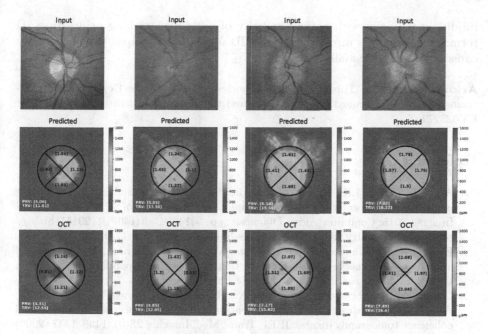

Fig. 5. Input cropped and registered fundus photographs with their accompanying thickness maps generated by SD-OCT (used as the ground truth) as well as the predicted thickness maps generated by the model with regional volumes shown in the grid, and peripapillary plus total volumes shown in the lower left hand corner (in mm³) shown in order of increasing swelling. Each column represents a different subject.

alternative to the SD-OCT in the primary care or telemedical settings in cases of assessment for optic disc swelling. Using features pertaining to textural and color information obtained directly from the fundus photographs, akin to properties neuro-ophthalmologists use, local thickness measurements can be estimated using a regression model trained on a variety of subjects with varying degrees of swelling. The ability to assess depth information at the swollen regions can help clinicians identify onset of diseases in earlier stages; for example, papilledema (a particular type of optic disc swelling due to increased intracranial pressure) often presents with swelling in the nasal quadrant relatively early [11]. In addition, the predicted thickness maps provide a future opportunity to create 3D retinal shape models directly from the 2D monocular fundus photographs. The retinal shape information is hypothetically helpful for distinguishing the different causes of the optic disc swelling.

The limitations of this work may include: (1) the lack of subjects with severe optic disc swelling causes the trained classifiers to slightly underestimate the retinal thickness at the extremely swollen regions, and (2) the thickness predictions at the regions with imperfect vessel inpainting are less accurate due to the inconsistent texture information between the swollen retinal tissue and the vessel residual. Potential future work can involve collecting more SD-OCT and fundus image pairs with various severity of optic disc swelling or creating 3D retinal shape models directly from the 2D fundus photographs to help identify causes of optic disc swelling.

Acknowledgments. This study was supported, in part, by the Department of Veterans Affairs Merit Award I01 RX001786 and the National Institutes of Health R01 EY023279.

References

1. Agne, J., Wang, J.K., Kardon, R.H., Garvin, M.K.: Determining degree of optic nerve edema from color fundus photography. In: Proceedings of SPIE 9414, Medical Imaging 2015: Computer-Aided Diagnosis, pp. 94140F_1–94140F_9 (2015). http://proceedings.spiedigitallibrary.org/proceeding.aspx?doi=10.1117/12.2081423
2. Echegaray, S., Zamora, G., Yu, H., Luo, W., Soliz, P., Kardon, R.: Automated analysis of optic nerve images for detection and staging of papilledema. Investig. Ophthalmol. Vis. Sci. **52**(10), 7470–7478 (2011)
3. Garvin, M.K., Abràmoff, M.D., Wu, X., Russell, S.R., Burns, T.L., Sonka, M.: Automated 3-D intraretinal layer segmentation of macular spectral-domain optical coherence tomography images. IEEE Trans. Med. Imaging **28**(9), 1436–1447 (2009)
4. Haralick, R.M., Dinstein, I., Shanmugam, K.: Textural features for image classification. IEEE Trans. Syst. Man Cybern. SMC **3**(6), 610–621 (1973)
5. Hayreh, S.S.: Optic disc edema in raised intracranial pressure - v. pathogenesis. Archiv. Ophthalmol. **95**(9), 1553–1565 (1977)
6. Ronneberger, O., Fischer, P., Brox, T.: U-Net: convolutional networks for biomedical image segmentation. In: Navab, N., Hornegger, J., Wells, W.M., Frangi, A.F. (eds.) MICCAI 2015 Part III. LNCS, vol. 9351, pp. 234–241. Springer, Cham (2015). https://doi.org/10.1007/978-3-319-24574-4_28
7. Roslan, R., Jamil, N.: Texture feature extraction using 2-D Gabor filters. In: 2012 International Symposium on Computer Applications and Industrial Electronics (ISCAIE), pp. 173–178, December 2012
8. Scott, C.J., Kardon, R.H., Lee, A.G., Frisén, L., Wall, M.: Diagnosis and grading of papilledema in patients with raised intracranial pressure using optical coherence tomography vs clinical expert assessment using a clinical staging scale. Archiv. Ophthalmol. **128**(6), 705–711 (2010)
9. Tang, L., Kardon, R.H., Wang, J.K., Garvin, M.K., Lee, K., Abràmoff, M.D.: Quantitative evaluation of papilledema from stereoscopic color fundus photographs. Investig. Ophthalmol. Vis. Sci. **53**(8), 4490–4497 (2012)
10. Wang, J.K., Kardon, R.H., Kupersmith, M.J., Garvin, M.K.: Automated quantification of volumetric optic disc swelling in papilledema using spectral-domain optical coherence tomography. Investig. Ophthalmol. Vis. Sci. **53**(7), 4069–4075 (2012)
11. Wang, J.K., Miri, M.S., Kardon, R.H., Garvin, M.K.: Automated 3-D region-based volumetric estimation of optic disc swelling in papilledema using spectral-domain optical coherence tomography. In: Proceedings of SPIE 8672, Medical Imaging 2013: Biomedical Applications in Molecular, Structural, and Functional Imaging, pp. 867214_1–867214_8 (2013)

Visual Field Based Automatic Diagnosis of Glaucoma Using Deep Convolutional Neural Network

Fei Li[1], Zhe Wang[3], Guoxiang Qu[2], Yu Qiao[2], and Xiulan Zhang[1(✉)]

[1] Zhongshan Ophthalmic Center, State Key Laboratory of Ophthalmology,
Sun Yat-Sen University, Guangzhou, China
zhangxl2@mail.sysu.edu.cn
[2] Guangdong Key Lab of Computer Vision and Virtual Reality,
Shenzhen Institutes of Advanced Technology, Chinese Academy of Sciences,
Beijing, China
[3] SenseTime Group Limited, Hong Kong, China

Abstract. In order to develop a deep neural network able to differentiate glaucoma from non-glaucoma patients based on visual filed (VF) test results, we collected VF tests from 3 different ophthalmic centers in mainland China. Visual fields (VFs) obtained by both Humphrey 30-2 and 24-2 tests were collected. Reliability criteria were established as fixation losses less than 2/13, false positive and false negative rates of less than 15%. All the VFs from both eyes of a single patient are assigned to either train or validation set to avoid data leakage. We split a total of 4012 PD images from 1352 patients into two sets, 3712 for training and another 300 for validation. On the validation set of 300 VFs, CNN achieves the accuracy of 0.876, while the specificity and sensitivity are 0.826 and 0.932, respectively. For ophthalmologists, the average accuracies are 0.607, 0.585 and 0.626 for resident ophthalmologists, attending ophthalmologists and glaucoma experts, respectively. AGIS and GSS2 achieved accuracy of 0.459 and 0.523 respectively. Three traditional machine learning algorithms, namely support vector machine (SVM), random forest (RF), and k-nearest neighbor (k-NN) were also implemented and evaluated in the experiments, which achieved accuracy of 0.670, 0.644, and 0.591 respectively. In glaucoma diagnosis based on VF, our algorithm based on CNN has achieved higher accuracy compared to human ophthalmologists and traditional rules (AGIS and GSS2). It will be a powerful tool to distinguish glaucoma from non-glaucoma VFs, and may help screening and diagnosis of glaucoma in the future.

Keywords: Glaucoma · Visual field · Machine learning

1 Introduction

Glaucoma is currently the second leading cause of irreversible blindness in the world [1], which is commonly characterized by sustained or temporary elevation of IOP and defects in visual field. Diagnosis of glaucoma depends on the information from various clinical examinations including visual field (VF), optical coherence tomography

© Springer Nature Switzerland AG 2018
D. Stoyanov et al. (Eds.): COMPAY 2018/OMIA 2018, LNCS 11039, pp. 285–293, 2018.
https://doi.org/10.1007/978-3-030-00949-6_34

(OCT) and fundus photo [1, 2]. Fundus photos are easy to capture and frequently used in glaucoma screening. Localization of optic cup and disc is the main clue for machines to make diagnosis [3, 4]. In clinical practice, VF is widely used as the gold standard to judge whether patients have typical glaucomatous damage. Specific patterns of defects such as nasal step and arcuate scotoma shown in visual field indicate existence of glaucoma [5, 6].

Researchers have developed several algorithms based on data from clinical studies, such as Advanced Glaucoma Intervention Study (AGIS) criteria and Glaucoma Staging System (GSS) criteria to grade glaucomatous VFs [5, 7–9]. However, it is hard to diagnose glaucoma depending on VF alone and for early stage glaucoma, even if retinal nerve fiber layer (RNFL) had been damaged there can be no obvious defect in VF. Therefore, it is necessary to develop new algorithm for glaucoma diagnosis. Thus, we designed this study to investigate the performance of deep neural network to identify glaucomatous VFs from non-glaucomatous VFs and to compare the performance of machine against human ophthalmologists.

2   Methods

2.1   Data Preparation

The study was approved by the Ethical Review Committee of the Zhongshan Oph-thalmic Center and was conducted in accordance with the Declaration of Helsinki for research involving human subjects. The study has been registered in clincaltrials.gov (**NCT: 03268031**). All the visual fields (VFs) were obtained by either Humphrey Field Analyzer 30-2 or 24-2 tests. To guarantee reliability, only VFs with fixation losses of less than 2/13, false positive and false negative rates of less than 15% were selected in the experiments. Representative examples of non-glaucoma and glaucoma PD plots are shown in Fig. 1.

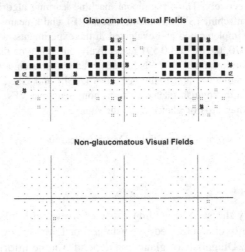

Fig. 1. Representative examples of pattern deviation figures in glaucomatous and non-glaucomatous visual fields.

The probability map of pattern deviation (PD image) is then cropped from the VF report and resized to 224×224 as the input of a deep CNN. All the VFs of both eyes of a single patient are assigned to either training or validation set to avoid data leakage. In this way, we split a total of 4012 PD images into two sets, 3712 for training and another 300 for validation. For data augmentation, we randomly flip the PD images in the training set horizontally to obtain final 7424 training samples. Cross validation is performed by randomly splitting the training and validation sets 3 times and no significant difference is observed. The validation set consists of 150 glaucomatous PD images and 150 non-glaucomatous PD images. The non-glaucomatous PD images include 50 images with only cataract and 150 images with no ocular disease, retinal diseases or neuro-ophthalmic diseases.

2.2 Diagnostic Criteria of Glaucoma

Glaucoma was diagnosed with similar criteria to UKGTS study [10]. VFs of patients who have glaucomatous damage to optic nerve head (ONH) and reproducible glaucomatous VF defects were included. A glaucomatous VF defect was defined as a reproducible reduction of sensitivity compared to the normative database in reliable tests at: (1) two or more contiguous locations with $P < 0.01$ loss or more, (2) three or more contiguous locations with $P < 0.05$ loss or more. ONH damage was defined as C/D ratio ≥ 0.7, thinning of RNFL or both, without a retinal or neurological cause of VF loss.

2.3 Deep CNN for Glaucoma Diagnosis

We adopted the powerful VGG [11, 12] as our network structure. The VGG network consists of 13 convolution layers and 3 fully connected layers. We modified the output dimension of the penultimate layer fc7 from 4096 to 200. And the last layer is modified to output a two-dimension vector which corresponds to the prediction scores of healthy VF and glaucoma VF. The network is first pre-trained on a large scale, natural image classification dataset ImageNet [13] to initialize its parameters. Then we modified the last two layers as mentioned above and initialized their parameters by drawing from a Gaussian distribution. All the parameters of the network were updated by the stochastic gradient descend algorithm with the softmax cross-entropy loss. The network structure is shown in Fig. 2.

2.4 Comparison Between CNN-Based Algorithm and Human Ophthalmologists in Glaucoma Diagnosis

We compared diagnostic accuracy between our algorithm based on deep neural network and ophthalmologists. We chose 9 ophthalmologists in 3 different levels (glaucoma experts: Professor YL-L, XC-D and SJ-F; attending ophthalmologists: Dr. T-S, WY-L and WY-Y; resident ophthalmologists: Dr. X-G, WJ-Z and YY-W), from 4 eye institutes (see details in acknowledgements). None of them has participated in the

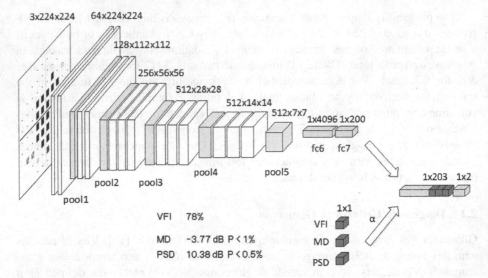

Fig. 2. VGG15 was adopted as our network structure. We modified the output dimension of the penultimate layer fc7 from 4096 to 200. And the last layer is modified to output a two-dimension vector which corresponds to the prediction scores of healthy VF and glaucoma VF. The network is first pre-trained on a large scale, natural image classification dataset ImageNet16 to initialize its parameters. Then we modified the last two layers as mentioned above and initialized their parameters by drawing from a Gaussian distribution.

current research. Attending ophthalmologists are doctors who have clinical training in ophthalmology for at least 5 years, while resident ophthalmologists are doctors who have clinical training in ophthalmology for 1–3 years. Ophthalmologists were shown the PD images alone and requested to assign one of five labels to each PD image, i.e., non-glaucoma, likely non-glaucoma, uncertain, likely glaucoma and glaucoma.

2.5   Traditional Methods for Glaucoma Diagnosis

As a comparison, we also evaluated several rule-based methods and traditional machine learning methods for glaucoma diagnosis.

Rule-based methods included AGIS and GSS methods. For AGIS, a VF is considered to be abnormal if three or more contiguous points in the TD plot are outside of normal limits [8]. GSS2 uses both MD and PSD values to classify VFs into 6 stages [9]. Only stage 0 is considered healthy and other stages are treated as glaucoma.

Moreover, we also compared our method with three other non-deep machine learning algorithms. Support Vector Machine (SVM) [14] maps training samples into high dimensional points that can be separated by a hyperplane as wide as possible. Random Forest (RF) [15] constructs a set of decision trees, and each sample is classified according to the number of training samples of different categories falling into the same leaf node. For k-Nearest-Neighbor (k-NN) [16] method, the sample is classified as healthy or glaucoma by majority voting from its k nearest training samples. Throughout these experiments, we used 52 PD values in VFs obtained in 24-2 test. For 30-2 test, 22 outermost values were discarded so that they can be treated equally. We optimized all the algorithms to improve their performance, e.g., we experimented whether to use Principal Component Analysis (PCA) for preprocessing, different kernel types in SVM, different numbers of trees in RF and various k values in k-NN.

3 Results

Baseline characteristics are shown in Table 1. We totally collected 4012 VF reports, including glaucoma and non-glaucoma reports. To compare the statistical difference between non- glaucoma group and glaucoma group, we run an unpaired test for numerical data and chi-square test for categorical data. It can be observed that there was no significant difference between left eye to right eye ratio ($P = 0.6211$, chi-square test), while age ($P = 0.0022$, unpaired t test), VFI ($P = 0.0001$, unpaired t test), MD ($P = 0.0039$, unpaired t test) and PSD ($P = 0.0001$, unpaired t test) exhibited obvious statistical differences.

Table 1. Baseline characteristics of participants

	Non-glaucoma group	Glaucoma group	P value
No. of images	1623	2389	–
Age (SD)	47.2 (17.4)	49.2 (16.3)	0.0022*
left/right	635/919	607/911	0.6211
VFI (SD)	0.917 (0.126)	0.847 (0.162)	0.0001*
MD (SD)	−5.0 (23.5)	−9.0 (44.8)	0.0039*
PSD (SD)	3.6 (3.3)	6.7 (22.2)	0.0001*

*shows results with a significant difference.
Comparison between non-glaucoma group and glaucoma group
(unpaired t test for numerical data and chi-square test for categorical
data).

To evaluate the effectiveness of the algorithm for automatic diagnosis of glaucoma, we summarized the performance of the proposed algorithm in Table 2.

On the validation set of 300 VFs, our algorithm based on CNN achieved an accuracy of 0.876, while the specificity and sensitivity was 0.826 and 0.932, respectively. In order to compare the results of ophthalmologists with machines, we also developed a software to collect evaluation results from ophthalmologists. Ophthalmologists were shown the PD images alone and requested to assign one of five labels to each image, i.e., non-glaucoma, likely non-glaucoma, uncertain, likely glaucoma and glaucoma. They were strongly advised not to choose the uncertain label. For final evaluation, the non-glaucoma and likely non-glaucoma labels were counted as normal, while the likely glaucoma and glaucoma labels were counted as glaucoma, and the

Table 2. Performance of the algorithm and the compared methods.

	Methods	Accuracy	Specificity	Sensitivity
Ophthalmologists	resident #1	0.640	0.767	0.513
	resident #2	0.593	0.680	0.507
	resident #3	0.587	0.630	0.540
	attending #1	0.533	0.213	0.853
	attending #2	0.570	0.670	0.473
	attending #3	0.653	0.547	0.760
	glaucoma expert #1	0.663	0.700	0.647
	glaucoma expert #2	0.607	0.527	0.687
	glaucoma expert #3	0.607	0.913	0.300
Rule based methods	AGIS	0.459	0.560	0.343
	GSS2	0.523	0.500	0.550
Traditional machine learning methods	SVM	0.670	0.618	0.733
	RF	0.644	0.453	0.863
	k-NN	0.591	0.347	0.870
Deep learning method	**CNN**	**0.876**	**0.826**	**0.932**

uncertain level is considered as a wrong answer. Although the ophthalmologists included three resident ophthalmologists, three attending ophthalmologists and three glaucoma experts, we did not observe significant differences among these three groups. The average accuracies are 0.607, 0.585 and 0.626 for resident ophthalmologists, attending ophthalmologists and glaucoma experts, respectively. However, there exists a huge performance gap between ophthalmologists and CNN, which indicates that CNN may have strong ability to identify the complex patterns presented in the PD images for glaucoma diagnosis. Two rule based methods, AGIS and GSS2, were also compared in the experiment. Both methods are not able to achieve satisfactory results. Interestingly, all the ophthalmologists performed better than GSS2 and AGIS, indicating the importance of human experience in the decision-making process. Three traditional machine learning algorithms were also included in the experiments. SVM performed best among these machine learning methods, but still much worse than CNN.

As shown in Fig. 3, we examined the receiver operating characteristic curve (ROC) of CNN and the compared methods. Our algorithm achieved an AUC of 0.966 (95%CI, 0.948-0.985). It outperformed all the ophthalmologists, rule based methods and traditional machine learning methods by a large margin.

We also studied the relative validation set accuracy as a function of the number of images in the training set. The training set is randomly chosen as a subset of the original training set at rates of (5%, 10%, …, 100%). Each set includes all the images in the smaller subset. As shown in Fig. 4, we can see the performance does not improve too much after the training set includes more than 3612 images.

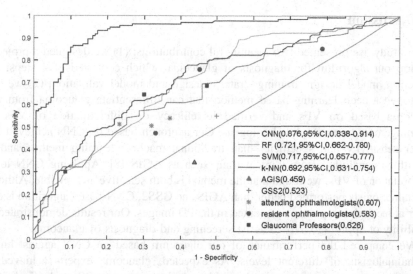

Fig. 3. Performance of CNN, ophthalmologists and traditional algorithms are presented. There were 9 ophthalmologists participating in evaluation of VFs. On the validation set of 300 VFs, CNN achieved an accuracy of 0.876, while the specificity and sensitivity was 0.826 and 0.932, respectively. The average accuracies are 0.607, 0.585 and 0.626 for resident ophthalmologists, attending ophthalmologists and glaucoma experts, respectively. Both AGIS and GSS2 are not able to achieve satisfactory results. Three traditional machine learning algorithms were also included in the experiments. SVM performed best among these machine learning methods, but still much worse than CNN. We also examined the receiver operating characteristic curve (ROC) of CNN and the compared methods. CNN achieved an AUC of 0.966 (95%CI, 0.948–0.985), which outperformed all the ophthalmologists, rule based methods and traditional machine learning methods by a large margin.

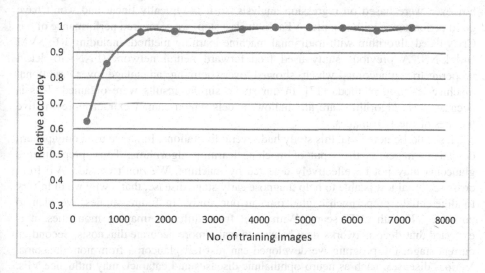

Fig. 4. We studied the relative validation set accuracy as a function of the number of images in the training set. The training set is randomly chosen as a subset of the original training set at rates of (5%, 10%, ..., 100%). Each set includes all the images in the smaller subset. As shown in the figure, the performance does not improve too much after the training set includes more than 3712 images.

4   Discussion

In our study, we presented two meaningful contributions: (1) we designed a project to develop our algorithm for diagnosis of glaucoma, which consisted of 4 steps: data collection, model design, training strategy design and model validation; (2) we have developed a deep learning-based method that can differentiate glaucoma from non-glaucoma based on VFs and verified its efficacy on differentiation of VFs and advantage over human ophthalmologists. Our approach based on CNN achieved both higher sensitivity and specificity than traditional machine learning method and the algorithms concluded from clinical trials such as AGIS [8]. Applying CNN to the interpretation of VFs, we found that the method is both sensitive and reliable. Although ophthalmologists performed better than AGIS and GSS2, CNN-based algorithm is even better at recognizing patterns presented in the PD images. Our results demonstrated the possibility of applying CNN to assist screening and diagnosis of glaucoma.

We compared the performance of our algorithm based on CNN against human ophthalmologists of different levels. As expected, glaucoma experts achieved the highest accuracy in VF interpretation, although there was just 2% and 4% different when compared to attending and resident doctors respectively. With accumulation of clinical experience, doctors tend to have higher specificity while lower sensitivity. Because doctors only have VFs as accessory examination to make a diagnosis, their diagnostic ability was restricted, and they would tend to be more careful about their decision. However, machines got the highest score in the test, achieving highest sensitivity while keeping high specificity. In our second step, we compared performance of our algorithm against 2 criteria summarized from clinical trials, AGIS and GSS2 [8, 9]. AGIS and GSS2 criteria were built to evaluate severity and staging of glaucoma based on VF. VF is divided into different areas with different weights. These algorithms, however, were based on regression analysis, so it is typically linear and won't have good performance with complex VFs. In the last step, we compared performance of our CNN-based algorithm with traditional machine learning method, including RF, SVM and k-NN. A previous study used feed forward neural network (FNN) to detect preperimetric glaucoma, which showed overwhelming advantage over traditional machine learning methods [17]. In our study, similar results were obtained. This is because these algorithms are all shallow models which cannot extract representative features of the PD images.

It should be noted that this study had several limitations. First, we used only pattern deviation images as the input of machine learning algorithms. Thus, preperimetric glaucoma may not be effectively detected by machine. We don't consider VF from cross-sectional test is able to help diagnose early stage disease, that's why we didn't try to differentiate preperimetric glaucoma in our study. In future studies, we plan to combine VF with OCT scans. With input from different imaging modalities, it is expected that deep networks may be able to make more accurate diagnosis. Second, at current stage, the program we developed can just tell glaucoma from non-glaucoma. Various diseases, such as neuro-ophthalmic diseases and cataract, may influence VFs. We hope to extend the function of our deep models to diagnose more ocular diseases.

5 Conclusion

In glaucoma diagnosis based on VF, our algorithm based on CNN has achieved higher accuracy compared to human ophthalmologists and traditional rules (AGIS and GSS2). The accuracy is 0.876, while the specificity and sensitivity are 0.826 and 0.932, respectively, indicating advantages of CNN-based algorithms over humans in diagnosis of glaucoma. It will be a powerful tool to distinguish glaucoma from non-glaucoma VFs, and may help screening and diagnosis of glaucoma in the future.

References

1. Quigley, H.A.: Glaucoma. Lancet **377**(9774), 1367–1377 (2011)
2. Jonas, J.B., Aung, T., Bourne, R.R., Bron, A.M., Ritch, R., Panda-Jonas, S.: Glaucoma. Lancet **390**, 2183–2193 (2017)
3. Fu, H., Cheng, J., Xu, Y., Wong, D.W.K., Liu, J., Cao, X.: Joint optic disc and cup segmentation based on multi-label deep network and polar transformation. arXiv preprint arXiv:180100926 (2018)
4. Fu, H., Cheng, J., Xu, Y., et al.: Disc-aware Ensemble Network for Glaucoma Screening from Fundus Image. IEEE Trans. Med. Imaging (2018)
5. The Advanced Glaucoma Intervention Study (AGIS): 7. The relationship between control of intraocular pressure and visual field deterioration.The AGIS Investigators. Am. J. Ophthalmol. **130**(4), 429–440 (2000)
6. Musch, D.C., Gillespie, B.W., Lichter, P.R., Niziol, L.M., Janz, N.K., Investigators, C.S.: Visual field progression in the collaborative initial glaucoma treatment study the impact of treatment and other baseline factors. Ophthalmology **116**(2), 200–207 (2009)
7. Nouri-Mahdavi, K., Hoffman, D., Gaasterland, D., Caprioli, J.: Prediction of visual field progression in glaucoma. Invest. Ophthalmol. Vis. Sci. **45**(12), 4346–4351 (2004)
8. Advanced Glaucoma Intervention Study. 2. Visual field test scoring and reliability. Ophthalmology **101**(8), 1445–1455 (1994)
9. Brusini, P., Filacorda, S.: Enhanced Glaucoma Staging System (GSS 2) for classifying functional damage in glaucoma. J. Glaucoma **15**(1), 40–46 (2006)
10. Garway-Heath, D.F., Crabb, D.P., Bunce, C., et al.: Latanoprost for open-angle glaucoma (UKGTS): a randomised, multicentre, placebo-controlled trial. Lancet **385**(9975), 1295–1304 (2015)
11. Chatfield, K., Simonyan, K., Vedaldi, A., Zisserman, A.: Return of the devil in the details: delving deep into convolutional nets. arXiv preprint arXiv:14053531 (2014)
12. Simonyan, K., Zisserman, A.: Very deep convolutional networks for large-scale image recognition. arXiv preprint arXiv:14091556 (2014)
13. Russakovsky, O., Deng, J., Su, H., et al.: Imagenet large scale visual recognition challenge. Int. J. Comput. Vision **115**(3), 211–252 (2015)
14. Cortes, C., Vapnik, V.: Support-vector networks. Mach. Learn. **20**(3), 273–297 (1995)
15. Ho, T.K.: Random decision forests. Paper presented at: Document Analysis and Recognition (1995). Proceedings of the Third International Conference
16. Altman, N.S.: An introduction to kernel and nearest-neighbor nonparametric regression. Am. Stat. **46**(3), 175–185 (1992)
17. Asaoka, R., Murata, H., Iwase, A., Araie, M.: Detecting preperimetric glaucoma with standard automated perimetry using a deep learning classifier. Ophthalmology **123**(9), 1974–1980 (2016)

Towards Standardization of Retinal Vascular Measurements: On the Effect of Image Centering

Muthu Rama Krishnan Mookiah[1], Sarah McGrory[2], Stephen Hogg[1],
Jackie Price[4], Rachel Forster[4], Thomas J. MacGillivray[3],
and Emanuele Trucco[1(✉)]

[1] VAMPIRE Project, CVIP, Computing (SSE), University of Dundee,
Dundee, UK
{m.r.k.mookiah,e.trucco}@dundee.ac.uk
[2] Centre for Cognitive Ageing and Cognitive Epidemiology,
University of Edinburgh, Edinburgh, UK
[3] VAMPIRE Project, Centre for Clinical Brain Sciences,
University of Edinburgh, Edinburgh, UK
[4] Usher Institute for Population Health Sciences and Informatics,
University of Edinburgh, Edinburgh, UK

Abstract. Within the general framework of consistent and reproducible morphometric measurements of the retinal vasculature in fundus images, we present a quantitative pilot study of the changes in measurements commonly used in retinal biomarker studies (e.g. caliber-related, tortuosity and fractal dimension of the vascular network) induced by centering fundus image acquisition on either the optic disc or on the macula. To our best knowledge, no such study has been reported so far. Analyzing 149 parameters computed from 80 retinal images (20 subjects, right and left eye, optic-disc and macula centered), we find strong variations and limited concordance in images of the two types. Although analysis of larger cohorts is obviously necessary, our results strengthen the need for a structured investigation into the uncertainty of retinal vasculature measurements, ideally in the framework of an international debate on standardization.

Keywords: Biomarkers · Retina · Microvasculature · VAMPIRE
Uncertainty

1 Introduction and Motivation

The eye is the only organ allowing direct, non-invasive and inexpensive observation of a rich portion of the human microvasculature. Ophthalmoscopic instruments include nowadays fundus cameras, optical coherence tomography (OCT), scanning laser ophthalmoscopes, ultra-widefield angiography, autofluorescence and OCT-angiography. Fundus camera imaging remains the most common modality, given its use in many decades of clinical practice and research. A number of software packages have been developed to quantitate the morphometry of the retinal vasculature efficiently in large numbers of images, e.g. IVAN [22], SIVA [23], QUARTZ [24] and VAMPIRE [25].

© Springer Nature Switzerland AG 2018
D. Stoyanov et al. (Eds.): COMPAY 2018/OMIA 2018, LNCS 11039, pp. 294–302, 2018.
https://doi.org/10.1007/978-3-030-00949-6_35

Coupled with the increasing availability of cross-linked clinical data repositories, the above has enabled a plethora of studies on retinal vascular biomarkers for a variety of conditions, among others diabetes, stroke, dementia, and cardiovascular disease. Morphometric vascular parameters commonly adopted include the central retinal arteriolar/venular equivalents (CRAE, CRVE) and their ratio, the arterio-venous ratio (AVR), all of which summarize measures of vessel calibers around the optic disc (OD); measures of tortuosity; bifurcation coefficients; and the fractal dimension (FD), assessing the complexity of the vascular network. Details of such measurements can be found in many publications; see e.g. MacGillivray *et al.* [1] for an introduction.

A crucial assumption of biomarker studies is that the *retinal vascular measurements are accurate, consistent and reliable.* Accuracy, consistency and reliability depend, in turn, on a considerable number of factors [2, 3], not all of which easily controllable. The effects of several of these factors have been reported in a few studies (Sect. 2). In this paper, we focus on a specific and little investigated factor, *the centering of fundus images.* Centering is commonly of two types: on the macula, named Type I, or on the OD, named Type II (Fig. 1). Clinical protocols require one or both, and for one eye or both (left and right), depending on the pathology of interest. Crucially for our discussion, *a standard ~30°–45° field-of-view (FOV) image centered on the macula does not capture vessels in nasal quadrants* (Fig. 1) *which would be visible in an OD-centered image.* Many studies on retinal vascular biomarkers have drawn on existing clinical repositories of retinal images, but it is not always specified whether images of different types have been analyzed separately. To our best knowledge, no reports exist on the quantitative difference in retinal fundus measurements of the same eye induced by different centering. This is what we present in this paper, contributing to the body of studies summarized in the next section.

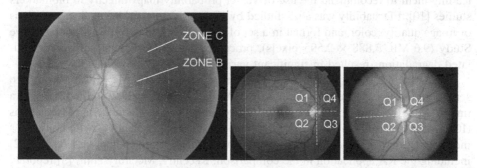

Fig. 1. Illustration of quadrants and retinal coordinates centered on the OD and circular zones used to compute retinal measurements in a left eye (left). The x axis goes through the estimated centers of OD and macula. Examples of Type I (center) and Type II (right) images of right eyes; Type I (macula-centered) images miss parts of the vasculature in nasal quadrants (Q3, Q4).

2 Related Work

The consistency and repeatability of retinal vascular measurements in fundus images, on which we focus (other imaging modalities have been considered, e.g. OCT [5]), has been investigated sparsely. The examples below, far from a systematic review, highlight concerns raised in the last decade. Notice that there is ample variability in the cohorts, numbers and statistical methods used in these studies themselves.

Chandler et al. [4] studied the variations of CRAE and CRVE, measured with IVAN in 30 fundus camera images (3 to 5 Mpixels from various cameras) of 3 subjects, applying systematically increasing blur. They found significant caliber broadening in blurred images, and almost twice as much on average for CRAE (20 µm) as for CRVE (10 µm). Lim et al. [6] analyzed the effect of variations of axial length and myopic refractive errors on the retinal vasculature in an Asian diabetic population (n = 2,882, Singapore Malay Eye Study), finding narrower retinal arterioles and venules, less tortuous arterioles, and increased branching coefficients in both arterioles and venules depending on axial length and refractive errors.

Knudtson et al. [7] analyzed the effect of the pulse cycle on width estimates in fundus images of 30 subjects. The retinal vessel diameter for one large arteriole/venule and one small arteriole/venule were measured by two trained graders. Results showed that the width of large retinal venules were less variable compared to that of arterioles. A related, more recent study was reported by Hao et al. [13].

Motivated by conflicting findings in the literature of the association of diseases with the FD, Huang et al. [8] performed a stability analysis of three FD measures (box, information and correlation coefficient) against different vessel segmentations from human annotators, automatic segmentation methods, threshold values, and regions of interest. Using 20 images from DRIVE [9], the authors observed substantial variations, leading them to recommend the use of vessel probability maps directly in biomarkers studies [10]. FD stability was also studied by Wainwright et al. [11] against variations of image quality, color, and format in a set of 30 images from the Blue Mountains Eye Study (9.6 Mb, 3,888 × 2,595 pixels), processed with IRIS-Fractal software. Simulated degradations resulted in significant variations of the FD coefficient.

Yip et al. [21] compared vessel width measurements (CRAE, CRVE) from two semi-automatic software applications, SIVA and IVAN, with 200 fundus camera images from the Singapore Chinese Eye Study. They found only moderate associations (ICC ∼ 0.5; for ICC see Sect. 3, Analysis) and discordant associations with body mass index and arterial blood pressure. The same authors report similar results in [22] including the RA application in the comparison. Recently, McGrory et al. [3] reported a similar comparison of SIVA and VAMPIRE with 655 images of participants in the Lothian Birth Cohort 1936 studies. They found ICC values indicating poor to limited agreement for all retinal parameters (0.159–0.410), but consistent associations with systemic variables relating to blood pressure, as well as significant differences in the magnitude of association between retinal and systemic variables for 7 of 77 comparisons. We omit some reports of small-scales comparison between software applications for reasons of space.

Other authors analyzed morphometric parameter variations induced by further factors (see [3] and references therein) including image resolution, operators and fundus cameras make and models.

In summary, several, independent authors have measured considerable variability of retinal vascular measurements. Understanding and reducing such variability seems crucial as statistical associations within biomarker studies rely on accurate and consistent measurements. To our best knowledge, we contribute the first quantitative pilot study on the effect of fundus image centering, an important part of any imaging protocol.

3 Materials and Methods

Data Set. 4 fundus-camera images of each of 20 subjects (2 per eye, macula and OD-centered, 80 images in total) were sourced from the Edinburgh Type 2 Diabetes Study (ET2DS), a population-based cohort study designed to investigate potentially modifiable risk factors for cognitive decrements in type 2 diabetes [12]. Images were acquired with a TOPCON TRC-50FX digital fundus camera at $35°$ FOV after pupil dilation using 1% tropicamide. Ethical approval for the ET2DS was granted by the Lothian Research Ethics Committee, and written informed consent was obtained from all participants; see Prince *et al.* [26] for details on the recruitment protocol. The images did not present diabetic lesions upsetting the detection and quantification of the vasculature, hence were considered a suitable sample for our purposes.

Retinal Measurements. All images were measured by a trained operator (author [1]) with VAMPIRE 3.1 (Universities of Dundee and Edinburgh), obtained from its authors [1, 3] following a standard protocol[1]. For each image, VAMPIRE computes 151 measurements (Sect. 1) and their basic statistics (mean, median, standard deviation, max, min). Measurements are computed by vessel type (arteriole or venule), by region (zone, whole image, quadrant) and vessel (path, generation). We considered the 149 measures describing vessel morphology: 39 widths and functions thereof (e.g. CRAE, CRVE, AVR, basic statistics, width gradients, different width estimation algorithms by artery and vein, average ratio length-diameter at branching points), 104 tortuosity measurements, and 6 FD coefficients (3 per vessel network type, arterial or venous).

Analysis. Two-way mixed model intra-class coefficients (ICC) were computed to evaluate the extent of correspondence between two measurements (e.g., right eye versus left eye, or OD centered versus macula centered) of the same parameter (e.g., CRAE). The ICC quantifies this agreement, combining a measure of correlation with a test of the difference in means correcting for systematic bias and agreement based on chance alone. ICCs are thought to be more appropriate for assessing whether two variations in measuring a quantitative parameter provide similar results than Pearson's r, which measures the extent to which two variables are linearly dependent [14]. Method-comparison studies have demonstrated that a perfect linear relationship does

---

[1] VAMPIRE operator training consists of a 3-day module followed by periodic intra- and inter-operator repeatability and consistency checks on sets of images analyzed by certified, experienced operators.

not necessarily reflect good or even moderate agreement as measured by ICC [15, 16]. ICC results are usually interpreted using 0.00–.49 = poor, 0.50–0.74 = moderate, and 0.75–1.00 = excellent [16]. Single-measure coefficients and 95% confidence intervals (CI) as well as correlations (raw, uncorrected Pearson's rs) were also computed. Tortuosity measurements were log-transformed to improve their distributions, which were positively skewed, as done elsewhere [3]. ICCs and Pearson's correlation were used to examine agreement between macula- and OD-centered images (right and left eye separately). In addition, we also analyzed measurement symmetry between right and left eye (macula and OD-centered images separately).

4 Results

Full-result tables are reported in the supplementary material[2] and summarized here. Following a well-established protocol (based on [17] and developments), VAMPIRE requires a minimum number of vessels visible in Zone B and C. These are however only partially visible in macula-centered images (e.g. Figure 1, center), as nasal quadrants are minimally or not at all visible, leading to higher rejection rates than in OD-centered ones (not enough vessels), or to an analysis based on fewer vessels in fewer quadrants. 5 macula-centered images of the right eye had major AVR segments missing in Q2, and one image in Q1. Similarly, 3 macula-centered images of the left eye had AVR segments missing in Q2. All four quadrants were visible in all OD-centered images (right and left eye).

In the right eye, 5 **width-related measurements** (of 39) showed at least moderate correlation, association and significance (defined for our purposes as $r > 0.5$, ICC > 0.6, $p < 0.1$) between OD- and macula-centered image, including CRAE, CRVE, arterial average ratio length-diameter in Zone C and the width gradient of the main artery in Q2 (LDR). Between 13 and 20 images supported these computations for OD- and macula centered images. In the left eye, only 3 width-related measures satisfied our conditions: CRAE (but not CRVE), AVR (not found in the right eye), and the venular (not arterial) LDR, with 12 to 20 images supporting the computation. Only the CRAE and arterial LDR satisfied our conditions in both eyes. For additional illustration, Bland-Altman graphs of two measures (right eyes) are shown in Fig. 2.

Tortuosity measures with at least moderate correlation and ICC (defined as above) between OD- and macula-centered images were only 17 (of 104) in the right and 20 in the left eyes. Of these, only 10 satisfied our conditions in both eyes: 8 arterial and 2 venular tortuosity measures, including 7 taken in Q1 and mean arterial tortuosity in Zone C. Of the 6 **FD measures** (3 for arteries, 3 for veins), only 2 of 20 images of the right eye supported full computation, leading to excellent but obviously not significant correlation and association; but in the left eyes, all 6 measures could be computed on the full set (20 images). Good and significant correlation ($r \sim 0.7$, $p < 0.01$) and moderate association (ICC ~ 0.7) was found for arterial measures only.

---

[2] The Supplementary Material for this article can be found online at: http://vampire.computing.dundee. ac.uk/upload/files/Mookiah-et-al-OMIA-2018-Supplementary-Material/.

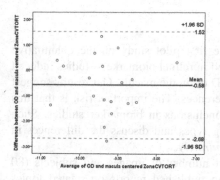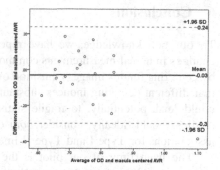

Fig. 2. Bland-Altman plots visualizing, for illustration, the association of venular tortuosity in Zone C (left) and AVR (right) between OD- and macula-centered images (right eyes).

5 Discussion

Number of Images and Vessels Measured. The absence or large occlusion of the nasal quadrants in macula-centered images implies higher image rejection rates (not enough vessels) or smaller number of vessels contributing to measurements compared to OD-centered images. Results show that *values for the same eye vary in the two cases*. What can we say of this variation (in our sample) is summarized below.

Width-Related Parameters. Given our results (Sect. 4), even imposing minimal requirements on r, ICC and significance, the effect of considering different sets of vessels for CRAE/CRVE and AVR calculations is considerable. This is supported by results reported by Heitmar *et al.* [18] in a related analysis.

Tortuosity. Again discrepancies between Type I and II images seem strong. In our sample, only 10 tortuosity measures satisfied our requirements simultaneously in both eyes. We notice that tortuosity values tend to be very small numbers, hence the numerical stability [19] of calculations involving them must be considered carefully.

Fractal Dimension. There was a marked discrepancy between OD- and macula centered images, with venular measures missing altogether in the right eye due to the exclusion of too many macula-centered images. Again this suggests that omitting substantial parts of the nasal quadrants induces substantial changes on FD measures compared to OD-centered images. This supports related findings and concerns by Huang *et al.* [8] on the stability of the FD of the retinal vasculature.

Symmetry. The right-left symmetry of morphometric measurements of the vascular network remains an object of study [20]. Our pilot strengthens the hypothesis that good symmetry levels must not be taken for granted. For instance, CRAE, CRVE, and AVR were poorly and not significantly correlated in macula-centered measurements of either eye; but in the OD-centered images, CRAE and CRVE showed strong correlation (r btw. 0.837 and 0.859, p < 0.001) and excellent agreement (ICC value range, 0.837 and 0.857, 95% CI); and good, significant correlation ($r = 0.673$, p < 0.01) was obtained for AVR. Similar discrepancies were found for tortuosity and FD (details omitted for conciseness).

6 Conclusions

To our best knowledge, we have reported the first pilot study on the quantitative changes in retinal measurements commonly used in retinal biomarkers studies induced by centering fundus image acquisition on the OD or on the macula. Our results suggest that different centering induces substantial differences. The important risk is that this could lead, potentially, to fragile statistical conclusions in biomarker studies. Such studies should, ideally, consider *both* centering types and discuss the differences in associations for Type I and Type II images *separately*.

The main limit of our pilot is the modest number of images and subjects (80 images, $n = 20$), larger however than those in published reports on related topics (Sect. 2). We notice that the question itself of what statistical analysis methods are resilient to what levels of uncertainty and errors requires attention. A second limit is the use of only two, if commonly used, statistics (r and ICC). We plan to extend our analysis to larger samples from independent populations to better understand the effects of centering on morphometric vascular measurements in the retina.

Ultimately, the many aspects of a protocol for reliable biomarker studies, of which centering is only one, require in our view an international collaborative standardization effort, which we strongly auspicate.

Acknowledgements. This work was partially supported by EPSRC grant EP/M005976/1 "Multi-modal retinal biomarkers for vascular dementia". MRK Mookiah is funded by a National Institute for Health Research Global Health Research (NIHR) award (Dundee-Chennai Unit on Diabetes Outcome Research). This paper presents independent research funded by the NIHR. The views expressed are those of the authors and not necessarily those of the NHS, the NIHR or the Department of Health. The support of NHS Lothian R&D, Edinburgh Imaging and the Edinburgh Clinical Research Facility at the University of Edinburgh is gratefully acknowledged.

ET2DS was supported by a grant from the UK Medical Research Council. We would like to thank participants of ET2DS and staff at the Wellcome Trust Clinical Research Facility and Princess Alexandra Eye Pavilion in Edinburgh.

References

1. MacGillivray, T., Trucco, E., Cameron, J., Dhillon, B., et al.: Retinal imaging as a source of biomarkers for diagnosis, characterization and prognosis of chronic illness or long-term conditions. Br. J. Radiol. **87**, 2014 (1040)
2. Trucco, E., Ruggeri, A., Karnowski, T., Giancardo, L., et al.: Validating retinal fundus image analysis algorithms: issues and a proposal. Invest. Ophthalmol. Vis. Sci. **54**, 3546–3559 (2013)
3. McGrory, S., Taylor, A., Pellegrini, E., Ballerini, L., et al.: Towards standardization of quantitative retinal vascular parameters: comparison of SIVA and VAMPIRE measurements in the Lothian Birth Cohort 1936. Transl. Vis. Sci. Tech. **7**(2), 12 (2018)
4. Chandler, C.S., Gangaputra, S., Hubbard, L.D., Ferrier, N.J., et al.: Suboptimal image focus broadens retinal vessel caliber measurement. Inv. Ophth. Vis. Sci. **52**(12), 8558–8561 (2011)

5. Shin, J.W., Shin, Y.U., Uhm, K.B., Sung, K.R., et al.: The effect of optic disc center displacement on retinal nerve fiber layer measurement determined by spectral domain optical coherence tomography. PLoS ONE **11**(10), e0165538 (2016). https://doi.org/10.1371/journal.pone.0165538
6. Lim, L.S., Cheung, C.Y., Lin, X., Mitchell, P., et al.: Influence of refractive error and axial length on retinal vessels geometric characteristics. Inv. Ophthalmol. Vis. Sci. **52**(2), 669–678 (2011)
7. Knudtson, M.D., et al.: Variation associated with measurement of retinal vessel diameters at different points in the pulse cycle. Br. J. Ophthalmol. **88**(1), 57–61 (2004)
8. Huang, F., Zhang, J., Bekkers, E.J., Dashtbozorg, B., et al.: Stability analysis of fractal dimension in retinal vasculature. In: Proceedings of the International MICCAI Workshop on Ophthalmic Medical Image Analysis (OMIA), Munich (2015)
9. Staal, J., Abramoff, M., Niemeijer, M., et al.: Ridge based vessel segmentation in color images of the retina. IEEE Trans. Med. Imag. **23**(4), 501–509 (2004)
10. Grauslund, J., Green, A., Kawasaki, R., et al.: Retinal vascular fractals and microvascular and macrovascular complications in type 1 diabetes. Ophthalmology **117**(7), 1400–1405 (2010)
11. Wainwright, A., Liew, G., Burlutsky, G., Rochtchina, E., et al.: Effect of image quality, color, and format on measurement of retinal vascular fractal dimension. Inv. Ophth. Vis. Sci. **51**(11), 5525–5529 (2011)
12. Ding, J., Strachan, M.W., Reynolds, R.M., Frier, B.M., et al.: Diabetic retinopathy and cognitive decline in older people with type 2 diabetes: the Edinburgh type 2 diabetes study. Diabetes **59**(11), 2883–2889 (2010)
13. Hao, H.A.O., Sasongko, M.B., Wong, T.Y.: Does retinal vascular geometry vary with cardiac cycle? Retinal vessel geometry during cardiac cycle. Inv. Ophthalmol. Vis. Sci. **53** (9), 5799–5805 (2012)
14. Lee, J., Koh, D., Ong, C.N.: Statistical evaluation of agreement between two methods for measuring a quantitative variable. Comput. Biol. Med. **19**, 61–70 (1989)
15. Bédard, M., Martin, N.J., Krueger, P., Brazil, K.: Assessing reproducibility of data obtained with instruments based on continuous measurements. Exp. Aging Res. **26**, 353–365 (2000)
16. Portney, L.G., Watkins, M.P.: Foundations of Clinical Research: Applications to Practice. Pearson/Prentice Hall, Upper Saddle River (2009)
17. Knudtson, M.D., Lee, K.E., Hubbard, L.D., Wong, T.Y., et al.: Revised formulas for summarizing retinal vessel diameters. Curr. Eye Res. **27**(3), 143–149 (2003)
18. Heitmar, R., Kalitzeos, A.A., Panesar, V.: Comparison of two formulas used to calculate summarized retinal vessel calibers. Optom. Vis. Sci. **92**(11), 1085–1091 (2015)
19. Strang, G.: Linear Algebra and Applications, 5th edn. Wesley-Cambridge Press, Cambridge (2016)
20. Cameron, J.R., Megaw, R.D., Tatham, A.J., McGrory, S., et al.: Lateral thinking: interocular symmetry and asymmetry in neurovascular patterning, in health and disease. Prog. Retinal Eye Res. **59**, 131–157 (2017)
21. Yip, W.F., Cheung, C.Y., Hamzah, H., Han, C., et al.: Are computer-assisted programs for measuring retinal vascular caliber interchangeable? Inv. Ophthmol. Vis. Sci. **53**, 4113 (2012)
22. Wong, T.Y., Islam, F.M., Klein, R., et al.: Retinal vascular caliber, cardiovascular risk factors and inflammation: the multi-ethnic study of atherosclerosis (MESA). Inv. Ophthmol. Vis Sci. **47**, 2341–2350 (2006)
23. Cheung, C.Y., Tay, W.T., Mitchell, P., et al.: Quantitative and qualitative retinal microvascular characteristics and blood pressure. J. Hypertens. **29**, 1380–1391 (2011)

24. Fraz, M.M., Welikala, R.A., Rudnicka, A.R., Owen, C.G., et al.: QUARTZ: quantitative analysis of retinal vessel topology and size. Expert Syst. Appl. **42**(20), 7221–7234 (2011)
25. Trucco, E., Giachetti, A., Ballerini, L., Relan, D., et al.: Morphometric measurements of the retinal vasculature in fundus images with VAMPIRE. In: Lim, J., Ong, S., Xiong, W. (eds.) Biomedical Image Understanding: Methods and Applications. Wiley, Hoboken (2015)
26. Prince, J.F., Reynolds, R.M., Mitchell, R.J., et al.: The Edinburgh type 2 diabetes study: study protocol. BMC Endocr. Disord. **18**(8), 18 (2008)

Feasibility Study of Subfoveal Choroidal Thickness Changes in Spectral-Domain Optical Coherence Tomography Measurements of Macular Telangiectasia Type 2

Tiziano Ronchetti[1,2,3]([✉]), Peter Maloca[2,4,6], Emanuel Ramos de Carvalho[6],
Tjebo F. C. Heeren[5], Konstantinos Balaskas[5,6], Adnan Tufail[6],
Catherine Egan[6], Mali Okada[6], Selim Orgül[4], Christoph Jud[1],
and Philippe C. Cattin[1]

[1] Center for medical Image Analysis and Navigation (CIAN),
Department of Biomedical Engineering, University of Basel, Allschwil, Switzerland
Tiziano.Ronchetti@unibas.ch
[2] OCTlab, Department of Ophthalmology, University of Basel, Basel, Switzerland
[3] HuCE-optoLab, Bern University of Applied Sciences, Biel, Switzerland
[4] University Hospital Basel, Basel, Switzerland
[5] Moorfields Ophthalmic Reading Centre, London, UK
[6] Moorfields Eye Hospital, London, UK

Abstract. Macular Telangiectasia Type 2 (MacTel2) is a disease of the retina leading to a gradual deterioration of central vision. At the onset of the disease a good visual acuity is present, which declines as the disease progresses to cause reading difficulties. In this paper, we present new insights on the vascular changes in MacTel2. We investigated whether MacTel2 progression correlates to changes in the thickness of the choroid. For this purpose, we apply a recently published registration-based approach to detect deviations in the choroid on a dataset of 45 MacTel2 patients. Between 2012 and 2016 these subjects and a control group were measured twice within variable intervals of time in the Moorfields Eye Hospital in the MacTel Natural History Observation and Registry Study. Our results show that in the MacTel2 group the thickness of the choroid increased while in the control group a decrease was noted. Manual expert segmentation and an automated state-of-the-art method were used to validate the results.

Keywords: Macular Telangiectasia Type 2
Choroidal thickness changes · Piecewise rigid registration

1 Introduction

Macular Telangiectasia Type 2 (MacTel2) [3] is a disease of the macula, causing loss of central vision. Typically, in MacTel2, perifoveal vessels leak and/or

© Springer Nature Switzerland AG 2018
D. Stoyanov et al. (Eds.): COMPAY 2018/OMIA 2018, LNCS 11039, pp. 303–309, 2018.
https://doi.org/10.1007/978-3-030-00949-6_36

become dilated. Fluid from leakage causes macular swelling [17], which can decrease visual acuity.

MacTel2 can be present without noticeable symptoms in the early stages. As the disease progresses, blurring, distorted vision, and loss of central vision can be detected and lead to a loss of central vision over a period of 10 to 20 years. MacTel2 does not affect side vision, nor does it usually lead to total blindness. Due to the lack of early symptoms and an adequate therapy for MacTel2, it is important to obtain early and regular eye exams. It may be an inherited disease and, especially in its later stage, it could be mistaken for age-related macular degeneration (AMD) [6], due to similar patterns of neovascularization. But all existing knowledge is restricted to advanced stages of the disease, while its beginnings and early stages are still unclear [3].

MacTel2 is diagnosed with the help of optical coherence tomography (OCT) [12] and specified in detail using fluorescein angiography (FA) [3]. There is no therapy of MacTel2. However, in the case of neovascular vessel growth in the course of MacTel2, the injection of anti-vascular growth agents has been shown to be effective.

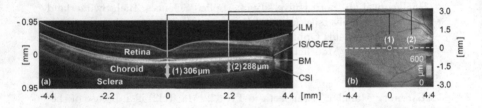

Fig. 1. (a) OCT B-scan with segmented layers corresponding to the white dotted line in (b). Inner Limiting Membrane (ILM), Bruch' s Membrane (BM) and Choroid-Sclera Interface (CSI). IS/OS/EZ denotes the Inner Segment/Outer Segment/Ellipsoid Zone. (b) Choroidal (BM-CSI) thickness map overlaid color-coded on fundus image. The indicated locations correspond to the central (1) and paracentral (2) choroidal thickness measurements, respectively.

Choroidal thickness (ChT) has been demonstrated to be an important bio-marker for MacTel2 [2,7,9] and other pathologies such as age-related macular degeneration (AMD), diabetic macular edema (DME) [17], glaucoma [8] and juvenile myopia [14]. Therefore, monitoring of the choroidal thickness delivers insight into the pathogenesis of such diseases and helps in the planning of their treatment. With this feasibility study, we want to evaluate the impact of the choroidal thickness over time in MacTel2.

State-of-the-Art Research
The authors of [11] examined microcystoid spaces in MacTel2 patients. Evidence was found in the pathogenesis of MacTel2 because the Müller cells dysfunction was reported. Not much research has been done to correlate choroidal thickness changes in the course of MacTel2. The authors of [1] examined choroidal thickness changes in eyes and concluded that the thickness of the choroid did not

vary between the eyes of MacTel2 patients and those of age-matched healthy subjects. In contrast, the authors of [10,15] compared subfoveal choroidal thickness in a similarly large group of MacTel2 patients to that of healthy subjects. They concluded that patients with MacTel2 show a clearly thicker choroid in the subfoveal area.

In this paper, we examine if a correlation between changes in the thickness of the choroid and the development of MacTel2 exists. We validate our results with the help of a state-of-the-art method [5] and manual segmentation by two experts.

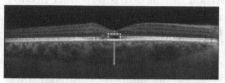 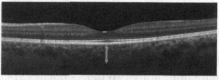

(a) Subfoveal ChT in MacTel2 (arrow)   (b) Subfoveal ChT in a healthy eye (arrow)

Fig. 2. (a) Right eye of a 53-year-old woman with MacTel2. The horizontal B-scan shows the increased subfoveal choroidal thickness of 422 μm (arrow). Typical findings of MacTel2 include an intraretinal outer retinal cavity and loss of the IS/OS/EZ indicated with the dashed box. (b) Healthy right eye of a 53-year-old woman from the control group.

2   Material and Method

The basis of this retrospective study was formed by forty-five MacTel2 patients who were enrolled at Moorfields Eye Hospital in the MacTel Natural History Observation and Registry Study from June 2012 until March 2016.

Their mean age was 57.1 (range 42 to 71). The study was approved by the local institutional review board and conducted according to the tenets of the Declaration of Helsinki. The volunteers were divided into three groups: the first group was measured a second time after 1 year, the second group after 2 years, the third after 4 years. Both eyes were measured.

Optical coherence tomography was acquired using the Heidelberg Spectralis, Human Reliability Analysis 2 (HRA2) system (Heidelberg Engineering, Heidelberg, Germany). The scan pattern for MacTel2 patients was between $3.8 \times 2.5 \times 1.9\,\mathrm{mm}^3$ and $4.4 \times 2.9 \times 1.9\,\mathrm{mm}^3$ including 49–261 B-scans per volume, imaging averaging 8–12 scans, interslice distance was 11–30 μm. The scan pattern for all healthy subjects was defined by raster lines of $4.5 \times 3.0 \times 1.9\,\mathrm{mm}^3$ (261 B-scans, averaging 8–12 scans, interslice distance was 11 μm).

The thickness of the choroid was calculated for each A-scan, defined as the Euclidean distance between the Bruch' s Membrane (BM) and the posterior surface of the choroid, delimited by the Choroid-Sclera Interface (CSI, see Fig. 1(a)).

Choroidal thickness maps for the $6 \times 6\,\mathrm{mm}^2$ macula-centered region imaged by spectral-domain (SD)-OCT scans were created (see Fig. 1(b)) and the average thicknesses were reported in micrometers. The changes in the choroidal thickness, measured between two visits, were automatically detected using CRAR [14], a registration-based algorithm, especially developed for longitudinal studies, which tackles the problems that occur in using image segmentation (i.e. low contrast, loss of signal and artifacts), the common approach to localize the exact position of the CSI.

Instead of segmentation, CRAR uses a piecewise rigid image registration-based approach, in which the changes in thickness in the region around the CSI are examined. In such a way the exact position of the CSI is no longer needed. In CRAR the uniform smoothness of the results is supported by a regularization which matches the anatomic structure of the eye [14].

In the next step, CRAR' s results were compared with those obtained using a graph search-based state-of-the-art segmentation method [5] and manual expert segmentation by two independent experts. Finally, the subfoveal choroidal thickness was determined. Since MacTel2 manifests itself in the juxtafoveal region, it is reasonable to assume that the subfoveal choroidal thickness is reliable for this kind of analysis (see Fig. 2(a)).

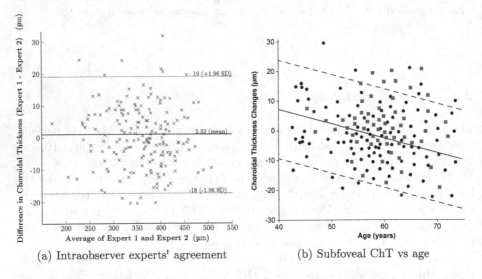

(a) Intraobserver experts' agreement   (b) Subfoveal ChT vs age

Fig. 3. (a) Bland-Altman plot representing the agreement between the choroidal thickness measured by the two experts. (b) Scatter plot comparing the relationship of subfoveal choroidal thickness and age in MacTel2 eyes (red squares) vs healthy control eyes (black dots). The dashed lines indicate 95% normal tolerance limits for choroidal thickness per year.

3   Result

Subfoveal choroidal thickness was manually measured independly by two experts, who showed an agreement with an interobserver correlation R^2 of 0.96 for both the MacTel2 affected and the healthy eyes. A Bland-Altman diagram in Fig. 3(a) shows a mean interobserver difference of 2.32 μm, while the 95% limits of agreement are −18 and 19 μm, respectively.

There was a predominance of females among both the MacTel2 subjects (32 ≅ 71%) and healthy subjects (28 ≅ 62%). However, there was no difference in the results between the groups with respect to gender ($P = 0.068$, Fisher' s exact test). Mean ages in the study and control groups were 58.4 ± 9.3 (≅ 49–68 years) and 52 ± 15.8 (≅ 36–68 years), respectively.

The relationship between changes in the subfoveal choroidal thickness and age in the MacTel2 and control group are shown in Fig. 3(b). All changes in the subfoveal choroidal thickness of both the MacTel2 and control group subdivided into time intervals between the two measurements are illustrated in Fig. 4(a).

On average, the subfoveal choroidal thickness of the MacTel2 patients increased in the time interval between the two measurements with 2.47 ± 7.89 μm (after 1 year), 4.32 ± 13.12 μm (2 years) and 10.85 ± 21.43 (4 years). In the case of healthy eyes, the same measurements showed a decrease of the subfoveal choroidal thickness of −0.71 ± 6.35 μm, −3.92 ± 10.76 μm, −8.37 ± 15.01 μm, respectively. The total mean increase of the MacTel2 group (over all time intervals and subjects) was 5.88 ± 14.15 μm, while in case of the control group this decrease was −4.33 ± 10.71 μm (see Fig. 4(b)). On average, the subfoveal choroidal thickness decreased by 2.55 μm per year.

There is a decrease in the subfoveal choroidal thickness in adults ranging from 14–54 μm every ten years [4,16]. On the other hand, the thickness increased in the MacTel2 group by 2.79 μm per year. The increase of the MacTel2 slope showed no correlation to the decrease of the control group slope, nor the other way round ($P = 0.54$, test of interaction).

For the whole MacTel 2 group, the state-of-the-art algorithm provided mean subfoveal choroidal thickness changes of 5.02 ± 10.55 μm, while both experts in average 6.84 ± 21.14 μm. For the whole control group the detected subfoveal choroidal thickness changes were −4.81 ± 12.87 μm and −6.32 ± 19.14 μm. After conducting a power analysis, we could attest a superior performance of CRAR vs the state-of-the-art-method ($P < 0.05$, medium effect size as Cohen' s d in the range $0.43 − 0.51$, paired t-test), as well as vs the manual expert segmentation ($P < 0.001$, large effect size as Cohen' s distance $d > 0.8$, paired t-test).

4   Discussion

As mentioned above, our main findings is the increase of the choroidal thickness in MacTel2 patients compared to healthy subjects. Similar results were mentioned in [10,15], where authors also came to the conclusion that a statistically significant positive correlation was found exist between subfoveal choroidal thickness and age in MacTel2 subjects, while subfoveal choroidal thickness and age are negatively correlated in healthy adults [4,16]. A thickening of the choroid may be an early manifestation of MacTel2 eyes and therefore be potentially useful for diagnosis and monitoring. In previous studies of MacTel2 eyes, there has been no mention of an abnormal choroid [13]. However, diurnal changes were not considered in this study what can be a limiting factor. Müller cell dysfunction might be behind the pathogenesis of MacTel2, because Müller' s cell loss is commonly associated with macular pigment depletion [13]. Müller cell dysfunction may potentially lead to changes not only in the retinal but also in the choroidal vessels.

Based on our findings of choroidal involvement, the thesis that MacTel2 is only linked to the retina needs to be reconsidered as there is a reasonable assumption that the choroid may also be involved. Another limitation regarding this study is that we only measured the subfoveal choroidal thickness, whereas the rest of the choroid was not evaluated. But because the disease activity is primarily present in the juxtafoveal region, it is nevertheless reasonable to assume that the subfoveal choroidal thickness is reliable for this kind of analysis.

Finally, the age disparity between patients with MacTel2 and healthy subjects should be taken more into consideration in future studies, although even when adjusted for age, using analysis of covariance as shown in [10], the differences in choroidal thickness measurements between MacTel2 and control eyes persisted.

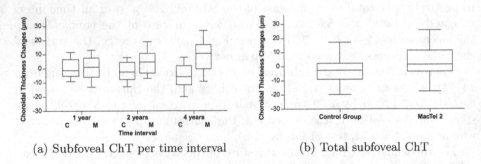

(a) Subfoveal ChT per time interval   (b) Total subfoveal ChT

Fig. 4. (a) Subfoveal ChT changes detected by CRAR in the control (C) and MacTel2 (M) group subdivided into the different time intervals between the two measurements. (b) Total average of subfoveal ChT changes for MacTel2 and control group.

5   Conclusion and Outlook

A thickened choroid may be a valuable diagnostic and monitoring clue in identifying MacTel2, and this may lead to better understand its pathogenesis. Further studies are needed to validate the prognostic value of these findings and to determine whether there are quantitative or qualitative changes in the choroid that predict the onset and progression of this disease.

References

1. Chhablani, J., et al.: Choroidal thickness in macular telangiectasia type 2. Retina **34**(9), 1819–1823 (2014)
2. Chhablani, J., Wong, I.Y., Kozak, I.: Choroidal imaging: a review. Saudi J. Ophthalmol. **28**(2), 123–128 (2014)
3. Issa, P.C., et al.: Macular telangiectasia type 2. Prog. Retin. Eye Res. **34**, 49–77 (2013)
4. Margolis, R., Spaide, R.F.: A pilot study of enhanced depth imaging optical coherence tomography of the choroid in normal eyes. Am. J. Ophthalmol. **147**(5), 811–815 (2009)
5. Mazzaferri, J., Beaton, L., Hounye, G., Sayah, D.N., Costantino, S.: Open-source algorithm for automatic choroid segmentation of oct volume reconstructions. Sci. Rep. **7**, 42112 (2017)
6. McLeod, D.S., Grebe, R., Bhutto, I., Merges, C., Baba, T., Lutty, G.A.: Relationship between RPE and choriocapillaris in age-related macular degeneration. Investig. Ophthalmol. Vis. Sci. **50**(10), 4982–4991 (2009)
7. Mutti, D.O., Gwiazda, J., Norton, T.T., Smith III, E.L., Schaeffel, F., To, C.: Myopia-yesterday, today, and tomorrow. Optom. Vis. Sci. **90**(11), 1161 (2013). Offcial publication of the American Academy of Optometry
8. Mwanza, J.C., Hochberg, J.T., Banitt, M.R., Feuer, W.J., Budenz, D.L.: Lack of association between glaucoma and macular choroidal thickness measured with enhanced depth-imaging optical coherence tomography. Investig. Ophthalmol. Vis. Sci. **52**(6), 3430–3435 (2011)
9. Nickla, D.L., Wallman, J.: The multifunctional choroid. Prog. Retin. Eye Res. **29**(2), 144–168 (2010)
10. Nunes, R.P., et al.: Spectral-domain optical coherence tomography measurements of choroidalthickness and outer retinal disruption in macular telangiectasia type 2. Ophthalmic Surg. Lasers Imaging Retin. **46**(2), 162–170 (2015)
11. Okada, M., Egan, C.A., Heeren, T.F., Tufail, A., Fruttiger, M., Maloca, P.M.: Macular telangiectasia type 2: quantitative analysis of a novel phenotype and implications for the pathobiology of the disease. Retina **38**, S97–S104 (2018)
12. Považay, B., Hermann, B., Unterhuber, A., Hofer, B.: Three-dimensional optical coherence tomography at 1050 nm versus 800 nm in retinal pathologies: enhanced performance and choroidal penetration in cataract patients. J. Biomed. Opt. **12**(4), 041211 (2007)
13. Powner, M.B., Gillies, M.C., Zhu, M., Vevis, K., Hunyor, A.P., Fruttiger, M.: Loss of müller's cells and photoreceptors in macular telangiectasia type 2. Ophthalmology **120**(11), 2344–2352 (2013)
14. Ronchetti, T., et al.: Detecting early choroidal changes using piecewise rigid image registration and eye-shape adherent regularization. In: Cardoso, M.J., et al. (eds.) FIFI/OMIA -2017. LNCS, vol. 10554, pp. 92–100. Springer, Cham (2017). https://doi.org/10.1007/978-3-319-67561-9_10
15. Shah, M., et al.: Subfoveal choroidal thicknessin macular telangiectasia type 2. Investig. Ophthalmol. Vis. Sci. **53**(14), 2138 (2012)
16. Wei, W.B., et al.: Subfoveal choroidal thickness: the Beijing eye study. Ophthalmology **120**(1), 175–180 (2013)
17. Yiu, G., Manjunath, V., Chiu, S.J., Farsiu, S., Mahmoud, T.H.: Effect of antivascular endothelial growth factor therapy on choroidal thickness in diabetic macular edema. Am. J. Ophthalmol. **158**(4), 745–751 (2014)

Segmentation of Retinal Layers in OCT Images of the Mouse Eye Utilizing Polarization Contrast

Marco Augustin[✉], Danielle J. Harper, Conrad W. Merkle,
Christoph K. Hitzenberger, and Bernhard Baumann

Center for Medical Physics and Biomedical Engineering,
Medical University of Vienna, Vienna, Austria
marco.augustin@meduniwien.ac.at

Abstract. Retinal layer segmentation is crucial for the interpretation and visualization of optical coherence tomography (OCT) image data. In this work we utilized a polarization-sensitive OCT system to enhance the segmentation of the retinal pigment epithelium in the mouse retina together with the segmentation of five additional retinal surfaces. Hereby, retinal layers are segmented on a tomogram basis using a graph-based approach in the reflectivity images as well as the cross-polarization images. Thickness changes in the superoxide dismutase 1 (SOD1) knock-out mouse model were assessed and compared to a control group and revealed a thinning of the total and outer retina. Pathological drusen-like lesions were identified in the outer retina. Incorporating additional image contrast offered by the functional extensions of OCT into traditional layer segmentation approaches proved to be valuable. The proposed approach might be extended with other contrast channels such as OCT angiography.

1 Introduction

Retinal thickness changes are associated with various ophthalmic diseases, for example glaucoma [9]. Hence, the development of segmentation algorithms to measure the retinal layers in optical coherence tomography (OCT) images was an early task in ophthalmic image analysis and nowadays clinical OCT systems are equipped with a retinal layer analysis software [1]. Segmentation of the retina does not only allow to assess thickness changes but has lately also been used to enhance data visualization and interpretation in OCT angiography (OCTA) [11]. Various approaches exist for layer segmentation and are summarized in detail elsewhere [1]. While many traditional methods are based on graph theory and dynamic programming, which can be performed in either two dimensions (2D) [5,16] or three dimensions (3D) [7], recent approaches use machine learning theory to label different layers of the retina [6,12]. The aforementioned methods work on conventional OCT images (intensity of the backscattered light). Within

© Springer Nature Switzerland AG 2018
D. Stoyanov et al. (Eds.): COMPAY 2018/OMIA 2018, LNCS 11039, pp. 310–318, 2018.
https://doi.org/10.1007/978-3-030-00949-6_37

the past three decades of OCT development various functional extensions to the conventional OCT emerged. For example OCT based angiography (OCTA) just lately found its way into clinical diagnostics [11]. Other functional extensions such as polarization-sensitive (PS)-OCT, proved to be valuable for identifying pathological hallmarks of ophthalmic diseases such as age-related macular degeneration (AMD) [14]. While these functional extensions are mainly used for gaining additional contrast of pathological structures e.g. fibrosis, neovascularizations or drusen, only a few approaches were presented, where the additional contrast is used for retinal layer segmentation. Segmentation of the RPE was reported in the past by using PS-OCT contrast [8]. Other groups reported the segmentation of the Bruch's membraned using OCTA image contrast [15]. An approach where the OCTA and PS-OCT information is combined to segment the RPE and choroidal stroma in patients was presented recently [3]. Animal models play an important role in preclinical ophthalmology for the understanding of pathological mechanisms and to foster the development of novel treatment strategies for eye diseases. OCT for preclinical imaging of the rodent eye enables longitudinal imaging of the same eye and therefore allows a detailed characterization of pathological processes such as neovascularizations [2,13]. OCTA and PS-OCT imaging were successfully adapted for the rodent eye and have demonstrated their potential in preclinical research. In this work, we developed a retinal layer segmentation algorithm for the mouse retina. Utilizing a PS-OCT system in combination with a graph-based approach for retinal layer segmentation [5] we performed retinal layer thickness analysis in the superoxide dismutase 1 (SOD1) knock-out mice as well as in controls. The SOD1 knock-out mice are used as a model for AMD [10]. Utilizing the segmented retina, hyper-reflective drusen-like deposits were successfully visualized.

2   Methods

2.1   Rodent PS-OCT Imaging

Imaging of the mouse retina was performed using a prototype polarization-sensitive OCT system. The scanning optics were optimized to image the posterior eye of mice and rats. The system featured a broadband lightsource (superluminiscent diode, $\lambda_c = 840\,\text{nm}$, $\Delta\lambda = 100\,\text{nm}$) offering an axial resolution of approximately $3.8\,\mu\text{m}$ in retinal tissue. In contrast to conventional OCT systems, the interferometer is equipped with polarization optics (polarizer, quarter-wave plates, polarizing beam splitter; depicted in green in Fig. 1(a)). The sample is illuminated with light having a defined polarization state and the backscattered light was detected using two cameras, one for the co-polarization state (CAM 1) and the other for the cross-polarization state (CAM2). This setup allows to not only detect the intensity of the backscattered interfered signal, but also to analyze its polarization state. Thereby additional information on the optical properties of the tissue sample can be assessed. For example pigmented tissue, which is found in the RPE or the choroid, scrambles a defined incident polarization state and can be differentiated from polarization preserving tissue, such as most of the inner retinal layers. While tissue which is polarization maintaining (co-polarizing) is detected in CAM1, other ocular tissue causing a change in the polarization of light can be captured with CAM2 (cross-polarization). A simplified sketch of such a differentiation is shown in Fig. 1(b), where only the pigmented layers, namely the RPE and the choroid, are visible with CAM2.

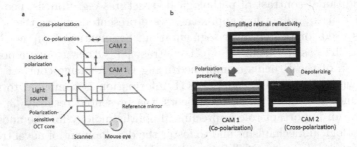

Fig. 1. Polarization-sensitive imaging of the mouse retina. (a) Simplified sketch of a PS-OCT system. (b) Depending on the optical polarization properties of the retina the input polarization state is altered and detected using a pair of spectrometers.

2.2 Image Dataset and Pre-processing

The system was operating at an 83 kHz A-scan rate and the spectral data was acquired with 3072 pixels per channel. A repeated raster scanning protocol (5 repetitions) was used for acquiring images with a field-of-view of $1 \times 1 \, mm^2$ centered at the optic nerve head (ONH). Each volume comprised 512 A-scans (x) and 2000 B-scans at 400 distinct B-scans positions (y). The acquisition time for one dataset was ≈ 15 s. After OCT raw-data processing, the intensity images I_i from CAM1 ($i = 1$) and CAM2 ($i = 2$) were automatically cropped in depth (z) and the repeated B-scans were averaged resulting in a dataset of $nz \times 512 \times 400$ voxels. All images were corrected for axial motion and the retina was flattened with respect to an approximation of the hyper-reflective retinal pigment epithelium (RPE)/choroidal complex before the retinal layers were segmented.

2.3 Segmentation Using Co- and Cross-Polarization Images

In this work the two channels I_1 and I_2 of the PS-OCT systems were used to segment the mouse retinal layers. In PS-OCT the reflectivity is defined as $I_R \propto I_1 + I_2$, where I_1 denotes the co-polarized and I_2 the cross-polarized channel [4]. The cross-polarized image I_2 was used to segment the anterior and posterior surface of the RPE, while inner retinal interfaces were segmented in the reflectivity images I_R. The segmentation approach is based on the algorithm described by Chiu et al. for retinal layer segmentation, where graph theory is used to iteratively segment retinal layers in OCT images [5]. The approach, initially published for human retinal OCT images [5], was later successfully applied to mouse retinal OCT images [16] having a comparable resolution to the system used throughout this work. Hereby, the minimum weighted path connecting the A-scans located at the image borders is found in each B-scan [5,16]. Figure 2 shows a sample reflectivity, co-polarization and cross-polarization B-scan. The images were filtered using a 2D median filter (size: 3×15 $(z \times x)$) where the filter size was chosen to enhance the contrast of the retinal layers in the x-direction but not deblurring thinner retinal layers such as the RPE in the z-direction. Consequently the image gradient was calculated along depth for each pixel and normalized to $[0, 1]$. Exemplary gradient images are shown in Fig. 2 (right column) with an overlay of the gradient profile along depth of the region highlighted in orange. While the gradient in the reflectivity and co-polarized image reveals multiple edges along z, the image gradient G_2 based on the cross-polarized image I_2 only shows substantial gradient changes at the RPE and choroidal interfaces. Hence, the RPE was segmented in the gradient image G_2 while the remaining layers were segmented using G_R. Hereby, a graph was constructed where each pixel corresponded to a node which was connected to neighboring nodes by an directional edge [5,16]. Each node can only have three neighbors (up-right, right, down-right) as the images were traversed from the first (very left) to the last (very right) A-scan. The weights w_{AB} of an edge connecting two nodes A and B at (x, z) were determined by

$$w_{AB} = 2 - G_i(A) + G_i(B) + w_v + w_{min}, \tag{1}$$

where w_v is an additional penalty for traversing up or down (set to 0.5 in this work) and w_{min} is minimum weight chosen to be $1e^{-5}$ [5,16]. As shown in Fig. 2 the RPE can be segmented using the proposed approach quite unambiguously in I_2. Hence, the upper and lower surface of the RPE are segmented first and are used as a reference for other layers. For each interface to be segmented a upper and lower reference border (urb and lrb), the positive G_i or negative $G_{i'}$ gradient image as well as relative (thickness) or absolute distance offset to the reference surfaces (d_{urb} and d_{lrb}) were defined to iteratively segment seven retinal layers. If the same surface was used for the upper and lower reference border, the surface was searched within the neighborhood of the reference border. The parameter settings were based on a pre-evaluation of thickness maps and are summarized in Table 1. The segmentation was carried out on every fourth B-scan (100 B-scans per volume) to reduce computation time, and interpolation was used for the remaining B-scans in a volume.

2.4 Animals

Homozygous superoxide dismutase 1 knock-out mice (SOD1^{m1BCM}) were purchased from The Jackson Laboratory (Stock No. 002972). Heterozygous breeding was established and wildtype $(+/+)$ as well as heterozygous $(+/-)$ litter mates were used as a control group. Four SOD1$^{-/-}$ mice (age range 159–368 days) and five control mice (age range 175–319 days) were used in this work. Eight datasets (four right, four left eyes) of each the knock-out and the control group were analyzed in total. Mice were immobilized during the measurement using inhalational anesthesia (isoflurane). All experiments were performed in accordance with the Association for Research in Vision and Ophthalmology Statement for the Use of Animals in Ophthalmic and Vision Research.

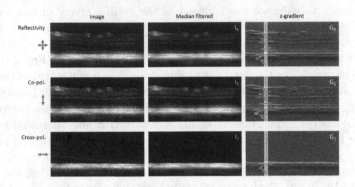

Fig. 2. Segmentation based on the co- and cross-polarization analysis. Left: Reflectivity, co-polarization and cross-polarization B-scans. Center: Images filtered using a 2D median filter. Right: Gradient for each image along z-direction (depth). Depth profile for A-scans are shown as an overlay.

Table 1. Parameter settings for segmentation of retinal layers in the mouse retina.

#	Retinal interface	G_i	urb	lrb	d_{urb}	d_{lrb}
1	RPE anterior	G_2	-	-	-	-
2	RPE posterior	$G_{2'}$	1	1	0	10
3	Inner limiting membrane (ILM)	G_R	-	1	10	85
4	Inner nuclear layer (INL)/outer plexiform layer (OPL)	G_R	3	1	0.3	0.3
5	OPL/outer nuclear layer (ONL)	$G_{R'}$	4	4	1	10
6	Inner plexiform layer (IPL)/INL	$G_{R'}$	3	4	0.5	0.05
7	Ganglion cell layer (GCL)/IPL	$G_{R'}$	3	6	0.05	0.05

3   Results

3.1   Segmentation of Seven Retinal Interfaces

A total of 1600 B-scans (16 datasets) were automatically segmented and visually inspected before further evaluation. An exemplary 3D volume of one dataset is

shown in Fig. 3(a) together with a surface plot of the seven segmented interfaces in Fig. 3(b). A reflectivity tomogram and the seven segmented surfaces are
shown in Fig. 3(c). The total retinal thickness was defined as the axial distance
between the inner limiting membrane and the posterior surface of the RPE. The
outer retina in this work was defined as the slab between the posterior outer
plexiform layer (OPL) and the posterior surface of the RPE. As an example, a
total retinal thickness map is shown in Fig. 3(d). For the evaluation of retinal
thickness changes, the area of the optic nerve head (ONH) was excluded and
only an annulus of 100 pixels $\approx$ 200 μm was evaluated.

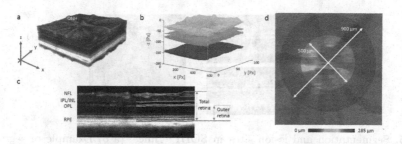

Fig. 3. Segmentation of retinal layers in the mouse eye. (a) 3D rendering of the reflectivity contrast. Datasets are acquired around the optic nerve head (ONH). (b) Seven
retinal interfaces are segmented and (c) used to define the nerve fiber layer (NFL),
the interface between the IPL/INL, the OPL and the RPE. (d) Total retinal thickness
map. An annulus of 200 μm centered around the ONH is defined for the evaluation.

3.2   Thickness Changes in the SOD1$^{-/-}$ Mouse Model

To investigate changes in the retina of SOD1$^{-/-}$ mice the total retinal thickness,
the outer retinal thickness as well as the thickness of the RPE was determined
and compared to control litter mates. Different mice at different ages were analyzed. The total and outer retina was thinner in the knock-out mice when compared to the control group. The RPE thickness was similar in both groups with
a slight trend to RPE thickening in the knock-out mice (Fig. 4).

3.3   Morphological Changes in the Outer Retina of SOD1$^{-/-}$ Mice

The SOD1$^{-/-}$ mice develop drusen-like deposits in the outer retina. Figure 5 shows
two B-scans with hyper-reflective lesions in the outer retina. Segmentation of the
RPE around the lesion site is unaffected at this stage of lesion-development. An
intensity-variance projection in the outer retinal slab reveals the presence of the
hyper-reflective spots around the optic nerve head.

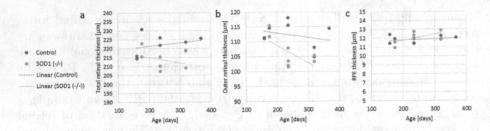

Fig. 4. Retinal thickness changes of the SOD1$^{-/-}$ mice compared with control litter mates. (a) Total retinal, (b) outer retinal and (c) RPE thickness changes.

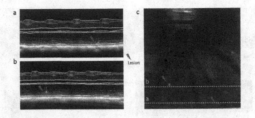

Fig. 5. Segmentation and lesion sites in SOD1$^{-/-}$ mice. (a-b) Example of segmentation around lesions in the SOD1 knock-out mice. (c) An intensity-variance in depth projection in the outer retina reveals hyper-reflective spots.

4   Discussion and Conclusion

In this work a novel framework for segmentation of seven retinal interfaces in the mouse eye was developed. Utilizing the cross-polarization channel demonstrated to be valuable for a robust segmentation of the RPE. Especially the anterior surface of the RPE, which sometimes is challenging to distinguish from the outer segments of the photoreceptors, was differentiated successfully. The remaining inner-retinal layers were segmented using the conventional OCT reflectivity contrast. The RPE plays a major role in several eye diseases, such as AMD, and hence a precise segmentation of the RPE is desired. The SOD1 knock-out mouse model is a popular mouse model in basic research. While younger mice have a regular appearance of the retinal layers, older SOD1$^{-/-}$ show retinal thickness changes, thickened Bruch's membrane, drusen-like structures and choroidal neovascularization [10]. Thickness changes between control and knock-out mice were investigated in an age range between 159 and 368 days. The total and outer retina was thinner in the knock-out mice. Analysis of the RPE thickness showed only minor differences between the groups, although it has to be mentioned that the number of animals and eyes used in this work was rather low (eight eyes for each group). A detailed comparison between the retinal layer segmentation with and without functional extensions (polarization contrast, OCTA) will be part of our future investigations to assess the benefits of using functional extensions of OCT for retinal layer segmentation. This work shows the potential of functional extensions of OCT to enhance retinal layer segmentation. The applied imaging protocol would further allow to incorporate OCTA contrast, which potentially could further improve the segmentation results. For example the retinal vessels in the ganglion cell layer influence the segmentation and cause an overestimation of the thickness. Furthermore, the proposed approach might be exchangeable with OCTA images, where a strong gradient change between the retina and the choroid is observed. A translation to human OCT images comprising polarization information would be feasible as the polarization characteristic of the RPE is similar [3,8]. The proposed approach is based on traditional image analysis methods. Extending the approach to machine learning based approaches or a combination of the two strategies, e.g. as proposed by Fang et al. [6], might further enhance the interpretation of the image data especially if severe pathological tissue changes are present.

References

1. Abràmoff, M.D., et al.: Retinal imaging and image analysis. IEEE Rev. Biomed. Eng. **3**, 169–208 (2010)
2. Augustin, M., et al.: In vivo characterization of spontaneous retinal neovascularization in the mouse eye by multifunctional optical coherence tomography. Invest. Ophthalmol. Vis. Sci. **59**(5), 2054–2068 (2018)
3. Azuma, S., et al.: Pixel-wise segmentation of severely pathologic retinal pigment epithelium and choroidal stroma using multi-contrast jones matrix optical coherence tomography. Biomed. Opt. Express **9**(7), 2955–2973 (2018)
4. Baumann, B.: Polarization sensitive optical coherence tomography: a review of technology and applications. Appl. Sci. **7**(5), 474 (2017)
5. Chiu, S.J., et al.: Automatic segmentation of seven retinal layers in SDOCT images congruent with expert manual segmentation. Opt. Express **18**(18), 19413–19428 (2010)
6. Fang, L., et al.: Automatic segmentation of nine retinal layer boundaries in OCT images of non-exudative AMD patients using deep learning and graph search. Biomed. Opt. Express **8**(5), 2732–2744 (2017)
7. Garvin, M.K., et al.: Automated 3-d intraretinal layer segmentation of macular spectral-domain optical coherence tomography images. IEEE Trans. Med. Imaging **28**(9), 1436–1447 (2009)
8. Götzinger, E., et al.: Retinal pigment epithelium segmentation by polarization sensitive optical coherence tomography. Opt. Express **16**(21), 16410–16422 (2008)
9. Guedes, V., et al.: Optical coherence tomography measurement of macular and nerve fiber layer thickness in normal and glaucomatous human eyes. Ophthalmology **110**(1), 177–189 (2003)
10. Imamura, Y., et al.: Drusen, choroidal neovascularization, and retinal pigment epithelium dysfunction in SOD1-deficient mice: a model of age-related macular degeneration. Proc. Natl. Acad. Sci. **103**(30), 11282–11287 (2006)
11. Kashani, A.H., et al.: Optical coherence tomography angiography: a comprehensive review of current methods and clinical applications. Prog. Retin. Eye. Res. **60**, 66–100 (2017)
12. McDonough, K., et al.: A neural network approach to retinal layer boundary identification from optical coherence tomography images. In: 2015 IEEE CIBCB, pp. 1–8 (2015)

13. Park, J.R., et al.: Imaging laser-induced choroidal neovascularization in the rodent retina using optical coherence tomography angiography. Invest. Ophthalmol. Vis. Sci. **57**(9), OCT331 (2016)
14. Pircher, M., et al.: Polarization sensitive optical coherence tomography in the human eye. Prog. Retin. Eye Res. **30**(6), 431–451 (2011)
15. Schottenhamml, J., et al.: Oct-octa segmentation: a novel framework and an application to segment bruch's membrane in the presence of drusen. Invest. Ophthalmol. Vis. Sci. **58**(8), 645 (2017)
16. Srinivasan, P.P., et al.: Automatic segmentation of up to ten layer boundaries in SD-OCT images of the mouse retina with and without missing layers due to pathology. Biomed. Opt. Express **5**(2), 348–365 (2014)

Glaucoma Diagnosis from Eye Fundus Images Based on Deep Morphometric Feature Estimation

Oscar Perdomo[1](✉), Vincent Andrearczyk[2], Fabrice Meriaudeau[3],
Henning Müller[2], and Fabio A. González[1]

[1] MindLab Research Group, Universidad Nacional de Colombia, Bogotá, Colombia
ojperdomoc@unal.edu.co
[2] University of Applied Sciences Western Switzerland (HES-SO),
Sierre, Switzerland
[3] Universiti Teknologi PETRONAS, Seri Iskandar, Malaysia

Abstract. Glaucoma is an ophthalmic disease related to damage in the optic nerve and it is without symptoms in its early stages. Left untreated, it can lead to vision limitation and blindness. Eye fundus images have been widely accepted by medical personnel to examine the morphology and texture of the optic nerve head and the physiologic cup but glaucoma diagnosis is still subjective and without clear consensus among experts. This paper presents a multi-stage deep learning model for glaucoma diagnosis based on a curriculum learning strategy. In curriculum learning, a model is sequentially trained to solve incrementally difficult tasks. Our proposed model includes the following stages: segmentation of the optic disc and physiological cup, prediction of morphometric features from segmentations, and prediction of disease level (healthy, suspicious and glaucoma). The experimental evaluation shows that our proposed method outperforms conventional convolutional deep learning models from the state of the art reported on the RIM-ONE-v1 and DRISHTI-GS1 datasets with an accuracy of 89.4% and an AUC of 0.82 respectively.

Keywords: Deep convolutional neural network · Curriculum learning
Morphometric features · Glaucoma diagnosis · Eye fundus images

1 Introduction

Glaucoma is one of the leading causes of vision loss and blindness worldwide [14]. It is defined as an increment of intraocular pressure producing morphological changes in the optic disc (OD) and physiological cup (PC) affecting the ability of the optic nerve to transmit images to the brain [15]. The main problem with glaucoma is associated with a delayed diagnosis causing an irreversible damage to the eye [14,15]. The examination of the optic disc, physiological cup and neuroretinal rim structures is important for an early detection and proper treatment [14].

© Springer Nature Switzerland AG 2018
D. Stoyanov et al. (Eds.): COMPAY 2018/OMIA 2018, LNCS 11039, pp. 319–327, 2018.
https://doi.org/10.1007/978-3-030-00949-6_38

The ocular tonometry or measurement of intraocular pressure does not quantify the damage or glaucoma progression [13]. Thus, a complete ophthalmoscopy examination through an eye fundus image is widely used to grade and monitor the disease [7,13]. Additionally, an accurate and objective diagnosis is required to avoid the minimal damage to the eye structure [7]. Thus, the design of computer-aided diagnosis models for automatic disease assessment is important to improve the glaucoma detection and minimize the subjectivity in the diagnosis.

The cup-to-disc ratio (CDR) is the most typical morphometric feature used in the diagnosis of glaucoma. However, locating and segmenting the OD or optic nerve head (ONH) and the physiological cup are not easy tasks. Septiarini et al. proposed an automatic glaucoma detection method extracting statistical features from the intensity in ONH: the mean, smoothness and 3rd moment, and using a k-nearest-neighbor algorithm as a classifier [10]. Pardha et al. reported a region-based active contour model using multiple image channels and gray level properties for optic disc and cup segmentations [7].

Deep learning models, such as Deep Convolutional Neural Networks (DCNNs) have been applied with success to different medical image analysis tasks and, in particular, to automatically discriminate between glaucoma and non-glaucoma patterns in eye fundus images. Al-Bander et al. presented an 8-layer CNN model to automatically extract features of the optic disc from the raw images and a linear Support Vector Machine (SVM) classifier to classify the images into normal or glaucoma subjects [2]. On the other hand, Chen et al. reported a deep learning model that contains four convolutional layers and two fully-connected layers, where a dropout layer and data augmentation strategies are used to improve the performance of glaucoma diagnosis [4]. In addition to this, Orlando et al. fine-tuned two deep learning approaches: OverFeat and VGG pre-trained from non-medical data for automated glaucoma detection on the DRISHTI-GS1 dataset [8]. Sevastopolsky reported a DCNN for automatic OD and PC segmentations, using a modification of the U-Net CNN tested on the DRIONS-DB, RIM-ONE v.3 and DRISHTI-GS1 database [11]. Finally, Abbas presented an unsupervised CNN architecture to extract the features and used a deep belief network model to select the most discriminative deep features with a softmax linear classifier to differentiate between glaucoma and non-glaucoma retinal fundus image [1]. Despite the results obtained, these studies have only been tested on a binary classification task, as many images are not clear cases but in a continuum between healthy and glaucoma.

This paper presents a novel model for automatic analysis of eye fundus images to support glaucoma diagnosis. The model is based on an end-to-end deep convolutional neural network. Deep convolutional neural networks have been highly successful in solving several image analysis tasks. However, they require a large number of labeled samples for training, which is not necessarily the case when dealing with medical images. To mitigate this problem, we devised a curriculum learning strategy that trains the network in stages that solve incrementally more complex tasks [3]. The division of the problem into subtasks allows to better train the different network modules even with a small number of samples. The sequence of tasks is motivated by the current practice for glaucoma diagnosis from eye fundus images by specialists who use morphometric measures estimated from the optical disc and physiological cup segmentations. Thus, the proposed deep learning method is composed of three stages: OD and PC segmentations, morphometric feature estimation and glaucoma detection, which are sequentially trained using a curriculum learning strategy.

The remainder of this paper is organized as follows: in Sect. 2, we give a detailed description of the proposed method including the DCNN used for automatic segmentation, the DCNN for extraction of Morphometric Features (MF), and the multilayer perceptron neural network used for glaucoma diagnosis. In Sect. 3 we define the experimental setup used to split the dataset and the baseline methods to evaluate our proposed method. The results are reported for the three tasks in Sect. 4. Finally, Sect. 5 discusses the results, presents the conclusions and future work.

2   Methods

Figure 1 shows the architecture of the deep neural network model for automatic analysis of eye fundus images to support glaucoma diagnosis. The model is organized in three consecutive stages that are sequentially trained using a curriculum learning approach, i.e. at each stage the training process focuses on different learning goals. This learning strategy regularizes the optimization process to converge faster, guiding the search towards better local minima [3]. The network stages were designed following a process analogous to the one followed by experts. The first stage performs the segmentation of the OD and PC using a 15-layer DCCN. The second stage uses as input the two segmentations generated by the first stage, stacking a third image mask corresponding to the union of the OD and PC segmentations to create a 3D-binary mask, which are fed to a 12-layers DCNN. The goal of this stage is to calculate different morphometric features, which are generally used by experts to diagnose glaucoma. Finally, the third stage applies a multilayer neural network to produce the final prediction that classifies the input image into three possible classes: normal, suspicious or glaucoma. The following subsections discuss the details of the three stages.

2.1   DCNN for Automatic Segmentation of Optic Disc and Physiological Cup

The first stage of the model corresponds to a DCNN that receives as input an RGB eye fundus image and calculates a segmentation of the OD and the PC. The DCNN is based on a deep retinal image understanding (DRIU) model using the last four sub-blocks called coarse feature maps [6], but with two additional convolutional layers. The DRIU model contains 13 convolutional layers with different filters sizes and 4 max-pooling layers [6]. The DRIU model is initialized with VGG weights pretrained on ImageNet. It is fine-tuned for 10,000 epochs with a learning rate of $1e - 6$, which is gradually decreased as the training

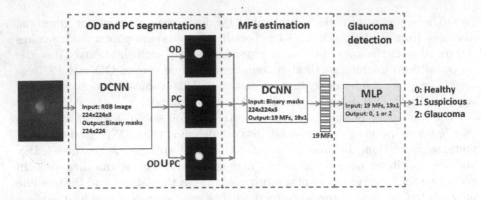

Fig. 1. Block diagram used to segment binary masks (first block), to extract morphometric features (second block), and to classify into healthy, suspicious and glaucoma classes (third block).

process proceeds. A real time data augmentation strategy is implemented to grow the training data. Class weights of 0.1 and 0.9 for background and foreground respectively are used to handle the imbalance of the number of background pixels (nor disc neither cup pixels) compared to foreground pixels (disc and cup pixels). The coarse DRIU feature maps or the 4th, 7th, 10th and 13th convolutional layers are extracted, stacked and up-sampled to generate a binary mask with size 224×224. A modification of the original DRIU was done adding two convolutional layers in cascade with one-padding and kernels size of 3×3 and 1×1 to improve resolution details at the optic and cup edges, without affecting the final size. The model is trained using binary cross-entropy as loss function with ground-truth segmentations of the OD and the PC provided by experts.

2.2 DCNN for Automatic Morphometric Estimation

The second stage takes as input the segmentations of the OD and the PC, along with a third image corresponding to the union, to calculate 19 morphometric features used by Perdomo et al. [9]. The 19 morphometric features can be divided into four subsets: geometric, distances, axis and ratio. The geometric subset contains areas and perimeters of OD and PC; the distance subset is composed of the superior, inferior, nasal and temporal distances between the OD and PC; the axis subset is defined by major and minor axis of the OD and the PC, finally, the ratio subset includes the eccentricity of the OD and PC and the five ratios between upper OD and PC parameters to seek disproportions and relationships between optic disc and the physiologic cup [9].

The DCNN designed for automatic morphometric feature estimation is composed of five convolutional, five pooling layers and two fully connected layers. The four max-pooling and the global average pooling layers are non-linear size reducers that are applied to reduce the spatial dimensions, to minimize overfitting and the number of parameters in the model. Finally, the first fully-connected layer connects all the neurons obtained from the global average-pooling layer with 512 neurons to the next fully-connected layer with the number of morphometric features to predict. The model is trained using mean absolute error as the loss and, morphometric features directly predicted from the OD, PC and union segmentations as ground-truth.

2.3 Multilayer Perceptron for Glaucoma Classification from Morphometric Features

The final stage corresponds to a multilayer perceptron (MLP) that receives as input the 19 morphometric features and classifies them into three possible classes: normal, suspicious or glaucoma. The MLP is composed of two fully-connected layers with 64 hidden and 3 output units. The batch size, number of epochs and initial learning rate used were explored using a grid search strategy, the best performing parameters found experimentally were 16, 500 and 0.01 respectively.

3 Experimental Setup

3.1 Eye Fundus Image Databases for Glaucoma Detection

The DRISHTI-GS1, RIM-ONE_v1 and RIM-ONE_v3 databases are used in this study [5,12]. The DRISHTI-GS1 dataset has been acquired and labeled as healthy and glaucomatous by Aravind Eye Hospital (India), and it contains 101 color fundus images distributed in two subsets: 50 images for training, and the 51 remaining for the testing subset [12]. RIM-ONE_v1 and RIM-ONE_v3 focus on optic nerve head segmentation for glaucoma detection with manual reference segmentations as gold standard with 455 and 159 images respectively, created by ophthalmologists from the Department of Ophthalmology at the Hospital Universitario de Canarias in Spain [5]. The RIM-ONE_v1 dataset was labeled according to a binary classification (healthy vs. glaucomatous), and RIM-ONE_v3 was labeled as a 3-class classification problem (healthy, suspicious and glaucoma).

The proposed method was evaluated in 2 setups: binary-classification task (DRISHTI-GS and RIM-ONE_v1) and 3-class classification (RIM-ONE_v3). Additionally, the proposed method used a stratified sampling to randomly divide two RIM-ONE datasets into three subsets with 60%, 10% and 30% for each class that correspond to training, validation and test sets respectively.

3.2 Evaluation

The proposed method used several performance metrics in each stage. The OD and PC segmentations stage were assessed with the Jaccard index and the Dice coefficient. The MF estimation stage was evaluated using the Mean Average Percentage Error (MAPE) among all the predicted MFs. Moreover, the MFs calculated from the OD and PC segmentations by the experts called Real Morphometric Features (RMFs) were compared to the Estimated Morphometric Features

(EMFs) or the MFs calculated from the OD and PC segmentations by the first stage as reported in Table 1. Finally, the complete proposed method for glaucoma detection was evaluated using accuracy, sensitivity, specificity, Area Under the Curve (AUC), precision, recall, f-score, Kappa coefficient, and Overall Accuracy (OA) performance metrics reported on the test sets.

For the binary task, a combination of an 8-layer CNN model and a linear SVM applied to RIM-ONE_v1 [2], and a fine-tuning of VGG pre-trained on ImageNet applied to DRISHTI-GS1 [8] were chosen as binary-classification baseline models, as reported for the test set in Table 2. The end-to-end data-fusion deep learning model that combines raw color fundus images and RMFs was chosen as the 3-class classification baseline [9] on the RIM-ONE_v3 dataset. Furthermore, a DCNN feeding with color fundus images and a 3D-binary mask described in Sect. 2 was compared to the proposed method. The proposed approach was implemented with Keras[1] using a GeForce GTX TITAN X from NVIDIA.

4   Results

The best performance of the proposed model for OD and PC segmentations (first stage) was obtained with a learning rate of $1e-6$, a batch size of 2, a number of samples per epoch of 300, and a number of epochs of 10.000. The Jaccard index (JI) and Dice coefficient (DC) were monitored during training in the first stage (not reported here), and these parameters were evaluated on the test set for OD segmentation with of $JI = 0.9975$ and $DC = 0.9987$ of, and PC segmentation with $JI = 0.9983$ and $DC = 0.9991$ respectively.

For the second stage, the EMFs presented a MAPE during training of 3.57%, and a MAPE in the test set of 6.30% compared to RMFs. Table 1 presents a comparison between the RMFs and EMFs for 3-class classification on the RIM-ONE_v3 dataset using SVM and Random Forest (RF) classifiers as reported in [9].

Table 1. Performance measures for the RMFs and EMFs on the test RIM-ONE_v3 dataset, bold values show the best score for each performance metric.

Method	Source	Precision	Recall	f-score	Kappa	OA
SVM [9]	RMFs	0.63	0.56	0.55	0.35	0.66
RF [9]	RMFs	**0.64**	0.57	0.58	**0.37**	**0.65**
SVM	EMFs	0.54	0.41	0.39	0.23	0.57
RF	EMFs	0.60	**0.64**	**0.61**	0.35	0.64

Finally, we evaluated the proposed method with baseline methods for binary and 3-class classifications (third stage) as reported in Tables 2 and 3 respectively. These tables present the comparison between the methods, information

---

[1] http://keras.io.

sources and performance metrics evaluated in classification tasks for the two experimental setups respectively.

Table 2. Binary-classification performance metrics for baseline models and the proposed method on the test data set. The bold values show the best score for each performance metric.

Method	Source	Accuracy	Sensitivity	Specificity	AUC
Al-Bander et al. [2]	RGB	0.882	0.85	**0.908**	–
Orlando et al. [8]	RGB	–	–	–	0.763
Proposed method	RGB	**0.894**	**0.895**	0.889	**0.82**

Table 3. Comparison of performance metrics for 3-class classification for baseline models and the proposed method on the test dataset. The bold values show the best score for each performance metric.

Method	Source	Precision	Recall	f-score	Kappa	OA
DCNN + RMF [9]	RMF, RGB	0.46	0.56	0.50	0.42	0.68
DCNN	RGB	0.48	0.55	0.51	0.20	0.55
DCNN	OD -PC-Union	0.60	0.60	0.59	0.29	0.60
Proposed method	RGB	**0.76**	**0.72**	**0.69**	**0.48**	**0.70**

5   Discussion and Conclusions

We present a novel method for automatic glaucoma assessment from eye fundus images based on DCNNs. The results show that our method is competitive with the best results reported for each dataset: RIM-ONE-v1 accuracy of 89.4% vs. 88.2% reported by [2], and DRISHTI-GS AUC of 0.82 vs 0.76 reported by [8], as shown in Table 2. The most remarkable characteristic of this model is its architecture and training strategy. The model is organized in stages that follow a conventional process for glaucoma diagnosis based on the calculus of morphometric features. The multistage architecture allows us to train the model using a curriculum learning approach, which gradually trains the model to accomplish subtasks with increasing complexity. This approach allows training a complex deep learning model with a reduced set of training samples, resulting in an improved performance of the model that was corroborated by the experimental evaluation. In particular, the experimental results showed that the multistage architecture along with the curriculum training, produces better results than conventional DCNNs, as reported in Table 3. The resulting model being end-to-end, it is able to directly produce a prediction from the input image without requiring a manual intermediate segmentation required by the conventional diagnosis protocol from eye fundus images.

The work shows that it is possible to involve domain knowledge in deep learning models. Additionally, intermediate results produced by the model (segmentations and morphometric features) can help the interpretability of the model predictions, making them more useful in support of the diagnosis process. We hypothesize that this approach can be extended to other medical image analysis applications and exploring this hypothesis will be the focus of our future work.

References

1. Abbas, Q.: Glaucoma-deep: detection of glaucoma eye disease on retinal fundus images using deep learning. Int. J. Adv. Comput. Sci. Appl. **8**(6), 41–45 (2017)
2. Al-Bander, B., Al-Nuaimy, W., Al-Taee, M.A., Zheng, Y.: Automated glaucoma diagnosis using deep learning approach. In: 14th International Multi-Conference on SSD, pp. 207–210. IEEE (2017)
3. Bengio, Y., Louradour, J., Collobert, R., Weston, J.: Curriculum learning. In: Proceedings of the 26th Annual International Conference on Machine Learning, pp. 41–48. ACM (2009)
4. Chen, X., Xu, Y., Wong, D.W.K., Wong, T.Y., Liu, J.: Glaucoma detection based on deep convolutional neural network. In: 37th Annual International Conference of the IEEE, EMBC, pp. 715–718. IEEE (2015)
5. Fumero, F., Alayón, S., Sanchez, J., Sigut, J., Gonzalez-Hernandez, M.: Rim-one: an open retinal image database for optic nerve evaluation. In: 24th International Symposium on Computer-Based Medical Systems (CBMS), pp. 1–6. IEEE (2011)
6. Maninis, K.-K., Pont-Tuset, J., Arbeláez, P., Van Gool, L.: Deep retinal image understanding. In: Ourselin, S., Joskowicz, L., Sabuncu, M.R., Unal, G., Wells, W. (eds.) MICCAI 2016. LNCS, vol. 9901, pp. 140–148. Springer, Cham (2016). https://doi.org/10.1007/978-3-319-46723-8_17
7. Mittapalli, P.S., Kande, G.B.: Segmentation of optic disk and optic cup from digital fundus images for the assessment of glaucoma. Biomed. Signal Process. Control. **24**, 34–46 (2016)
8. Orlando, J.I., Prokofyeva, E., del Fresno, M., Blaschko, M.B.: Convolutional neural network transfer for automated glaucoma identification. In: 12th International Symposium on Medical Information Processing and Analysis. vol. 10160, p. 101600U. International Society for Optics and Photonics (2017)
9. Perdomo, O., Arevalo, J., González, F.A.: Combining morphometric features and convolutional networks fusion for glaucoma diagnosis. In: 13th International Conference on Medical Information Processing and Analysis. vol. 10572, p. 105721G. International Society for Optics and Photonics (2017)
10. Septiarini, A., Khairina, D.M., Kridalaksana, A.H., Hamdani, H.: Automatic glaucoma detection method applying a statistical approach to fundus images. Healthc. Inform. Res. **24**(1), 53–60 (2018)
11. Sevastopolsky, A.: Optic disc and cup segmentation methods for glaucoma detection with modification of u-net convolutional neural network. Pattern Recognit. Image Anal. **27**(3), 618–624 (2017)
12. Sivaswamy, J., Krishnadas, S., Chakravarty, A., Joshi, G., Tabish, A.S.: A comprehensive retinal image dataset for the assessment of glaucoma from the optic nerve head analysis. JSM Biomed. Imaging Data Pap. **2**(1), 1004 (2015)

13. Stein, D.M., Wollstein, G., Schuman, J.S.: Imaging in glaucoma. Ophthalmol. Clin. N. Am. **17**(1), 33 (2004)
14. Tham, Y.C., Li, X., Wong, T.Y., Quigley, H.A., Aung, T., Cheng, C.Y.: Global prevalence of glaucoma and projections of glaucoma burden through 2040: a systematic review and meta-analysis. Ophthalmology **121**(11), 2081–2090 (2014)
15. Weinreb, R.N., Aung, T., Medeiros, F.A.: The pathophysiology and treatment of glaucoma: a review. Jama **311**(18), 1901–1911 (2014)

2D Modeling and Correction of Fan-Beam Scan Geometry in OCT

Min Chen[1(✉)], James C. Gee[1], Jerry L. Prince[2], and Geoffrey K. Aguirre[3]

[1] Department of Radiology, University of Pennsylvania, Philadelphia, PA 19104, USA
minchen1@upenn.edu
[2] Department of ECE, Johns Hopkins University, Baltimore, MD 21218, USA
[3] Department of Neurology, University of Pennsylvania,
Philadelphia, PA 19104, USA

Abstract. A-scan acquisitions in OCT images are acquired in a fan-beam pattern, but saved and displayed in a rectangular space. This results in an inaccurate representation of the scan geometry of OCT images, which introduces systematic distortions that can greatly impact shape and morphology based analysis of the retina. Correction of OCT scan geometry has proven to be a challenging task due to a lack of information regarding the true angle of entry of each A-scan through the pupil and the location of the A-scan nodal points. In this work, we present a preliminary model that solves for the OCT scan geometry in a restricted 2D setting. Our approach uses two repeat scans with corresponding landmarks to estimate the necessary parameters to correctly restore the fan-beam geometry of the input B-scans. Our results show accurate estimation of the ground truth geometry from simulated B-scans, and we found qualitatively promising result when the correction was applied to longitudinal B-scans of the same subject. We establish a robust 2D framework that can potentially be expanded for full 3D estimation and correction of OCT scan geometries.

Keywords: Distortion correction · Retina · OCT

1 Introduction

Optical coherence tomography (OCT) imaging has gained widespread popularity in the past decade for the analysis of retinal health and disease in both clinical and research settings. However, a longstanding limitation of OCT imaging is the absence of scan geometry correction to accurately match the image with its physical acquisition space [2,6,9]. A-scan acquisitions in OCT images are acquired in a fan-beam pattern, however they are saved and visualized in a rectangular pattern. This distorts the resulting image in two ways. First, due to the spread of the fan-beam, the arc distance between A-scans increases with the depth in the scan. When represented as a rectangular pattern, this results in structures deeper in the image to appear compressed. Second, the A-scans on

© Springer Nature Switzerland AG 2018
D. Stoyanov et al. (Eds.): COMPAY 2018/OMIA 2018, LNCS 11039, pp. 328–335, 2018.
https://doi.org/10.1007/978-3-030-00949-6_39

the edges of a fan-beam must travel further than the A-scans at the center of the scan. This causes objects in the periphery of the OCT image to be located deeper in the A-scan. When placed in a rectangular pattern, this results in the appearance that objects are curved downwards when moving away from the center of the scan. Figure 1 shows two examples of distortions due to uncorrected scan geometry on a flat object.

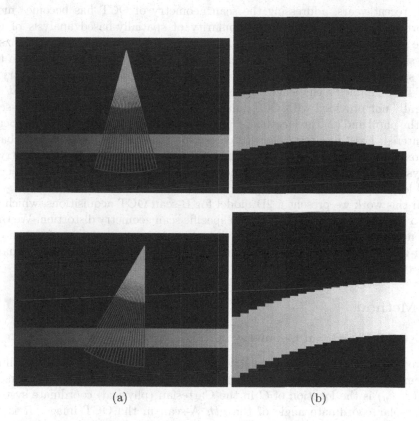

(a)   (b)

Fig. 1. Two examples of simulated OCT scans and the effect of the scan geometry distortion on a flat object. (a) shows the relative positioning between the fan-beam and the object. Blue represents the scan area acquired in the OCT image, and green is the non-acquired region between the nodal point and OCT image. (b) shows the resulting OCT B-scan of the object, where each column in the image corresponds with a blue A-scan line in the respective image in (a). (Color figure online)

Due to the presence of scan geometry distortions, OCT images do not fully represent the correct underlying physical structure of the retina [6]. However, it is interesting to note that errors from this misrepresentation are largely mitigated in contemporary analysis of OCT which focuses on observing regional means of retinal layer thickness. This is because the thickness of a layer is typically measured as the direct distance within each A-scan in the OCT [7]. Neither the depth dependent compression of the fan-beam nor the shift of the A-scans in the periphery influences this thickness measure. The main effect that scan geometry has on thickness values is in the non-uniform path each A-scan takes across a layer, resulting in thicker measurements away from the fovea. However this effect is often masked by the natural concentric increase of layer thickness with distance from the fovea, and the use of mean layer thickness for most analysis.

In recent years, addressing the scan geometry of OCT has becomes more important due to the emerging popularity of spatially-based analysis of the shape [5,8] and morphology [1,3,4] of the retina. In contrast to thickness-based analysis, these methods are highly sensitive to structural distortions due to the local specificity of the techniques. However, correcting for the scan geometry of OCT has remained a challenge due to two missing pieces of information that are generally not provided with OCT images: (1) the angle of entry of each A-scan into the pupil and (2) the location of the A-scan nodal point, where the fan-beam is centered. Existing approaches for correcting the scan geometry rely on basic approximations regarding these parameters, such as assuming the beam entry is always centered or approximating the nodal point location using the subject's axial length [6].

In this work we present a 2D model for B-scan OCT acquisitions, which we use to generate simulated images with specific scan geometry distortion. We then present an analytic approach for solving the missing parameters of the model using trigonometric relationships established between corresponding landmarks from two different scans.

2   Method

2.1   OCT Fan-Beam Geometry

We represent the fan-beam geometry of the OCT image as a polar coordinate system where the origin, C, is the nodal point where the A-scans intersect, and (c_x, c_y) is the location of C in the Cartesian (physical) coordinate system. ϕ_n is polar coordinate angle of the nth A-scan in the OCT image. R is the distance from the nodal point to the top of the OCT scan, and r_n is the A-scan measurement from the top of the OCT scan to the surface of the retina. Figure 2 shows a diagram of this model. For a physical location (x_n, y_n) on the surface of the retina, we can describe it in terms of the image coordinate system using standard conversion between polar and Cartesian coordinates:

$$x_n = (R + r_n)\cos(\phi_n) + c_x \tag{1}$$
$$y_n = (R + r_n)\sin(\phi_n) + c_y. \tag{2}$$

2.2   Two Scan Solution

Given the geometry described in Sect. 2.1, we introduce an approach for solving for the length of R and relative location of C when given two scans (with nodal

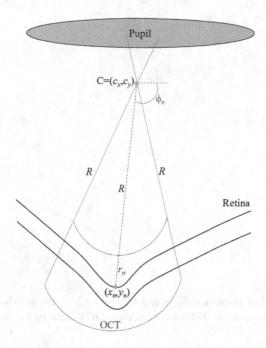

Fig. 2. Diagram of the OCT fan-beam geometry in our model. C is the location of the nodal point, R is a constant distance from C to the beginning of the OCT scan. For the nth A-scan, r_n is the distance from the retina to the beginning of the scan and ϕ_n is the polar angle of the A-scan.

points C_A and C_B) with corresponding landmarks (x_1, y_1) and (x_2, y_2). Figure 3 shows the geometric setup for the problem. We see from the diagram that the nodal points and corresponding landmarks form two triangles with a shared side L. $r_1^A, r_2^A, r_1^B, r_2^B$ are the distance from each A-scan in images A and B to (x_1, y_1) and (x_2, y_2), respectively. $\alpha_{1,2}^A$ and $\alpha_{1,2}^B$ are the angles between the two A-scans associated with the landmarks in each image. These angles can be found as a fraction of the total angular field of view of each scan. Using the law of cosines on both triangles, we can establish the equivalency:

$$L^2 = (R + r_1^A)^2 + (R + r_2^A)^2 - 2(R + r_1^A)(R + r_2^A)\cos(\alpha_{1,2}^A) \qquad (3)$$
$$L^2 = (R + r_1^B)^2 + (R + r_2^B)^2 - 2(R + r_1^B)(R + r_2^B)\cos(\alpha_{1,2}^B). \qquad (4)$$

This allows us to solve for R as a quadratic equation:

$$R = \frac{-b \pm \sqrt{(b^2 - 4ac)}}{2a} \qquad (5)$$

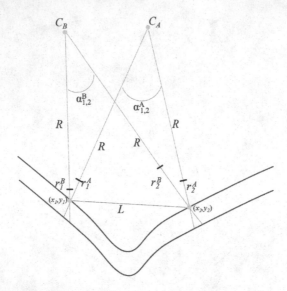

Fig. 3. Diagram of our approach for using two scans to solve for the nodal points C_A and C_B and, R, the constant distance from each OCT scan to its nodal point.

where,

$$a = 2(\cos(\alpha_{1,2}^A) - \cos(\alpha_{1,2}^B)) \tag{6}$$

$$b = 2(r_1^B r_2^B (1 - \cos(\alpha_{1,2}^B)) - r_1^A r_2^A (1 - \cos(\alpha_{1,2}^A))) \tag{7}$$

$$c = (r_1^B)^2 + (r_2^B)^2 - (r_1^A)^2 + (r_2^A)^2 - 2(r_1^B r_2^B \cos(\alpha_{1,2}^B) - r_1^A r_2^A \cos(\alpha_{1,2}^A)). \tag{8}$$

Taking the positive solution for R, we can then use Eqs. 1 and 2 to place each A-scan from each image into the physical coordinate space. This is done by first setting the nodal point as the origin $(c_x, c_y) = (0, 0)$ and placing the first A-scan at zero degrees ($\phi_1 = 0$). This allows both images to be converted from their polar coordinate systems into Cartesian coordinate systems. Once in Cartesian space, we can then apply a rigid body rotation and translation such that the thinnest line that crosses the fovea center is aligned with the visual axis of the eye (see Fig. 5b).

3   Evaluation and Results

3.1   Simulated Reconstruction

One challenge in modeling and correcting for the scan geometry in retinal OCT is the lack of a ground truth representation for the retina being imaged. Thus, it is difficult to assess if the scan geometry was accurately corrected. To evaluate our approach, we use our model to generate simulated OCT images from specific scan geometries (see Fig. 1a). Each voxel of the object in the simulation is tagged with a unique identifier, which can be observed as the gradient in Fig. 1a. Using these identifier, this allows us to establish voxel-wise correspondences between two simulated OCT images.

Our correction method was applied to these simulated OCT images using two randomly chosen locations in the images as landmarks to estimate R. This process was repeated 10000 times to acquire a distribution of the R estimated from the simulated OCT images. Figure 4 shows a histogram of the estimated R from the 10000 trials. From the figure we see that majority of the estimated R were clustered around the true distance of $R = 500$ pixels. There were however several degenerate estimations of zero and infinity for R which resulted when the two randomly chosen landmarks were too close together or co-linear, causing the equation to have zero or infinite solutions. Ignoring these degenerate cases, the RMSE of the estimated R compared to the ground truth in this experiment was 84.9 pixels, which is a mean spread of 16.98% relative to the ground truth R length being estimated.

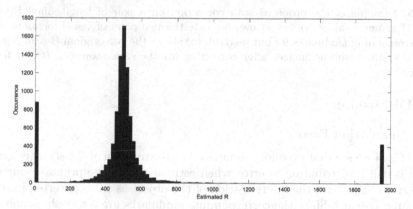

Fig. 4. Histogram of the estimated R across 10000 random trials using simulated OCT images with a ground truth $R = 500$.

3.2   Application to Real Data

We also applied our approach for correcting the scan geometry of real OCT data. A corresponding pair of B-scans from longitudinal OCT scans of the same subject was manually selected. A pair of vessel locations in each B-scan was then selected and used as corresponding landmarks, and our approach was applied to the images to correct for the scan geometry in both images simultaneously. Both corrected images were then realigned such that the fovea is centered. Figure 5 shows the original and corrected images. Since the true distortion and the shape of the underlying retina is unknown, this primarily served as a preliminary demonstration of our technique on real data. Qualitatively, we observe that the shape and structure of the retina became better aligned after correcting for the scan geometry. However, unlike an image registration technique where one image is used as a reference and the other image is transformed to match, our approach solves and corrects for the underlying shared geometry of the retina in both images. Thus, our approach increased the similarity between the images without being biased towards a chosen reference.

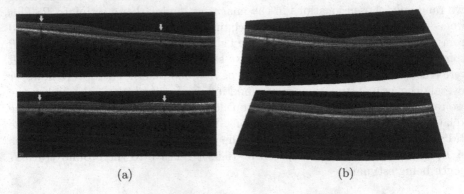

(a)   (b)

Fig. 5. Example of the proposed scan correction on a pair of longitudinal B-scans from the same subject. Yellow arrows indicate the two pairs of vessel locations used as corresponding landmarks for our method. (a) shows the two original B-scan images. (b) shows the resulting images after correcting for the scan geometry. (Color figure online)

4   Discussion

4.1   Simulation Errors

From Fig. 4 we see that even for a simulated reconstruction of the scan geometry, there is still a distribution of error when estimating R. The primary source of this error comes from the discretization of the object as it is converted into the simulated B-scans. Since the corresponding landmarks are establish as pairwise voxels, the precise location of a corresponding feature can be lost due to the voxel resolution. This error then translates to inaccuracies in the reconstruction of the scan geometry.

4.2   Limitations

We observe two primary limitations to our proposed approach. First, the 2D model currently limits the practicality of the method for use with real 3D OCT images. While in Sect. 3.2 we were able to find two corresponding B-scans to apply our method, often corresponding landmark locations between OCT images will not necessary land on the same B-scan. To address this, we are currently working to extend the model to a 3D solution, which will account for the scan geometry along both the A-scan and B-scan directions. Second, we recognize that a repeat scan of the same subject may not always be available to be used with our approach. Thus we are looking into adaptations of this method that will be able to use different types of regularly acquired scans (e.g. horizontal and vertical scans) to reduce this limitation.

5   Conclusion

In this work we introduced a 2D model for correcting the fan-beam geometry of OCT B-scans. From the model, we derived an analytic solution for solving the unknown location and distance of the nodal point from a pair of B-scans, which showed promising results in simulated and real data. Our goal is to establish a robust 2D framework that will be expanded in the future for full 3D estimation and correction of OCT scan geometries. This correction will allow for more accurate analysis of retinal shape and morphology for the study of retinal health and disease.

Acknowledgment. This work was supported by our funding sources NEI/NIH grants P30EY001583, 1R01EY024655 and U01EY025864.

References

1. Antony, B.J., et al.: Voxel based morphometry in optical coherence tomography: validation and core findings. In: Medical Imaging 2016: Biomedical Applications in Molecular, Structural, and Functional Imaging, vol. 9788, p. 97880P. International Society for Optics and Photonics (2016)
2. Asami, T., et al.: Development of a fiber-optic optical coherence tomography probe for intraocular use. Investig. Ophthalmol. Vis. Sci. **57**(9), OCT568–OCT574 (2016)
3. Brar, M., et al.: Correlation between morphological features on spectral domain optical coherence tomography and angiographic leakage patterns in macular edema. Retina (Philadelphia, Pa.) **30**(3), 383 (2010)
4. Chen, M., Lang, A., Ying, H.S., Calabresi, P.A., Prince, J.L., Carass, A.: Analysis of macular oct images using deformable registration. Biomed. Opt. Express **5**(7), 2196–2214 (2014)
5. Kajić, V., et al.: Robust segmentation of intraretinal layers in the normal human fovea using a novel statistical model based on texture and shape analysis. Opt. Express **18**(14), 14730–14744 (2010)
6. Podoleanu, A., Charalambous, I., Plesea, L., Dogariu, A., Rosen, R.: Correction of distortions in optical coherence tomography imaging of the eye. Phys. Med. Biol. **49**(7), 1277 (2004)
7. Schuman, J.S., et al.: Reproducibility of nerve fiber layer thickness measurements using optical coherence tomography. Ophthalmology **103**(11), 1889–1898 (1996)
8. Sibony, P., Kupersmith, M.J., Rohlf, F.J.: Shape analysis of the peripapillary rpe layer in papilledema and ischemic optic neuropathy. Investig. Ophthalmol. Vis. Sci. **52**(11), 7987–7995 (2011)
9. Westphal, V., Rollins, A.M., Radhakrishnan, S., Izatt, J.A.: Correction of geometric and refractive image distortions in optical coherence tomography applying fermat's principle. Opt. Express **10**(9), 397–404 (2002)

A Bottom-Up Saliency Estimation
Approach for Neonatal Retinal Images

Sharath M. Shankaranarayana[1(✉)], Keerthi Ram[2], Anand Vinekar[3],
Kaushik Mitra[1], and Mohanasankar Sivaprakasam[1,2]

[1] Indian Institute of Technology Madras (IITM), Chennai, India
ee15s050@ee.iitm.ac.in
[2] Healthcare Technology Innovation Centre, IITM, Chennai, India
[3] Narayana Nethralaya, Bengaluru, India

Abstract. Retinopathy of Prematurity (ROP) is a potentially blinding disease occurring primarily in prematurely born neonates. Staging or classification of ROP into various stages is mainly dependant on the presence of ridge or demarcation line and its distance with respect to optic disc. Thus, computer aided diagnosis of ROP requires method to automatically detect the ridge. To this end, a new bottom up saliency estimation method for neonatal retinal images is proposed. The method consists of first obtaining a depth map of neonatal retinal image via an image restoration scheme based on a physical model. The obtain depth is then converted to a saliency map. Then the image is further processed to even out illumination and contrast variations and the border artifacts. Next, two additional saliency maps are estimated from the processed image using gradient and appearance cues. The obtained saliency maps are then fused by pixel-wise multiplication and addition operators. The obtained final saliency map facilitates the detection of demarcation line and is qualitatively shown to be more suitable for neonatal retinal images compared to the state of the art saliency estimation techniques. This method could thus serve as tool for improved and faster diagnosis. Additionally, we also explore the usefulness of saliency maps for the task of classification of ROP into four stages.

Keywords: Retinopathy of prematurity · Saliency estimation
Deep Learning

1 Introduction

Retinopathy of prematurity (ROP) is a sight-threatening disease occurring primarily in prematurely born neonates and is seen as one of the leading causes of childhood blindness. The diagnosis is usually performed upon the examination of retinal images from a wide-field fundus camera. It is necessary to perform screening at regular intervals for detecting ROP and staging the progression and also planning interventions. A report published by an international consensus panel

© Springer Nature Switzerland AG 2018
D. Stoyanov et al. (Eds.): COMPAY 2018/OMIA 2018, LNCS 11039, pp. 336–343, 2018.
https://doi.org/10.1007/978-3-030-00949-6_40

[1] provides guidelines for staging of the disease. The staging of the disease is performed based on the presence and location of a ridge-like structure which occurs
due to abnormal development of retinal vasculature. The other important signs
include venous dilation and arteriolar tortuosity which indicate the presence of
plus disease [1]. Even though ROP is considered difficult to detect, plus-disease
is considered to be easier to detect. In-fact, the detection of plus disease has
received considerably more attention than the staging of ROP by detecting the
presence of ridge, since the latter is considered harder. Rising incidences of ROP
due to improvements in neonatology, coupled with the lack of ophthalmologists
proficient in diagnosing ROP necessitates the need for automated and computer
aided diagnosis.

Compared to the numerous works on adult retinal image analyses, there are
very few works on neonatal retinal image analyses. Majority of the works in ROP
focus on the identification and quantification of plus disease in neonatal retinal
images. A standard pipeline employed in such works consists of two major steps -
to extract vessels as a first step using vessel segmentation techniques and extract
tortuousity information as the next step. Jomier et al. [2] use a neural network for
identification of plus disease. The authors extract features in the form of width
and tortuousity-index information from four quadrants of vessel segmentation
map and feed the extracted features as the input to the neural network for
classification. Other works [3, 4] also extract similar width and tortuousity based
information for the quantification of the disease. Recently, Worrall et al. [6]
proposed a deep learning framework for detection of ROP. They fine-tune a
pretrained convolutional neural network (CNN) as an ROP detector. Similar to
the previous works, the authors focus only on the presence or absence of plus
disease and also consider only a small region of interest around the optic disc.

Thus, we see that almost all the works on ROP focus on identification of
plus disease. Another important cue for ROP- the demarcation line or the ridge
is not included in any of the prior works. The ridge is an important factor in
ROP analyses since it occurs in the majority of neonatal retinal images, and
moreover, the staging or classification of ROP into various stages is dependent
the distance of the ridge from the optic disc. But automated ridge detection
is a difficult task because the neonatal retinal images are inherently of poorer
quality due to limited dilation, heavy fundus pigmentation, corneal and vitreous
haze, which render lower information content in images. This can be addressed
to an extent using restoration schemes as proposed in [5] which uses a physical
model similar to dehazing model. In addition to this, ridge detection is also
hard because of variations color, texture, size and appearances of the ridges. In
this paper, we explore the possibility of using saliency based models [7–10] for
detecting ridge regions. Saliency is one of the well studied concepts in computer
vision and is related to human visual perception and processing of visual stimuli.
A saliency map gives a map which indicates relative importance of attributes in
an image. Saliency is usually estimated from attributes such as color, texture,
edges and boundaries. We see that even ridges in neonatal retinal images can be
characterized by such attributes.

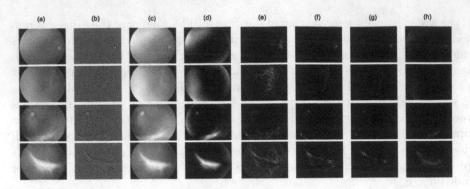

Fig. 1. Output of intermediate steps - (a) Original Image, (b) Processed Image, (c) Depth Map D, (d) Sal_d, (e) Sal_{grad}, (f) Sal_{app}, (g) Sal_{fused}, (h) Sal_{final}

Saliency based methods have been employed in retinal image analysis. The work in [13] employs Spectral residual (SR) saliency [10] for bright lesion detection in retinal images. Recently, Zhao et al. [14] proposed a saliency estimation techniques for fluorescein angiography (FA) images for leakage detection in diabetic and malarial retinopathy. Saliency based techniques have also been applied to other medical images. To the best of our knowledge, this is the first work exploring the use of saliency models for neonatal retinal images. The main contributions of our work can be summarized as follows:

1. We propose a new bottom-up unsupervised saliency estimation approach for neonatal retinal images to highlight the presence of ridge. The saliency estimation method takes into account important characteristic features of ridge.
2. We qualitatively compare our estimated saliency maps with other state of the art saliency estimation techniques and show that our method is most suitable for highlighting the ridge.
3. Additionally we explore the usability of saliency maps estimated by different techniques for the task of staging or classification of ROP into four stages using deep learning.

2 Methods

2.1 Image Restoration and Preprocessing

The neonatal retinal images tend to be corrupted due to vitreous haze. We employ the restoration scheme provided in [5] as the first step to obtain a restored image. In addition to the restored image, the method gives us a relative depth-map which we later use in our saliency model. Even after restoration, the images suffer from effects such as uneven color distribution, uneven illumination and artifacts on the retinal periphery due to the nature of the camera. These variations could hamper the performances of saliency models. We simply perform high pass filtering and remove some of the pixels in retinal periphery and fill the otherwise black background with average color values of the retinal image (see Fig. 1 (b)).

2.2   Saliency Detection

In the context of neonatal retinal images, the most important cue for staging is the presence and location of ridge or the demarcation line. Thus the most salient region in this case happens to be the ridge. Thus, the proposed saliency estimation technique employs three most characteristic features of the ridge.

1. Since the ridge is generally characterized by sharp changes, we first estimate saliency using the gradient information of the image.
2. The ridge, in most cases, is seen to have a different appearance with respect to the background in terms of intensity and color. Therefore we also estimate saliency using luminance and color information.
3. As reported in the medical literature [1], the retinal detachment occurs near the ridge which cause ridge to protrude out of the retinal plane and hence the depth also serves as one of the cues to characterize the ridge.

Gradient Based Saliency. Edge being a characteristic feature in determining saliency, we employ image gradient information for the estimation of saliency map. Given the image I, we first blur the image with Gaussian blur at five different levels giving $I_{\sigma_1 - \sigma_5}$. Next we calculate the magnitude of gradient for all the Gaussian blurred images-

$$L_{\sigma_i} = Ig_{x\sigma_i}{}^2 + Ig_{y\sigma_i}{}^2 \tag{1}$$

where Ig_x and Ig_y are gradients along x and y directions respectively and L_{σ_i} is the magnitude for blur level σ_i. The final saliency map is calculated by-

$$Sal_{grad} = \frac{1}{5} \sum_{i=1}^{5} L_{\sigma_i} \tag{2}$$

where Sal_{grad} is the final gradient based saliency.

Appearance Based Saliency. Since the ridge also differs in appearance when compared to the background and vasculature, we employ the luminance and color features to estimate saliency based on appearance. The given color image I is first transformed to Lab color space and then the following operation is performed to find the saliency map.

$$Sal_{app} = (L - L_\mu).^2 + (a - a_\mu).^2 + (b - b_\mu).^2 \tag{3}$$

where Sal_{app} is the appearance based saliency and L, a, b are the corresponding channels in Lab color space and L_μ, a_μ, b_μ are the mean values of the corresponding channels.

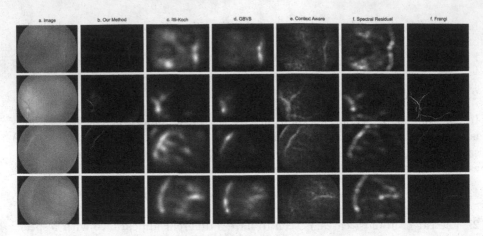

Fig. 2. Comparison with different saliency estimation techniques.

Depth Based Saliency. The depth information obtained as an extra during the restoration operation [5] is also used for estimating the saliency. Let D be the normalized depth map of the image I, the saliency map Sal_d is calculated using the following-

$$Sal_d = (D - D_\mu)^2 \qquad (4)$$

where D_μ is the mean value of the depth map. The above saliency map contains high values in the regions which are far away from the mean. This might also lead in high values for regions with low values of depth. Since the ridge regions tend to have high values os depth, we perform a pixel-wise product of the above obtained saliency map with the depth map to obtain final depth based saliency map S_{deoth}

$$Sal_{depth} = Sal_d.D \qquad (5)$$

The final saliency map is estimated by combining the individual saliency maps. We first fuse the gradient based saliency map and appearance based saliency map by

$$Sal_{fused} = Sal_{grad}.Sal_{app} \qquad (6)$$

where Sal_{fused} is the fused saliency map and this saliency map is then normalized. Finally, this normalized saliency map is combined with depth based saliency map using the following operation-

$$Sal_{final} = \lambda_1 Sal_{fused} + \lambda_2 Sal_{depth} \qquad (7)$$

where Sal_{final} is the final saliency maps and λ_1 and λ_2 are parameters which are set empirically to 0.7 and 0.3 respectively.

The outputs obtained at various stages is shown in Fig. 1. We thus incorporate clinically relevant features in the estimation of saliency map.

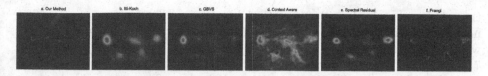

Fig. 3. Comparison with different saliency estimation techniques.

3   Experiments and Results

We evaluate the different saliency models on a privately collected dataset having a total of 647 (having multiple images from the same patients) and the retinal scans of each patient having been assigned on of the four stages from $1-4$. We visually compare the proposed saliency technique with the other state of the art techniques- Itti-Koch (IT) [7], graph based visual saliency (GBVS) [8], context aware saliency (CA) [9], spectral residual saliency (SR) [10] and Frangi filter [11]. We compare the normalized saliency maps obtain by each of the methods. The results are shown in Fig. 2. It can be seen from the figure that the proposed method yields superior results in highlighting the presence of ridge when compared to the other state of the art. The other methods contain high values in their saliency maps even for the non-ridge regions. Although, this could be potentially masked out by using vessel detection and optic disc detection, the other methods do not yield finely delineated saliency maps for the masking to work effectively.

Since the saliency models used are unsupervised and since we also lack the ground truth data in the form pixel-wise markings of the ridge, it is not possible to quantitatively evaluate the different saliency models. As another preliminary study, which could also potentially serve as a method to evaluate the saliency models, we employ the saliency maps obtained from the different methods in a classification task. For this task, we first fuse the processed image and the corresponding saliency maps from different methods to generate an overlaid image as shown in Fig. 3. These fused images are then used for the task of classifying the images into various stages using a deep network [15]. We use the whole dataset of 647 images having 233 images for *Stage*1, 217 images for *Stage*2, 121 images for *Stage*3 and 76 images for *Stage*4 for the classification task. We perform the classification experiment for the seven cases-

  (i) train with raw RGB images
 (ii) train with preprocessed images
(iii) train with fused images with saliency obtained from proposed method
(iv) train with fused images with saliency obtained from IT [7] method
 (v) train with fused images with saliency obtained from GBVS [8] method
(vi) train with fused images with saliency obtained from CA [9] method
(vii) train with fused images with saliency obtained from SR [10] method
(viii) train with fused images with saliency obtained from Frangi [11] method.

We use the architecture RESNET18 [15] with 18 layers as the network of choice. The network is initialized with imagenet pretrained weights. We finetune the whole network after changing final layer to predict for four classes. The network is trained with crossentropy loss using stochastic gradient descent (SGD) with initial learning rate of 0.001 which is decayed by a factor of 0.7 after every 7 epochs. We compute the average accuracies for all the above experiments upon five-fold cross-validation using the whole data. The obtained values are listed in the Table 1. It can be seen that naive fusion of saliency map with the image does not result in improved performance. But viewed from the point of view comparison of different saliency detection methods, it can be seen that classification with fused images from the proposed saliency estimation technique yields better results when compared to the other state of the art techniques with context aware [9] saliency being an exception. Again, it should be noted that the classification of ROP using deep learning was performed as a study to explore the usefulness of estimated saliency maps. A more rigorous classification pipeline needs to explored for ROP, given the large variability in the images present in the dataset. This is also evident from the overall classification accuracies obtained which are very low.

Table 1. Performance

Method	Accuracy
Raw RGB images	0.538
Processed images	0.587
Proposed saliency	0.576
Itti Koch	0.571
GBVS	0.532
Spectral residual	0.561
Context aware	0.596
Frangi	0.570

4   Conclusion

In this paper, we proposed a novel method to estimate saliency for neonatal retinal images. We employed clinically important cues for the estimation of saliency and by qualitative comparison, we showed that proposed method gives superior results compared to the state of the art when it comes to highlighting the ridge. We also explored the usability of saliency models in deep learning framework and saw that naive fusion does not yield an improved performance. Since the proposed work lacks concrete quantitative evaluation, we would work on it in future by possibly creating a dataset with pixel-wise annotated ridges. Also in future, we would like to explore a better method to incorporate features from saliency maps in a learning framework.

References

1. Gole, G.A., et al.: The international classification of retinopathy of prematurity revisited. JAMA Ophthalmol. **123**(7), 991–999 (2005)
2. Jomier, J., Wallace, D.K., Aylward, S.R.: Quantification of retinopathy of prematurity via vessel segmentation. In: Ellis, R.E., Peters, T.M. (eds.) MICCAI 2003. LNCS, vol. 2879, pp. 620–626. Springer, Heidelberg (2003). https://doi.org/10.1007/978-3-540-39903-2_76
3. Swanson, C., Cocker, K., Parker, K., Moseley, M., Fielder, A.: Semiautomated computer analysis of vessel growth in preterm infants without and with ROP. Br. J. Ophthalmol. **87**(12), 1474–1477 (2003)
4. Aslam, T., Fleck, B., Patton, N., Trucco, M., Azegrouz, H.: Digital image analysis of plus disease in retinopathy of prematurity. Acta Ophthalmol. **87**(4), 368–377 (2009)
5. Shankaranarayana, S.M., Ram, K., Vinekar, A., Mitra, K., Sivaprakasam, M.: Restoration of neonatal retinal images. In: Chen, X., Garvin, M.K., Liu, J., Trucco, E., Xu, Y. (eds.) Proceedings of the Ophthalmic Medical Image Analysis Third International Workshop, OMIA 2016, MICCAI 2016, pp. 49–56 (2016)
6. Worrall, D.E., Wilson, C.M., Brostow, G.J.: Automated retinopathy of prematurity case detection with convolutional neural networks. In: Carneiro, G., et al. (eds.) LABELS/DLMIA -2016. LNCS, vol. 10008, pp. 68–76. Springer, Cham (2016). https://doi.org/10.1007/978-3-319-46976-8_8
7. Itti, L., Koch, C., Niebur, E.: A model of saliency-based visual attention for rapid scene analysis. IEEE Trans. Pattern Anal. Mach. Intell. **20**(11), 1254–1259 (1998)
8. Harel, J., Koch, C., Perona, P.: Graph-based visual saliency. In: Advances in Neural Information Processing Systems, pp. 545–552 (2007)
9. Goferman, S., Zelnik-Manor, L., Tal, A.: Context-aware saliency detection. IEEE Trans. Pattern Anal. Mach. Intell. **34**(10), 1915–1926 (2012)
10. Hou, X., Zhang, L.: Saliency detection: a spectral residual approach. In: IEEE Conference on Computer Vision and Pattern Recognition, CVPR 2007, pp. 1–8. IEEE, June 2007
11. Frangi, A.F., Niessen, W.J., Vincken, K.L., Viergever, M.A.: Multiscale vessel enhancement filtering. In: Wells, W.M., Colchester, A., Delp, S. (eds.) MICCAI 1998. LNCS, vol. 1496, pp. 130–137. Springer, Heidelberg (1998). https://doi.org/10.1007/BFb0056195
12. Lpez, A.M., Lumbreras, F., Serrat, J., Villanueva, J.J.: Evaluation of methods for ridge and valley detection. IEEE Trans. Pattern Anal. Mach. Intell. **21**(4), 327–335 (1999)
13. Deepak, K.S., Chakravarty, A., Sivaswamy, J.: Visual saliency based bright lesion detection and discrimination in retinal images. In: 2013 IEEE 10th International Symposium on Biomedical Imaging (ISBI), pp. 1436–1439. IEEE, April 2013
14. Zhao, Y., Zheng, Y., Liu, Y., Yang, J., Zhao, Y., Chen, D., Wang, Y.: Intensity and compactness enabled saliency estimation for leakage detection in diabetic and malarial retinopathy. IEEE Trans. Med. Imaging **36**(1), 51–63 (2017)
15. He, K., Zhang, X., Ren, S., Sun, J.: Deep residual learning for image recognition. In: Proceedings of the IEEE Conference on Computer Vision and Pattern Recognition, pp. 770–778 (2016)

Author Index

Printed in the United States
By Bookmasters